NEC®Answers

OTHER BOOKS IN THE McGRAW-HILL NEC® SERIES

NEC®Answers

Michael A. Anthony, P.E.
University of Michigan

McGraw-Hill

New York • San Francisco • Washington, D.C. • Auckland • Bogotá
Caracas • Lisbon • Madrid • Mexico City • Milan
Montreal • New Delhi • San Juan • Singapore
Sydney • Tokyo • Toronto

Library of Congress Cataloging-in-Publication Data

Anthony, Mike.
  NEC answers / Michael A. Anthony
     p.   cm.
  ISBN0-07-1344494-2
  1.National Fire Protection Association. National Electrical Code (1992)   2.
  Building—Contracts and specifications.   3. Electric engineering—United States—
  Insurance requirements.   I. Title.
  TK260.A68   1999
  621.319'24'0218—dc21   99-35502
                                                                                    CIP

McGraw-Hill

A Division of The **McGraw·Hill** Companies

1 2 3 4 5 6 7 8 9 0 DOC/DOC 9 0 4 3 2 1 0 9

ISBN 0-07-134494-2

*The sponsoring editor for this book was Zoe Foundotos, the editing supervisor was
Frank Kotowski, Jr., and the production supervisor was Sheri Souffrance. It was set
in Sabon per the NM2 design by Kim Sheran of McGraw-Hill's Professional Book
Group composition unit, Hightstown, N.J.*

Printed and bound by R. R. Donnelley & Sons.

McGraw-Hill books are available at special quantity discounts to use as premiums
and sales promotions, or for use in corporate training programs. For more information, please write to the Director of Special Sales, McGraw-Hill, 11 West 19th Street,
New York, NY 10011. Or contact your local bookstore.

 This book is printed on recycled, acid-free paper containing
a minimum of 50% recycled, de-inked fiber.

Contents

Preface

This book establishes a new platform for putting the 1999 National Electric Code to work. Until now, all of the companion literature written for the **NEC** has targeted electrical professionals who are familiar with the manner in which the nine chapters of the **NEC** progress, complement, and cross-reference. *NEC® Answers* provides other building industry professionals—architects; civil, mechanical, computer, and telecommunication engineers, for example—with a book to help them gain sufficient familiarity with the Code fundamentals to participate in informed discussion of its implications for design and requirements for construction.

We hope to bring new life to the supporting literature of the Code by arranging the presentation of these core requirements in the Construction Specifications Institute *MasterSpec* format. The *MasterSpec* format is the *de facto* industry standard for thousands of short-run printings of construction documents in use at job sites every working day. It provides the framework of the construction contract itself and is common ground for all the building trades.

NEC® Answers is for readers who want familiarity with the broad, underlying principles of the code rather than sharp focus upon wiring practice. It has practical value for facility planners and engineers whose influence in the early design stage of a building typically eclipses the influence of electricians and inspection authorities, but who, just as typically, know the least about the code. While this book covers many of the 1999 changes to the **NEC**, it is not a code change book per se and should not be regarded as a text for **NEC** instructors and/or electricians seeking renewal of trade licenses. The electrician looking for such would be advised to consult a title from any one of the titles indicated in the core references.

The National Electric Code as Literature

Building construction in the United States today is governed by a constellation of standards from which the National Electric Code stands apart. As a document whose first pages were put to print in 1896 it is the oldest standard of them all: predating the founding of the American National Standards Institute by some twenty years, and having its inspiration originate in Thomas Edison's concern for safety in the public lighting systems of New York City in the 1880s.

While its content may apply to only a minor part of a building project, the process by which it is updated is a standard for other disciplines in the construction industry. The democratic process by which the code is amended periodically is similar to the manner in which laws are amended in governments at all levels. As such, not a few of its pages form the basis of safety legislation enforced by governmental agencies of all levels to ensure the practical safeguarding in the use of electricity. In the best sense of the word it is a consensus standard. In the electrical trades, it is the North Star.

The National Electric Code is not a simple document, nor can it be. It is easy to forget what a small miracle building construction is in the first place; that near-strangers can come to a job site and, with plans and specifications, actually have a language in common and share common rules for safety and profit. It is easy to forget that electricity is a fearsome form of energy, with destructive capacity equal to its capacity to improve our lives.

You never get a true sense of how remarkable a document the **NEC** is until you undertake research for a book like this and observe how so many standards converge within it, and how few other technologies are governed by anything like it. The comfort of the general public with electrical energy in buildings is due, in large part, to the success of the National Electric Code as its widespread acceptance. It is big. Its narrative is far from inspirational, and requires interpretation by oracles. Its finest aesthetic qualities seem to be subverted by technical ambiguity, redundancy, and quasilegal arcania. No document was ever less conducive to pleasure or understanding than is the curious and complicated panoply in which the **NEC** has come down to us. Nothing but a work of collective genius could have survived such a handicap at all.

Remarks on the Organization of this Book

This book has been largely a task of re-formatting existing information. Considerable effort has been put into getting economic information on the same page with safety information to help designers, planners, and building owners reconcile their often competing requirements. The use of the word "shall," in most cases, is replaced by language that is much softer. We will occasionally strip away some of the formal typography of the **NEC** in order to get to the important points. We hope that this will remind the reader that this book is no substitute for the **NEC** itself.

Traditional companion literature of the **NEC**, written for electricians, usually mimics the numerical order in which topics appear in the code. For better or worse, this book contains three numbering regimes: its own 26 chapter format, the **NEC** format (Chapters 1 through 9), and the CSI *Masterspec* format (Divisions 0 through 16). By tracking the *Masterspec* division system (albeit, somewhat loosely) we discuss electrical topics in the sequence in which construction actually occurs: from utility site work to equipment acceptance testing.

Wherever possible we have tried to eliminate repetition and crossover. This has not been an easy job. Some issues have several topical characteristics that make it necessary that they appear in more than one place in the code and in this textbook. To a certain degree, we have also tried to make the answers as independent from one another as possible. The pitfall in meeting this requirement is that there are passages that will seem somewhat repetitious for the user reading several related questions in sequence. Some things in electrotechnology cannot be made simple, and that is why there is a market for electrical professionals who are comfortable with fine points and necessary ambiguities.

While many of the answers restate the first principles of the science, art, and economics of electrotechnology that appear in other books, we also hope that you will find something in this book that you have not seen elsewhere. A book becomes a standard by being tested in the public domain, and hammered-upon—not unlike the **NEC** itself. First editions of any technical book have errors in them, practically by definition. At best, a first edition will establish a platform, a basis for further building in which reader feedback is indispensible. Comments to the publisher will be appreciated and answered personally by the author.

Mike Anthony
11 April 1999

Acknowledgments

An old adage claims that the best research lies in the number of authors from whom you steal. Thus the author makes no claim to originality in this book other than choosing the selected material by the point of view of someone who has read widely and who has worked for nearly 20 years from both sides of the decimal point that divides electronics from electrical power. He prefers to be regarded as a standards "popularizer" rather than a standards expert. To the NEC experts who have written textbooks and articles for the trade press belongs underlying credit, and no author has been and will continue to be under the deepest obligations to their labors.

Thanks to all the electricians that I have known and have had the particular pleasure of working with at the University of Michigan. It is impossible to have a career in engineering without their candor and their insight. Thanks to my reviewers: Doug Hanna, Jim Hoogersteger, Fred Mayer, and Dave Stockson of the University of Michigan, Rivan Frazee of Michigan State University, Bill Serantoni of UCLA, Roy Warwick of the University of Tennessee, Knoxville, John Laetz of the University of North Carolina–Chapel Hill, Paul Kempf of the University of Notre Dame, Brian Goss of the Joint Apprenticeship Training Council, and Herb Alfaro at The Taubman Company. Though they are in no way responsible for any errors of fact or judgement in these pages, there is no part of the book which has not been improved by their remarks.

I wish to express gratitude for the support of the executive committee of our trade association: *The Association of Higher Education Facilities Officers* (APPA): Jim Christenson of the University of Michigan, Ron Flynn of Michigan State University, John Harrod of the University of Wisconsin, and Jack Colby at the University of North Carolina–Greensborough.

Thanks to the speedsters at North Market Street Graphics: Mark Ammerman, Ginny Caroll, Jill Lynch, and Dick Stipe. Thanks to Frank Kotowski, Jr. of McGraw-Hill and their production staff for their good nature and focus on results. Special thanks to Harold Crawford, my first editor, who retired after decades of service to McGraw-Hill and the building industry. The electrical world has been a little safer for the books he ushered into the light of print. Ultimate thanks to Zoe Foundotos for clearing the way for me to continue and making the new path so easy to follow.

Great improvements have been made in making the **NEC** more usable in recent years. This book takes another step in the same direction by attempting to present the **NEC** in as enjoyable and understandable form as possible. The FAQ (frequently asked questions) format leavens the Code, and helps Code users who learn best through Socratic interplay. If the present edition attains in any degree its high and humble aim of fostering appreciation of the **NEC**, or it merely inspires others to do a better job of it, it will have been well worth while.

Mike Anthony
11 April 1999

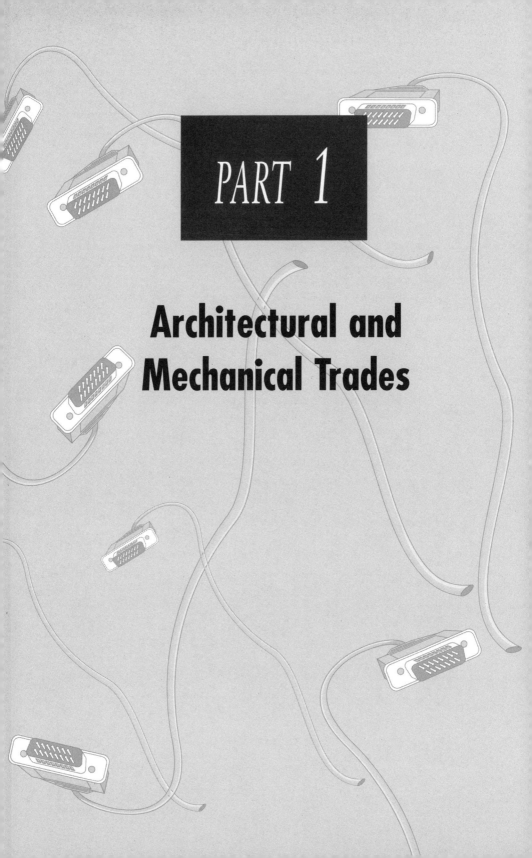

PART 1

Architectural and Mechanical Trades

Division 00000—Standard General Conditions

Remarks. A contract for the installation of electric systems consists of a written document and drawings. These become part of the construction contract, which contains what is more commonly known as "specs and blueprints." The legal, nontechnical document that precedes the specifications is sometimes called the standard general conditions and contains the general terms of agreement between contractor and owner and a description of the routine of business. We will refer to this document as Division 0.

Some variants of Division 0 also contain "boilerplate" material required by the federal and local governments to ensure fair bidding and equal opportunity. There are several milestones in the routine of business that refer to the process of securing permits and passing inspection. The NEC—coupled, perhaps, with a local, more rigorous, reciprocal version of it—establishes the conditions for passing electrical inspections. Division 0 may tell you where the more rigorous construction standards appear.

The practice of engineers and architects addressing the same subject matter in different locations in their respective bidding and construction documents frequently leads to confusion and unanticipated legal consequences. For example, whether or not the minimum standards established by the NEC are to be superseded by more rigorous requirements indicated on either the blueprints or in the specifications is always a touchy issue. It is often a matter of architect and engineer stylistic efficacy, dependent upon such factors as the size of the job or whether or not it is cheaper to pay a drafter to put it on the blueprint or pay a secretary to key it into the specifications. The American Society of Civil Engineers have a series of documents, one of which is titled, "Uniform Location of Subject Matter," that attempts to solve this problem.

Q1.   *What NEC requirements commonly appear in Division 00000?*

A.   The standard general conditions indicates the contractor's basic duties and responsibilities and covers such areas as bonds and insurance; progress and final payments; substantial completion; mutual responsibility of contractors, status of the architect-engineer during construction; owner's responsibilities; subsurface and hidden site conditions; temporary elevators, fire precautions; and the like. It may contain a list of contract definitions and perhaps a general "Who's Who" and "Who Does What" of the enterprise. One of the most important issues has to do with the manner in which the authority having jurisdiction (AHJ) is approached, how the temporary construction power will be supplied and metered. In deference to the many variations in standard general conditions, the subject of temporary electric power will be covered here.

Q2.   *Who determines the facility classification of a building?*

A.   The architect and the owner, usually with some input from local authorities. It is usually obvious except in the case of mixed-use, multipurpose facilities. The local building code will be used as a guideline. The NFPA defines type I and type II buildings in NFPA 220-1995, Standard on Types of Building Construction.

Q3.   *What role does the electrical engineer have in meeting NEC requirements?*

A.   Electrical engineers have an obligation to protect the interests of the public at large as well as protect the interests of the client. The NEC has evolved over the years as a document that rarely put these two interests at odds. Engineers, in addition to the duty to apply their technical judgment to protect a client from the consequences of the client's lack of building knowledge, must be guided by a firm sense of fairness in their relationships with contractors and with other professionals and must provide society in general with buildings which are safe.

If engineers have the requisite skill and do not use it, they are chargeable with negligence. If they do not possess the requisite standard of skill, they are liable for the lack of it. Engineers are not held to absolute accuracy in performing their professional duties, however; nor do they warrant the perfection of their plans and specifications. The law requires only the exercise of ordinary skill and care in light of present-day knowledge. Electrical engineers may be charged with the consequence of errors only where such errors have occurred for want of reasonable skill or reasonable diligence.

Q4. *Can an engineer be held responsible for violations of the NEC?*

A. The mere fact that an engineer makes an error is not in itself evidence of negligence, so long as he or she has exercised reasonable care. The preponderance of court decisions in a majority of states indicates that engineers do not guarantee perfection in their plans and specifications, nor do they guarantee that a safe and economical power system will result from their work. Although they may explicitly make such a guarantee—and if they do, the courts hold them to the promise—such a promise will not be read into a contract by implication. A long history of precedent upholds the opinion that architects or engineers are not ordinarily liable for defects, even if they have made a mistake, if they possess the degree of skill and knowledge usually possessed by a member of their profession, and if they exercise reasonable care and diligence in applying their skill and knowledge to the situation at hand. However, if their skill and care fall below this standard, they may be adjudged liable for negligence.

Q5. What is a utility?

A. This is not as easy to answer as it used to be in the days of central station power and telecommunications. With traditional investor-owned utilities getting into the commercial arena and branches of government exhibiting private-enterprise behavior, it may be more difficult to define with each code cycle. Many industrial plants and universities have substantial cogeneration "behind the fence" and operate very much like their own utility company. Thus they become their own supplier and customer. The definition of "on-site" generation may become more complicated, as the case may be, for many universities and industrial plants with cogeneration. The same holds true for the telecommunications systems of multifacility organizations, many of which operate their own data and telephone exchanges.

TABLE 1.1 WHO DOES WHAT? COORDINATION BETWEEN THE TRADES

Item	Responsibility		Comments
Division 16A Electrical			
Power wiring for motors for plumbing, mechanical, and fire protection equipment	16A Electrical	P	Motor provided by plumbing, mechanical, or fire protection trade
Motor controls and starters for motors for plumbing, mechanical, and fire protection equipment	16A Electrical 15A Plumbing 15B HVAC 15C Fire Protection	I F F F	Factory-installed motor starters to be furnished and installed by trade providing motor
Electric baseboard heat with line voltage thermostats	16A Electrical	P	16A Electrical to provide entire system
Electric baseboard heat, cabinet heaters, and unit heaters with integral thermostats and/or fans	16A Electrical	P	16A Electrical to provide entire system
Duct smoke detectors	16A Electrical 15B HVAC	F I	16A Electrical to make all connections to fire alarm system, 15B HVAC to mount detector in ductwork and make connections to HVAC controls
Electric heater cables for pipe tracing with local thermostat control	16A Electrical	P	Coordinate installation with pipe insulation
Electric heater cables for pipe tracing with remote or DDC control	16A Electrical	P	Coordinate installation with pipe insulation. 15B HVAC to provide control sensors and connections

TABLE 1.1 WHO DOES WHAT? COORDINATION BETWEEN THE TRADES (*CONTINUED*)

Item	Responsibility		Comments
Division 16A Electrical (*Continued*)			
Electric snow melting equipment with local sensor control	16A Electrical	P	Coordinate installation with 3A Concrete
Electric snow melting equipment with remote or DDC control	16A Electric	P	Coordinate installation with 3A Concrete 15B HVAC to provide control sensors and connections
Division 15B HVAC			
Electric duct heaters	15B HVAC	P	16A Electrical to provide power wiring to and including disconnect
Electric duct heater controls	15B HVAC	P	15B HVAC to provide power wiring from disconnect to unit and all controls
Through-wall A/C and electric heating units	15B HVAC	P	Sleeve installation by trade building wall. Sleeve furnished by 15B HVAC. Power wiring by 16A Electrical
Miscellaneous			
Access doors	Trade requiring access	F	
	Trade constructing wall or ceiling	I	

TABLE 1.1 WHO DOES WHAT? COORDINATION BETWEEN THE TRADES (*CONTINUED*)

Item	Responsibility	Comments
Miscellaneous (*Continued*)		
Masonry shafts, drywall shafts, tunnels utilized for air ducts		It is mandatory to assure the airtightness of all joints, holes, and other openings to make the air conveyors acceptable for their function
Thermal and acoustic insulation for mechanical room walls and ceilings	7B Insulation	P

P = provide; F = furnish; I = install.

From R. S. Means, *HVAC Handbook*; used with permission.

A great deal of electrical work appears in nonelectrical divisions of the specifications. Table 1.1 shows how the electrical trades are allied with other trades with respect to electric power, communication, and signaling systems.

Q6. *Does the NEC have anything to say about application and certification of inspections?*

A. No, the Code is silent, even though the document itself is at the center of the process. As the authority having jurisdiction (AHJ), the electrical inspector will be charged with assessing conformance of the new electrical system with the accepted edition of the NEC. Appendix item A2 is a sample building inspection form of a well-known United States city that has allowed electrical engineers to do wonderful things with electricity. It is very similar to the permit applications for the majority of United States municipalities.

Q7.   *Does the NEC have anything to say about shop drawings?*

A.   No. There is a running assumption throughout the NEC that electrical equipment shall be specified for the purpose and will be installed consistent with manufacturer instructions. A knowledgeable AHJ will understand how the bidding process may allow a piece of unlisted equipment to "slip through" and will want to review shop drawings to confirm that all apparatus is listed (and not counterfeit). In many cases, the shop drawings are pulled together from sales literature and then used as the basis for an operations and maintenance manual that would, for example, be the only guide to an electrician needing to repair a VSD or an elevator.

Q8.   *Shouldn't the architect-engineer be responsible for approval of shop drawings?*

A.   It may seem counterintuitive to route all shop drawings through the architect and then have him or her reject all claims of responsibility, but it happens. Practice in the field varies widely. A legal opinion on this matter is not within the scope of this book, but a few things are worth remembering: an electrical equipment manufacturer will meet many NEC requirements as a precondition of having its products classified by NEMA. Most of the problems occur in the field installation phase. Nevertheless, a complicated piece of equipment (such as a substation) should have the drawings reviewed carefully by engineers and tradespersons equally. Each electrical professional will tend to see something different in the drawings which, if communicated to and implemented by the manufacturer, will result in a safe installation.

Q9.   *Shouldn't the architect-engineer be responsible for performing essential design calculations to confirm compliance with NEC requirements for short-circuit withstand and voltage drop?*

A.   One of the most lamentable practices in the electrical design profession today is the practice of letting the lowest bidder farm out the engineering. The classical arrangement has the engineer doing the engineering *up front*—before the bidding documents hit the street. Ultimately, a registered professional engineer is retained to do these calculations (and usually overcurrent coordination studies as well), but an owner who will not compensate the architect-engineer sufficiently to pay for up-front engineering risks a great deal. The cost of good electrical engineering is frequently a single-digit percentage of a building project budget and ought not to be the place where corners are cut.

Q10.   *What does the NEC have to say about protection of work and property?*

A.   Nothing explicit, though the ideal of responsible workmanship can be read in every passage of the **NEC**. It seems safe to assume that if a contractor owns the equipment (such as a new substation) he or she is obligated, by contract, to make sure that the apparatus is protected. Contractors will need to protect the apparatus according to the terms of Art. 300, Temporary Wiring. For example, during the course of construction, a contractor will own and operate a substation. Before the building is commissioned and turned over to the owner, the electrical contractor must assure that the breaker's trip units are set appropriately for the conditions of use. That may mean that the protection strategy during construction may differ from the protection strategy after the building is turned over to the owner.

Q11.   *Does the NEC assert requirements for testing and inspection?*

A.   Yes, and so do OSHA, NETA, NEMA, and ANSI/IEEE. First, when a new system is being commissioned; afterward, NFPA 70E—the National Electrical Maintenance Code—asserts requirements for testing and inspection. The **NEC** cites requirements for testing of Emergency Systems, 700-4; Ground fault protection 230-95c; insulation resistance. In emergency and legally required standby systems (Sections 700-4 and 701-5), "The AHJ shall conduct or witness a test of the complete system upon installation." Typically, this involves on-site generators, battery systems, elevators, and the like.

While a system that is not adequately engineered, designed, and constructed will not provide reliable service, regardless of how good or how much preventive maintenance is accomplished, it is worth investigating how preventive maintenance oriented design affects system performance. Table 1.2 was assembled by the

Reliability Subcommittee of the IEEE Industrial and Commercial Power Systems Committee in 1974. Of the 1469 failures reported from all causes, inadequate maintenance was blamed for 240, or over 16 percent of all the failures.

TABLE 1.2 NUMBER OF FAILURES VERSUS MAINTENANCE QUALITY FOR ALL EQUIPMENT CLASSES COMBINED

Number of failures			
Maintenance quality	All causes	Inadequate maintenance	Percent of failures due to inadequate maintenance
Excellent	311	36	11.6
Fair	853	154	18.1
Poor	67	22	32.8
None	238	28	11.8
Total	1469	240	16.4

Q12. *Does the **NEC** contain the procedure for drawing a permit and complying with location regulations?*

A. No. Unless the electrical trade is the only trade involved in a construction project, the general contractor will obtain the permit. The electrical contractor, however, is responsible for communicating with the AHJ to schedule an inspection of the electrical system. Very often, inspections are done continuously as work progresses so that violations can be corrected before it costs too much to remove and rebuild.

Q13. *Does the **NEC** have anything to say about the use of premises and utility disruptions?*

A. Nothing beyond the usual common sense. Some states post prohibitions against the transport of switchgear and concrete during months when the roads would be damaged by heavy transport machinery. OSHA will require telephone service. A crane installed on a utility circuit may cause voltage to drop, so the local utility should be consulted. Water supply may be necessary for a fire pump even before the construction is complete.

Q14. *Does the NEC have requirements for temporary elevators?*

A.   Yes (see Arts. 305 and 620). There is virtually no difference between a permanent and temporary elevator used for construction. Apart from the given permissions for variation from rules on permanent wiring, all temporary systems are required to comply in all other respects with code rules covering permanent wiring.

TEMPORARY WIRING AND METERING

Q15. *What does the NEC require for temporary utility service to a construction project?*

A.   All the general rules for wiring methods that appear in Chap. 3 apply to temporary utility service. This is the only case in which the general rules of the early chapters of the Code are modified by an article within the early chapters. Wiring methods installed as temporary are required to be installed in accordance with the rules of the article governing that wiring method except where modified by the rules in Art. 305. Temporary wiring methods must be approved based on criteria such as length of time in service, likelihood of physical damage, exposure to weather, and other special requirements. All rules on temporary wiring must be observed wherever maintenance or repair work is in process. This expands the applicability of temporary wiring beyond new construction, remodeling work, or demolition. The intent is aimed at assuring that the equipment and circuits installed under this article are really "temporary" and not a back door to low-quality permanent installations. Temporary wiring must ultimately be taken down.

Q16. *How does a temporary wiring system differ from a permanent system?*

A.   The specific rules of Art. 305 cover the only ways in which a temporary wiring system may differ from a permanent system. An example might be where there is a transformer failure and conduit is laid on the ground from a nearby source. The conduit on the ground does not violate the Code, but the installation is an eyesore and sometimes this is enough to make it stop, safety issues notwithstanding. Open or individual conductor feeders are permitted only during emergencies or tests, not during special events. Refer to Division 16450 of Chap. 21 for particulars on the subject of ground-fault protection.

Q17.   *What kind of equipment is permitted in temporary wiring regimes?*

A.   In general, a temporary electrical system does not have to be put together with the detail that characterizes permanent wiring systems. Flexible cords are in more liberal use, for example. Hard usage or extra hard usage cords are permitted to be laid on the floor. Temporary wiring typically looks like fewer junction boxes, pigtail connections, and ruggedized modular receptacles to avoid damage due to pinching, abrasion, cutting, or other abuse. Taps are permitted to be made in temporary wiring circuits without the use of a junction box or other enclosure at the point of splice or tap but only with nonmetallic cords and cables. There are many vendors of so-called manufactured temporary systems. These temporary systems consist of harnesses, cord sets, and power centers. A variety of cord sets are available for use with GFCI plug-in units to supply temporary lighting and receptacle outlets. Many varieties of portable power-distribution centers (some on wheels) and modules have been developed with integral GFCI breakers.

Q18.   *Are OSHA requirements for temporary wiring any different from NEC requirements?*

A.   Yes, they tend to be more rigorous since OSHA covers workplace safety. For example, lamps must be protected from accidental contact or breakage. OSHA requires a suitable metal or plastic guard on each lamp, mandatory grounding of metal lamp sockets due to the high exposure to shock hazard, and the use of metal shell sockets. The **NEC** prohibits used of both receptacles and lighting in the same temporary branch circuit on construction sites. The purpose is to provide complete separation of lighting so that operation of an overcurrent device or a GFCI due to a fault or overload of cord-connected tools will not simultaneously disconnect lighting. In renovation projects the branch circuits should be partitioned from the circuits that supply power to other parts of the building which are not undergoing renovation. Thus pigtail connections are traditionally associated with temporary power on the jobsite. As long as portable tools are being used in damp locations in close proximity with grounded building steel and other conductive surfaces, the possibility of shock exists from faulty equipment whether it is energized from temporary or from permanent circuits.

Q19.   *How long is "temporary" wiring considered temporary?*

A.   Generally speaking, 90 days. The **NEC** recognizes two conditions under which wiring is installed that is not intended to remain a permanent installation: special events and during construction. See Sec. 305-3, where the **NEC** indicates that temporary electrical power and lighting installations are permitted during the period of construction, remodeling, maintenance, repair, or demolition of buildings, structures, equipment, or similar activities. Ultimately, temporary wiring must be removed.

Q20.   *Does temporary wiring have to be approved by an electrical inspector?*

A.   Yes. In Sec. 305-2b it is indicated that the AHJ must approve temporary wiring methods based on the conditions of use and any special requirements of the temporary installation. The AHJ is involved because an approval, by definition, can be given only by the AHJ. The 90-day limit as in Sec. 305-3b is only for seasonal uses such as Christmas lighting; there is no such limit on construction sites.

Q21.   *Does temporary wiring need to be tested?*

A.   Yes. All cord sets and receptacles that are not part of the permanent wiring of the building or structure. Each receptacle and attachment plug must be tested for correct attachment of the equipment grounding conductor. All required tests must be performed:

1. Before first use on the construction site

2. When there is evidence of damage

3. Before equipment is returned to service following any repairs

4. At intervals not exceeding 3 months

The tests must be recorded and made available to the authority having jurisdiction. All equipment grounding conductors must be tested for continuity. Cord- and plug-connected equipment is required to be grounded.

Q22.   *What does the Code have to say about temporary metering for construction work sites?*

A.   Construction of service equipment to the trailers shall not be of a lower standard. Again, apart from the given permissions for variation from rules on permanent wiring, all temporary systems are required to comply in all other respects with Code rules covering permanent wiring. Refer to the questions in Division 16400.

2

Division 01000—General Requirements

Remarks. In this chapter we discuss standards related to the NEC and the generally accepted role of each of the allied trades that apply them. We conclude with a few questions that deal with the issue of power system reliability. Reliable power underlies many of the NEC requirements for safety, and quantitative information on the subject is sparse. Answers to these questions are most helpful at the early stages of conceptual design.

Q1.   *What NEC requirements commonly appear in Division 01000?*

A.   This division typically contains a summary of the work and an inventory of the drawings, specifications, and related documents. It will cite all the relevant standards which apply, of which the NEC is one among many. It may duplicate—or cover in greater detail—some of the material that appeared in the Standard General Conditions with respect to project commissioning, temporary facilities and controls, materials, and substitutions.

The NEC does not contain any rule that requires consideration for future expansion of electrical use. The NEC is concerned solely with safety: but the electrical designer must be concerned with safety, efficiency, convenience, good service, and future expansion. Often, electrical systems are designed and installed that exceed NEC requirements. However, the inspector does not have the authority to require installations to exceed the NEC requirements, unless adopted by local ordinance.

Q2.   *What is the purpose of the NEC?*

A.   The purpose of the Code is the practical safeguarding of persons and property from hazards arising from the use of electricity. The Code contains provisions considered necessary for safety. Compliance therewith and proper maintenance will result in an installation essentially free from hazard but not necessarily efficient, convenient, or adequate for good service or future expansion of electrical use. The NEC is not a "how-to" book, and it was not intended to be used as a design specification or an instruction manual [90-1(c)]. Because the Code is developed by a consensus of special-interest groups, the rules can be confusing, frequently conflict, or omit too much.

Q3.   *Where does the authority for enforcement of the National Electric Code originate?*

A.   In the amendments of the U.S. Constitution which grant policing powers to all states. It has no legal standing of its own until it is adopted as law by a jurisdiction, which may be a city, county, or state. Most jurisdictions adopt the NEC in its entirety; some adopt it with variations, usually more rigid, to suit local conditions. A few large cities have their own electrical codes which are basically similar to the **NEC**. Construction professionals must determine which code applies in the area of a specific project. Designers must not only provide that a system meets minimum **NEC** requirements but that it meets the *intent* of the **NEC**.

Q4.   *By what authority is the NEC a national standard?*

A.   The **NEC** is one of a constellation of standards underwritten by the *American National Standards Institute*. The American National Standards Institute (ANSI) has served in its capacity as administrator and coordinator of the United States private sector voluntary standardization system for 80 years. Founded in 1918 by five engineering societies and three government agencies, ANSI remains a private nonprofit membership organization supported by a diverse constituency of private and public sector organizations.

Q5.   *Does the ANSI organization itself develop standards?*

A.   No; it sets the standards for making standards by facilitating their development by establishing consensus among qualified groups. The institute ensures that its guiding principles—consensus, due process, and openness—are followed by the more than 175 distinct entities currently accredited under one of the federation's three methods of accreditation (organization, committee, or canvass). In 1996 alone the number of American National Standards increased by nearly 4 percent to a new total of 13,056 approved by ANSI. ANSI-accredited developers—such as the National Fire Protection Agency—are committed to supporting the development of national and in many cases international standards addressing the critical trends of technological innovation, marketplace globalization, and regulatory reform.

Q6.   *What is the scope of the NEC?*

A.   Principally, building premises wiring. It applies to electrical apparatus and systems in their manner of installation by construction professionals as well as in their use by the general public. Building premises wiring can be broken down further into three classifications: *services, feeders,* and *branch circuits.* Service wiring involves interconnection with the utility power source; branch circuit wiring typically involves getting power to 120-V single-phase lights and receptacles; and feeders involve all the wiring *between* the service and the branch circuits. By comparison, electrical apparatus farther upstream from the building service will be built according to National Electric Safety Code (NESC), which covers provisions for safeguarding of persons from hazards associated with electric supply stations. Figure 2.1 offers an overview of the systems to which the NESC applies. The **NEC** covers building premises wiring. The NESC covers utility wiring. OSHA 1926 Subparts V; OSHA 1926 Subpart K; OSHA 1910.269 Subpart R; OSHA Subpart S covers workplace safety no matter which side of the service point the work takes place.

Figure 2.1 Facilities where the NESC applies. (*Adapted from State of California Public Utilities Commission report, "Electric Power Reliability."*)

Premises Wiring.

That interior and exterior wiring, including power, lighting, control, and signal circuit wiring together with all of their associated hardware, fittings, and wiring devices, both permanently and temporarily installed, that extends from the service point of utility conductors or source of a separately derived system to the outlet(s). Such wiring does not include wiring internal to appliances, fixtures, motors, controllers, motor control centers, and similar equipment. An example is an on-premises privately owned transformer supplying all building loads. The transformer would be considered part of the premises wiring system.

SEPARATELY DERIVED SYSTEM.

A premises wiring system whose power is derived from a battery, a solar photovoltaic system, or a generator, transformer, or converter windings, and that has no direct electrical connection, including a solidly connected grounded circuit conductor, to supply conductors originating in another system. Its theoretical basis is the Kirchhoff equations which define a closed electrical energy system. All current supplied by a source must return to that source.

Q7. *What is the difference between the NEC and the NESC?*

A. While the **NEC** and the NESC both stand alone and primarily cover different areas, they have much in common and both refer to one another. The NESC originates with the electric power and telecommunication industry. The NFPA originates with the fire safety community. As shown in Fig. 2.1, the NESC covers what might be called municipal wiring systems (street lights, bulk distribution grids, PBX wiring, etc.), and the NFPA covers the wiring within buildings.

Q8. *What does the NEC not cover?*

A. Generally speaking, the **NEC** does not cover things that move. Installations in ships, watercraft other than floating buildings, railway rolling stock, aircraft, or automotive vehicles (other than mobile homes and recreational vehicles) have standards of their own.

The **NEC** does not cover installations under the exclusive control of electric utilities for the purpose of communications, metering, generation, control, transformation, transmission, or distribution of electric energy. Such installations are typically located in buildings used exclusively by utilities for such purposes; outdoors on property owned or leased by the utility; on or along public highways, streets, roads, etc.; or outdoors on private property by established rights such as easements.

Another class of system that is not covered is the apparatus and systems of transportation, telecommunication, and electric power companies. Railroad yard warehouses will not be covered installations of railways for generation, transformation, transmission, or distribution of power used exclusively for operation of installations used exclusively for signaling and communications purposes.

Obviously, it does not cover those installations that have adopted another safety standard. The U.S. military, for example, has a group of standards of its own. A few major municipalities in the United States have electric codes of their own which bear a remarkable resemblance to the code produced by the NFPA and unwritten by ANSI.

Q9. *Where are the gray areas?*

A. In the boundary between the utility and the general public. While the physical region of this boundary may be getting narrower, the practical limit of the boundary is getting wider. This is largely due to the controlled divestiture of traditional utilities of their monopoly status. One of the many reasons the **NEC** needs to be updated every 3 years is that economic incentives are now in place that allow private enterprise to install electrical systems virtually identical to the electrical systems of traditional utilities. So-called behind the fence generation is one of them, large-scale medium-voltage bulk distribution systems being another.

Street lighting is a good example. You may now own your own lighting fixture and put it on the utility power pole and use utility electricity. It *does* cover installations in buildings used by the electric utility, such as office buildings, warehouses, garages, machine shops, and recreational buildings that are not an integral part of a generating plant, substation, or control center.

Q10. *Is the **NEC** a design standard?*

A. It tries not to be, though arguably it dominates design because of what it says about what a designer *cannot* do. When considering the design of an electrical system, the engineer must consider alternate design approaches which best fit the following overall goals: safety, minimum first cost and minimum maintenance, maximum reliability, flexibility, and expandability, and power quality. It is rare that designers have all the information they need to design a system or part of a system. As design professionals they must expand available information on the basis of experience with similar problems. The engineer should know whether certain loads function separately or as an assembled unit, the demand profile separately and in aggregate, the rated voltage of the devices, their physical location with respect to each other and with respect to the source, and the possibility of relocation of the load devices and addition of loads in the future.

Q11.   *Who interprets the NEC?*

A.   It is the responsibility of the local authority enforcing the Code to interpret the specific rules of the Code. The authority having jurisdiction may waive specific requirements in this Code or permit alternate methods where it is assured that equivalent objectives can be achieved by establishing and maintaining effective safety. The second paragraph of Sec. 90-4 is included to allow the AHJ the option of permitting alternative methods where specific rules are not established in the Code. For example, this allows the local authority to waive specific requirements in industrial occupancies, research and testing laboratories, and other occupancies where the specified type of installation is not contemplated in the Code rules. Some localities do not adopt the **NEC**, but even in those localities installations meeting the current Code are *prima facie* evidence that the electrical installation is safe.

Q12.   *How is the NEC enforced?*

A.   The **NEC** is intended to be suitable for mandatory application by governmental bodies exercising legal jurisdiction over electrical installations and for use by insurance inspectors. The authority having jurisdiction for enforcement of the Code will have the responsibility for making interpretations of the rules, for deciding upon the approval of equipment and materials, and for granting the special permission contemplated in a number of the rules. Section 90-4 advises that an authority must grant approval for all materials and equipment used under the requirements of the

Code in its area of jurisdiction. The terms "approved," "identified," "listed," and "labeled" are intended to provide a basis for the authority having jurisdiction to make the judgments that fall within its area of responsibility.

APPROVED.

Acceptable to the authority having jurisdiction. The phrase "authority having jurisdiction" is used in NFPA documents in a broad manner since jurisdictions and "approval" agencies vary, as do their responsibilities. Where public safety is primary, the authority having jurisdiction may be a federal, state, local, or other regional department or individual such as a fire chief, fire marshal, chief of a fire prevention bureau, labor or health department official, building official, electrical inspector, or other having statutory authority. For instance, an insurance inspection department, rating bureau, or other insurance company representative may be the authority having jurisdiction. In many circumstances, the property owner or owner's designated agent assumes the role of the authority having jurisdiction; at government installations, the commanding officer or departmental official may be the authority having jurisdiction.

IDENTIFIED (AS APPLIED TO EQUIPMENT).

Recognizable as suitable for the specific purpose, function, use, environment, application, etc., where described in a particular Code requirement. Suitability of equipment for a specific purpose, environment, or application may be determined by a qualified testing laboratory, inspection agency, or other organization concerned with product evaluation. Such identification may include *labeling* or *listing*.

Q13.   *Who is the authority having jurisdiction?*

A.   Some common titles for the AHJ are electrical inspector, chief electrical inspector, fire marshal. The **NEC** committee is not responsible for subsequent actions by authorities enforcing the **NEC** that accept or reject its findings. The AHJ has responsibility for interpreting the Code rules and should attempt to resolve all disagreements at the local level. There are remarkably few problems with the **NEC** as written but not so few that meeting every 3 years is not justified.

Q14.   *What role does the electrical inspector have?*

A.   The following material, adapted from the trade journals of the *International Association of Electrical Inspectors,* sums it up best. An inspector's authority and responsibilities include:

1.   Interpretation of the **NEC** rules. Electrical inspectors do not have the authority to require installations to exceed the requirements of the **NEC**.

2.   Approval of equipment and materials

3.   Granting of special permission.

4.   Waiver of specific requirements of the Code or permit alternate methods.

5.   Waiver of new Code requirements on materials.

6.   Ensure that equipment is installed properly.

7.   Detect any modification of the equipment by the electrician.

Q15.   *What are issues over which the AHJ may typically grant a variance in* **NEC** *requirements?*

A.   Frequently, where a local application of common sense will allow designers to use different demand factors. In Sec. 430-26, for example, the AHJ is allowed to grant permission to allow a demand factor of less than 100 percent where operational procedures, production demands, or the nature of the work is such that not all the motors are running at one time. Granting the AHJ permission to do this will often result in a reduction of construction costs.

Q16.   *Where do I find the final authority in matters of interpretation of the Code?*

A.   First, you must be careful to distinguish between needing a final authority on interpretation of the intent of the Code and needing a determination about whether a particular installation meets that intent.

To promote uniformity of interpretation and application of the provisions of the **NEC,** the NFPA has established formal interpretation procedures. These procedures are found in the "NFPA Regulations Governing Committee Projects," which may be obtained from the NFPA office in Quincy, MA.

Two general forms of formal interpretations are recognized: (1) those making an interpretation of the literal text, (2) those making an interpretation of the intent of the committee at the time the particular text was issued. Formal interpretations of the Code rules are published in the *NFPA Fire News, the NFPA Electrical Section News Bulletin (Current Flashes)*, and the *National Fire Codes Subscription Service,* and are sent to interested trade publications.

Most interpretations of the **NEC** are rendered as the personal opinion of the NFPA electrical department staff or of the involved member of the **NEC** committee, because the request for interpretation does not qualify for processing as a formal interpretation in accordance with the Regulations Governing Committee Projects for the **NEC** committee policy. Such opinions are rendered in writing only in response to written requests. The responses are identified as personal opinions and indicate that the opinion shall not be considered the official position of the NFPA or of the **NEC** committee and shall not be considered to be nor be relied upon as a formal interpretation.

Q17. *What is the difference between a chapter, an article, and a section?*

A. Many sections make up an article. Many articles make up a chapter. The **NEC** is divided into the Introduction and nine chapters. Chapters 1, 2, 3, and 4 apply generally; Chaps. 5, 6, and 7 apply to special occupancies, special equipment, or other special conditions. Any provisions in Chaps. 5 to 9 take precedence over contrary provisions in the first four chapters of the **NEC** per Sec. 90-3. The latter chapters supplement or modify the general rules. In other words, Chaps. 1 through 4 apply to all electrical systems unless they are amended by requirements that appear in Chaps. 5,

6, and 7. Chapter 8 covers communications systems and is independent of the other chapters except where they are specifically referenced therein. Chapter 9 consists of tables and examples.

Q18.   *Which edition of the NEC applies?*

A.   The Code may require new products, constructions, or materials that may not yet be available at the time the Code is adopted. In such event, the authority having jurisdiction may permit the use of the products, constructions, or materials that comply with the most recent previous edition of this Code adopted by the jurisdiction. This paragraph permits the AHJ to waive a new Code requirement during the interim period between the acceptance of a new edition of the NEC and the availability of a new product, construction, or material redesigned to comply with the increased safety required by the new Code edition.

It is difficult to establish a viable future effective date in the NEC because the time needed to change the existing products and standards and to develop new materials and test methods is not usually known at the time of adoption of the new requirement in the NEC. When the prevailing edition of the NEC is not the edition of the NEC cited in the bidding documents, follow the more rigorous version of the Code. Often bidding documents will say that the electrical contractor must confirm the version of the NEC in effect at the time of bidding.

Q19.   *How are midcycle changes to the NEC handled?*

A.   There are not very many of these, though the need to amend a part of a standard is a normal part of business in the production of a national consensus document. When a change in the middle of the 3-year production cycle of the NEC needs to be made, the NFPA will publish a *Tentative Interim Amendment* (TIA) in its own journals, in its newsletters, on its web site, and in the trade journals. When a standard needs to be released while appeals on the content of the standard are in process, the first page will contain a notice explaining the nature of the appeal. Errata to the first printing of the 1999 NEC appear in the Appendix. The first printing of the NEC is identified by the number 1 on the far right in the line of numbers at the bottom of the copyright page.

Q20.   *What are the principal organizations involved in setting standards for electrical systems in buildings?*

A.   It may seem as if there are many more than there actually are. The NFPA is only one of the principal "players" in the U.S. safety community; Factory Mutual and Underwriter's Laboratories are among others. As a leading developer of safety standards, UL publishes its standards for the benefit of the entire safety community. Of the more than 726 UL Standards for Safety currently in use, nearly 80 percent have been approved as American National Standards by the American National Standards Institute (ANSI). Factory Mutual, with its roots in the early years of U.S. industrial history when many industrial accidents were due to criminal negligence, is another key player. Along with the IEEE and the IEE, these organizations form the technical literature of our trade.

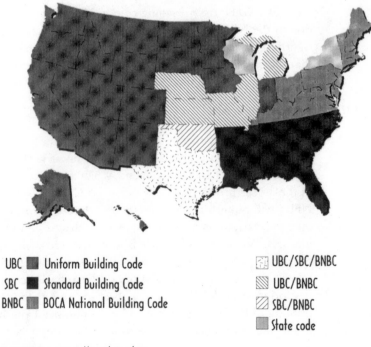

UBC ■ Uniform Building Code

SBC ■ Standard Building Code

BNBC ■ BOCA National Building Code

▨ UBC/SBC/BNBC

▨ UBC/BNBC

▨ SBC/BNBC

■ State code

Figure 2.2 Current U.S. Building Code Boundaries.

Building codes throughout the United States are remarkably similar in their concentration on fire safety. Where they differ is on the basis of geography, climate, available building materials, and economic and cultural conditions (see Fig. 2.2).

Q21. *Why is it so important to understand how the NEC defines a building?*

A. Because it is essential to defining the service drop and thereby the amount of on-site generation that may be necessary and whether feeders from an adjacent building may be brought over. The firewall is the key thing to determine. The **NEC** has a specific definition, but in some facilities this may be difficult to determine. Industrial and multibuilding complexes (for example, some universities) often include substations and other installations employing construction wiring similar to electric utility installations. Although such nonutility installations are within the scope of the **NEC,** its requirements may not always be sufficient or complete, for example, in such cases as clearances of conductors or clearance from building or structures for nominal voltages over 600 V. The identification of service entrance and independent power source may be further complicated with the presence of cogeneration facilities.

BUILDING.

A structure that stands alone or that is cut off from adjoining structures by fire walls with all openings therein protected by approved fire doors. A building is a structure that may be used or intended for supporting or sheltering any use or occupancy. It may also be a separate structure such as a pole, billboard sign, or water tower. Definitions of the terms "fire walls" and "fire doors" are the responsibility of the municipal and/or state building codes, and interpretations of building terms have been avoided by **NEC** committees. "Fire resistance rating" is defined as the time, in minutes or hours, that materials or assemblies have withstood a fire exposure.

Q22.   *How do OSHA requirements link with NEC requirements?*

A.   The requirements for electric apparatus in a workplace environment are more rigorous than in the Code. Much of the OSHA rules have been adapted from the **NEC**. This makes sense because a private standard can be changed to meet technological and legal developments faster than a federal standard (OSHA-29CFR-Part 1910-SUBPART "S" 1910.399).

Since a construction site is a workplace—and a hazardous location—the OSHA requirements bear heavily upon temporary wiring and accessibility issues cited in the **NEC**.

Q23.   *Why is the U.S. national standard for electrical system safety published by a fire safety organization?*

A.   Because of the manner in which illumination systems developed in the last few decades of the 1800s. These systems were based upon arc and gas technologies and were notorious for causing fires. With the incandescent electric light made practical and affordable by Swan and Edison, a great improvement was made in fire safety. Edison convinced the New York Board of Fire Underwriters of the safety of the incandescent lamp in 1881, and 6 years after the National Electric Code was born, originally published by the same board.

Q24.   *How many fires in the United States are caused by electricity?*

A.   One hundred years after the first edition of the **NEC** there are still about 56,200 building structure fires in the United States that can be attributed to electrical distribution systems. That ranks fifth among the 12 leading causes of fires (at about 9 percent of the total). It would be fair to conclude that without the widespread acceptance of the **NEC** there would be many more fires, because in the late 1890s there were many fires associated with electrical systems—"new" technology at the time.

Q25.   *What is the difference between "listed" and "labeled"?*

A.   An important distinction which requires a somewhat lengthy answer.

Labeled equipment is equipment or materials to which has been attached a label, symbol, or other identifying mark of an organization that is acceptable to the AHJ and is concerned with product evaluation that maintains periodic inspection of production of labeled equipment or materials and by whose labeling the manufacturer indicates compliance with appropriate standards or performance in a specified manner. Equipment and conductors required or permitted by the Code are acceptable only where approved for a specific environment or application by the AHJ. Listing or labeling by a qualified testing laboratory will provide a basis for approval.

Listed equipment is equipment or materials included in a list published by an organization acceptable to the authority having jurisdiction and concerned with product evaluation, which maintains periodic inspection of production of listed equipment or materials, and whose listing states either that the equipment or material meets appropriate designated standards or has been tested and found suitable for use in a specified manner. The means for identifying listed equipment may vary for each organization concerned with product evaluation, some of which do not recognize equipment as listed unless it is also labeled. The authority having jurisdiction should utilize the system employed by the listing organization to identify a listed product.

Q26.   *How do I get my equipment listed and labeled?*

A.   There are consulting companies that may be retained to determine whether a given apparatus or installation is listed by a testing laboratory. Installation instructions are usually supplied with equipment by the manufacturer for use by the general contractor and others concerned with the installation. For example, hot tubs may have controllers that need to be installed according to manufacturer instructions in order for them to perform as designed. It is the field installation that must be coordinated with the equipment. Section 110-3 does not in itself require listing or labeling. It does, however, require considerable evaluation of equipment and it must be acceptable to the AHJ. Before issuing the permit, the AHJ may require evidence of compliance with manufacturer field installation requirements. The most common form of this evidence is a listing or labeling by a third party. In some cases a third party consultant is retained.

Q27.   What is Underwriter's Lab?

A.   A major player in the U.S. safety community. UL Standards contain the technical requirements used to evaluate products for conformity with consensus performance standards. UL works with U.S. Customs to eliminate counterfeiting of the UL mark and protect consumers from unsafe products and UL clients from illegal competition. Products that have already been found to bear counterfeit marks include cord sets and power supply cords, current taps, ceiling fans, electric fans, incandescent fixtures, recessed incandescent fixtures, portable lamps, night lights, work lights, receptacles, ground-fault circuit interrupters (GFCIs), power taps, transformers, transient voltage surge suppressors, flexible cords, and decorative light strings.

UL Standards are developed with the input of many other affected parties including manufacturers, jurisdictional authorities, code developers, and others. This open process of standards development creates effective product safety requirements which manufacturers use in designing and producing safer products. UL Standards are designed to be compatible with the U.S. National Electrical Code and other nationally recognized installation, building, and safety codes.

Q28.   Can conformity assessment be provided in the field?

A.   Yes. Organizations such as UL and ETL evaluate installed products in the field that don't have a UL listing mark or classification marking on them or have been significantly modified since manufacture. Field inspection service applies specifically to products that were eligible to carry a UL mark at the time of manufacture, but for some reason a UL mark was not applied. In this case, the manufacturer of the product makes a request for a field inspection. A UL field representative visits the equipment site and determines if the product meets applicable requirements. If it does, the manufacturer can immediately apply the authorized listing mark or classification marking to the product—under the supervision of the UL field representative. The appropriate regulatory authority is notified of the results.

Q29.   *Does the NEC have anything to say about power system reliability?*

A.   Nothing explicit, though it does to the extent that the practical safeguarding of persons involved with electrical energy depends upon reliable power. Safety requirements are rooted in good design, and good design takes the reliability into consideration.

The level of debate on power system reliability has been raised in recent years now that so much of our economic health depends upon power delivery systems. For many years the reliability problem was solved, more or less, in an intuitive manner. It wasn't too difficult to understand that two services to a building would make power more available than a single service.

Complex electrical systems require a more quantitative treatment. A glimpse of quantitative treatment is given in the Solved Problem that appears at the end of this chapter. It affirms, in a somewhat more crude fashion, the data gathered from 41 plants in the United States and Canada. See Tables 2.1 through 2.4 which have been taken from the IEEE Standard 493-1990, Design of Reliable Industrial and Commercial Power Systems. We treat the issue of power system reliability here because the topic of reliability should prevail throughout the design, construction, and operation of a power system. Related questions appear in Chap. 19, in particular, the comparative reliabilities of on-site generators versus utility service drops. Appendix Item A-5 cites IEEE reliability data for various types of electrical equipment installed in industrial plants. More recent data is available from IEEE in updated versions of IEEE Standard 493.

TABLE 2.1 AVERAGE COST OF POWER INTERRUPTIONS FOR INDUSTRIAL PLANTS

All plants	$4.69/kW + $6.65/kWh
Plants > 1000 kW max demand	$2.60/kW + $2.33/kWh
Plants < 1000 kW max demand	$11.38/kW + $20.11/kWh

TABLE 2.2 MEDIAN COST OF POWER INTERRUPTIONS FOR INDUSTRIAL PLANTS

All plants	$1.71/kW + $2.06/kWh
Plants > 1000 kW max demand	$0.79/kW + $0.89/kWh
Plants < 1000 kW max demand	$9.13/kW + $10.96/kWh

TABLE 2.3 AVERAGE COST OF POWER INTERRUPTIONS FOR COMMERCIAL BUILDINGS

All commercial buildings	$14.65/kWh not delivered
Office buildings only	$17.99/kWh not delivered

TABLE 2.4 COST OF POWER INTERRUPTIONS AS A FUNCTION OF DURATION FOR OFFICE BUILDINGS (WITH COMPUTERS)

| Power interruptions | Sample size | Maximum | Cost/peak kWh not delivered | |
			Minimum	Average
15 min duration	14	$45.11	$3.82	$18.05
1 h duration	16	$50.61	$3,82	$16.85
Duration > 1 h	10	$137.35	$0.32	$19.91

Q30.   *How can we evaluate the cost of a power interruption?*

A.   The most comprehensive study of the cost of power interruptions was undertaken by the IEEE in 1973. Even then, the data was hard to come by. Tables 2.1 through 2.4 offer an overview. The cost figures date back to July 1987, and reader is cautioned that because these figures are based upon a limited sample set they should be used for power of 10 or "order of magnitude" cost comparisons only. If the data do nothing but identify a need for every organization to keep the records of power outages, then the challenge of the IEEE will have been met.

One method for estimating the cost of a power outage involves an assumption about the kilowatt demand for each person using a facility. You might assume, for example, that each person demands about 1 kW of electricity, a figure that resembles the peak U.S. electrical demand divided by a population of 250 million, but one that will vary regionally and according to facility class. In a commercial or industrial setting it is possible to make similar ballpark estimates about the cost of lost labor based upon the duration of the outage. The use of kilowatt demand is a handy quantity because it can easily be read from a meter at the moment the outage occurs.

Miscellaneous

Q31.   *Does the NEC have a requirement for marking underground cable after it has been buried?*

A.   Not after it is in the ground, only as you are putting it in the ground. Refer to the generic rules for color coding that appear in Arts. 200, 210, and 215. Another color code rule appears in Sec. 230-56 when you are dealing with a four-wire delta-connected service where the midpoint of one phase winding is grounded. (You will need to mark the high leg with some kind of permanent brown paint or tape.)

The **NEC** requires warning ribbon to be fastened to the top of the bank before the trench is filled in. After the trench is filled, you take your chances. There are no rules that say you must permanently mark a cable route so that it can be seen aboveground. After all, hiding the cable is one of the reasons you put it in the ground in the first place.

Some European countries, such as the United Kingdom, post signs above buried cables to warn gardeners, for example, of the danger of direct-buried circuits. The signs are just like the yellow signs used for traffic control and they have arrows pointing downward. Some local governments in the United States, recognizing the hazard and losses associated with people taking such chances in excavation, have passed legislation to prevent such misfortunes.

The sidebar on this page is an example of one state's effort. The state of Virginia, for example, levies fairly serious penalties against excavators who fail to follow the notification procedure and cause damage to underground infrastructure. Readers should consult with local authorities to determine if a similar color coding method had been adopted in your neighborhood or the neighborhood of your client.

Color Coding of Underground Utilities

Private utility and type of product	Identifying color or equivalent
Electric power distribution and transmission	Safety red
Municipal electric systems	Safety red
Gas distribution and transmission	High-visibility safety yellow
Oil and petroleum products distribution and transmission	High-visibility safety yellow
Dangerous materials, product lines, steam lines	High-visibility safety yellow
Telecommunications systems	Safety alert orange
Police and fire communications	Safety alert orange
Cable television	Safety alert orange
Water systems	Safety precaution blue
Slurry systems	Safety precaution blue
Sewer systems	Safety green
Proposed excavation	White

Excerpt from the state of Virginia's *"Underground Utility Damage Prevention Act."*

Q32. *What are the key changes to the 1999 NEC that could have an effect upon the design of a building site?*

A. Here are a few items that are likely to have the most profound effects upon site design and construction:

- Where more than one building or other structure is on the same property and under single management, each building serviced shall be supplied by one feeder or branch circuit unless permitted in exceptions that appear in new Part B of Art. 225. Both Part B and Part C of Art. 225 contain many modifications and clarifications of **NEC** rules for low- and medium-voltage systems for multibuilding facilities with underground distribution circuits (covered in more detail in this book in Chap. 22).

- Outside feeder taps do not have to be protected at the supply as long as they comply with conditions stated in Sec. 240-21(b)(5). Although this may seem to be an exclusively electrical design problem, it may have an effect on whether or not switchgear needs to be located on a multibuilding facility site.

- Below-grade wiring that is subject to movement by settlement or frost must be arranged to prevent damage to conductors and equipment [Sec. 300-50(b)].

- It is now permissible to install electrical nonmetallic tubing (ENT) at the bottom of a floor slab as long as it is fully embedded in the slab (Sec. 331-3).

- A new table has been provided for determining the bending radius of nonmetallic underground conduit. This may affect the dimensions and orientation of a trench (Sec. 343-10).

- Some new rules and clarification of older rules appear in Arts. 640 and 680 for wiring near pools. These rules may affect how you lay out the pool perimeter amenities and where you put the panelboard.

- The protection of exterior network-powered broadband communication cables by installing them in raceway at least 8 ft above grade may have an effect upon how you design the service entrance unless you use listed fault-protection devices (Secs. 830-10 and 11).

■   The new requirement for the telecommunication primary protector to be as close as practicable to the point of entrance may have significant effect on the siting of electric and telecommunication services (Sec. 830-30).

For details on each of the foregoing, the reader is referred to the core references.

Solved Problem

SITUATION

A 10,000 sq ft office building has an average demand of 50 kW. Power is lost for 1 h in the middle of the work day.

REQUIREMENTS

What is the cost of this power outage in terms of lost labor alone? Assume the nature of business requires an average labor rate of $25 per hour for every employee.

SOLUTION

The 50-kW demand is based upon an average of 5 W/sq ft, slightly higher than the design demand figures that appear in Chap. 2 of the **NEC.** To get an idea of how large a 10,000 sq ft facility is, a typical fast-food restaurant leaves a footprint of about 5000 sq ft; thus the facility under consideration has a footprint of roughly double this size.

While it is true that a given building will always have a fixed minimum electrical demand, the local building code will have already established occupancy rules for the number of people that may occupy a 10,000 sq ft building (ranging from 10 to 100 sq ft per person depending upon occupancy class). This admittedly loose correlation may allow us to produce a power of 10 estimate for the cost of the outage using an assumed demand of 1 kW per person. Thus, using the data given, the cost of lost labor alone is

$$50 \text{ persons} \times \$25 \text{ per hour} = \$1250$$

Of course, the underlying assumption here is that no recovery time is required, that no data has been lost such that the building occupants need a full day to recover data.

By way of comparison with the IEEE methodology, using the 1987 data in Table 1.4 we have $17.99/kWh for the 50 kW not delivered for 1 h.

$17.99 × 50 kw × 1 h = $899.50 in 1987 dollars. If you assume that 1999 dollars are worth at least 1250/899 → 40 percent more than 1987 dollars, then the cost of the power outage computed by either method is identical. It would be reasonable to conclude that the cost of an 8-h power outage to this business is on the order of $10,000 per day.

REMARKS

The best that the IEEE reliability work group could come up with is a methodology that would yield a power of 10 estimate. This fact alone should help build some tolerance for the loose network of underlying assumptions in this example. In the finest sense of the term, it is a quick and dirty estimate of an important engineering economic problem.

The 1 kW per person (or 0.5 or 1.5 or 3.0 kW per person) method illustrated here is an effective method for adding a human element into the power loss cost equations. The method may be even more effective at the utility distribution level, where you might want to know how many people are affected by a medium-voltage feeder that was supplying 10,000 kW. By inference, 10,000 people are affected by the outage. Conversely, when it is reported that 100,000 people are without power, you may assume that the load loss is on the order of 100 MW.

At the very least, this example should demonstrate the importance of having good data on power outage history when undertaking the design of an electrical system. A related discussion on the comparative reliabilities of on-site generators and standby utility service drops is in Chap. 19.

SERVICE EQUIPMENT.

The necessary equipment, usually consisting of a circuit breaker or switch and fuses, and their accessories, located near the point of entrance of supply conductors to a building or other structure, or an otherwise defined area, and intended to constitute the main control and means of cutoff of the supply.

3

Division 02000—Site Requirements

Remarks. The CSI specification format is orga-
nized in a fashion that resembles the order in
which actual building construction proceeds.
Having covered contract, regulatory, and general business issues in Divisions 00000 and
01000, we come to the first division with technical substance. Our focus will be upon anything electrical that affects the jobsite: overhead
and underground power and communications;
duct and switchgear construction; and landscaping. We cover the subject of transformers
just enough to make a determination about
whether the service should be located on the
site or in the building interior. Other **NEC**
requirements for transformers will be covered
in Division 16400.

Q1.   *What NEC requirements commonly appear in Division 02000?*

A.   Division 02000 contains requirements for safety in working with existing and new utilities, site preparation, demolition, excavation, etc. In some variants of the CSI specification format, you may see a reprise of temporary electric service and temporary elevators that is covered in Division 02000. In a typical electrical construction contract the requirements for placement of the transformer, switchgear, supply conduit or poles, site lighting, and signage will be located here. There may be modifying or related requirements in Division 03000 and Division 15000 and certainly in Division 16000. Some contracts place the construction of duct banks in Division 03000 (Concrete) and Division 16000. Underground circuits operating at less than 600 V are covered in Sec. 300-5; circuits operating at 600 V and above are covered in new Sec. 300-50. Transformer vaults are covered in Art. 450.

Q2.   *Do any NEC rules apply during demolition?*

A.   Yes. You can extend the general rules for the "practical safeguarding of persons and property from the hazards arising from the use of electricity" to the demolition phase of a project. A construction site is a hazardous location. As such you need to verify that utilities have been disconnected and sealed before demolition commences. You might also look into whether mechanical vibrations from a wrecking ball might trip nearby electromechanical relays or whether cranes are safe from swinging into overhead power lines or whether a power line will not fall upon a temporary construction fence.

Q3.   *Do any NEC rules apply during excavation?*

A.   Yes. Again we cite the stated purpose of the NEC in Art. 90-2. If trenches intended for underground duct systems are not protected appropriately, contractors can be assessed heavy penalties by OSHA and others. The general idea is to keep excavated material far enough away from the trenches so that it does not cause the surrounding compacted earth to crumble while electricians are still working in them. Per Sec. 370-29, conduit bodies, junction, pull, and outlet boxes must be so installed that the wiring contained in them can be rendered accessible without removing any part of the building or, in underground circuits, without excavating sidewalks, paving, earth, or other substance that is to be used to establish the finished grade. Pull boxes must be handled in the same fashion according to Sec. 426-41, Pull Boxes. Neither of these may affect your particular job unless the last contractor failed to comply with this requirement.

UNDERGROUND ELECTRIC POWER TRANSMISSION

Q4. *What are the basic* **NEC** *rules for underground circuit installations?*

A.   The bulk of them appear in Sec. 300-5 and new Sec. 300-50. A distinction is made between cables buried directly in the ground and cables pulled through underground raceway (with or without concrete encasement). Table 3.1 lays out the rules for direct buried cable or conduit or other raceways. These are minimums. In many jurisdictions, the owner has no choice but to install new power lines underground. Much depends upon the urban development philosophy of the host community. Engineers and contractors should confirm that reciprocal codes do not require reciprocal standards that are more rigorous than the minimums cited in this section of the **NEC**. It should be plain from the tables and figures accompanying this question that much depends upon the voltage level, the cable type, and the location itself. The general rules are easier to understand with diagrams (Figs. 3.1, 3.2).

Table 3.1 has been expanded in the 1999 **NEC** to include tabulations of burial dimensions that formerly appeared as text elsewhere.

Q5. *Can an underground cable be installed under a building?*

A.   Yes. Section 300-5 says that underground cable installed under a building must be in a raceway that is extended beyond the outside walls of the building. Conductors entering a building must be protected to the point of entrance. Where the enclosure or raceway is subject to physical damage, the conductors must be installed in rigid metal conduit, intermediate metal conduit, Schedule 80 rigid nonmetallic conduit, or equivalent.

TABLE 3.1 Minimum Cover Requirements, 0 to 600 V, Nominal, Burial in Inches

Location of wiring method or circuit	Type of wiring method or circuit				
	Column 1, direct burial cables or conductors	Column 2, rigid metal conduit or intermediate metal conduit	Column 3, nonmetallic raceways listed for direct burial without concrete encasement or other approved raceways	Column 4, residential branch circuits rated 120 V or less with GFCI protection and maximum overcurrent protection of 20 A	Column 5, circuits for control of irrigation and landscape lighting limited to not more than 30 V and installed with type UF or in other identified cable or raceway
All locations not specified below	24	6	18	12	6
In trench below 2-in-thick concrete or equivalent	18	6	12	6	6
Under a building	0 (in raceway only)	0	0	0 (in raceway only)	0 (in raceway only)
Under minimum of 4-in-thick concrete exterior slab with no vehicular traffic and the slab extending not less than 6 in beyond the underground installation	18	4	4	6 (direct burial) 4 (in raceway)	6 (direct burial) 4 (in raceway)

Location of Wiring Method or Circuit	Column 1	Column 2	Column 3	Column 4	Column 5
Under streets, highways, roads, alleys, driveways, and parking lots	24	24	24	24	24
One- and two-family dwelling driveways and outdoor parking areas, and used only for dwelling-related purposes	18	18	18	12	18
In or under airport runways, including adjacent areas where trespassing prohibited	18	18	18	18	18

Cover is defined as the shortest distance in inches measured between a point on the top surface of any direct-buried conductor, cable, conduit, or other raceway and the top surface of finished grade, concrete, or similar cover.

Notes:

1. For SI units, 1 in = 25.4 mm.
2. Raceways approved for burial only where concrete encased shall require concrete envelope not less than 2 in thick.
3. Lesser depths shall be permitted where cables and conductors rise for terminations or splices or where access is otherwise required.
4. Where one of the wiring method types listed in columns 1 to 3 is used for one of the circuit types in columns 4 and 5, the shallower depth of burial shall be permitted.
5. Where solid rock prevents compliance with the cover depths specified in this table, the wiring shall be installed in metal or nonmetallic raceway permitted for direct burial. The raceways shall be covered by a minimum of 2 in of concrete extending down to rock.

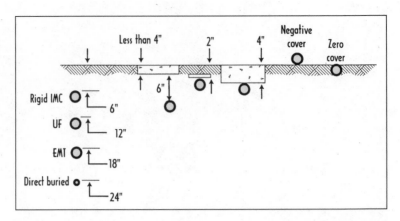

Figure 3.1 Minimum cover requirements, low voltage. Pictorial description of **NEC** requirements without concrete encasement.

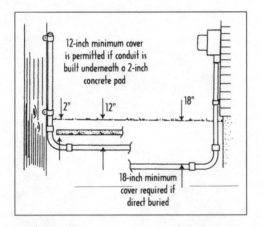

Figure 3.2 Minimum cover requirements for residential service. (Adapted from **NEC** Handbook.)

Q6.   *Can wires be run underneath a concrete driveway or a sidewalk?*

A.   Yes. A common example is the branch circuit to a residential site lighting (Fig. 3.3). Conductors under residential driveways must be at least 18 in below grade. If the conductors are protected by an overcurrent device rated at not more than 20 A and provided with ground-fault protection for personnel, the burial depth may be reduced to 12 in. Figure 3.3 shows minimum requirements for branch circuit wiring on a residential lot. More general rules for running branch circuit and feeder wiring underground appear in Art. 225.

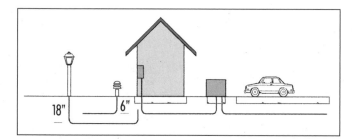

Figure 3.3 Underground branch circuit wiring on residential lots. Rules for low-voltage cable run to landscape lighting and lawn sprinkler controls.

Q7.   *Can we tap an existing direct buried branch circuit?*

A.   Yes. Not only that, the **NEC** allows cables to be spliced or tapped without the use of splice boxes as long as they follow the rule for electrical connections that appears in Sec. 110-14. Make sure that devices such as pressure terminal or pressure splicing connectors and soldering lugs are listed for the purpose. Materials such as solder, fluxes, inhibitors, and compounds should be of a type that will not adversely affect the conductors, installation, or equipment.

Q8.   *When an underground direct burial circuit is made up of conductors in multiple, must all the conductors be installed in the same trench?*

A.   Yes. All conductors making up a direct burial circuit of single conductors in parallel must be run in the same trench and must be "in close proximity." This rule just doesn't apply to underground circuits but is a general rule for all electrical wiring that will be discussed in detail in Division 16400.

Table 3.2 is new to the 1999 **NEC.** You may recall seeing it as the old Table 710-4(b) located in the old Art. 710. That article has since been deleted from its previous location and has reemerged as new Art. 490 in the 1999 **NEC.**

Q9.   *Can we pull low-voltage conductors through a medium-voltage underground raceway?*

A.   Yes, as long as the conductors do not exceed the percentage conduit fill requirements. This condition is quite common in network

TABLE 3.2 MINIMUM COVER REQUIREMENTS

Circuit voltage	Direct-buried cables	Rigid nonmetallic conduit approved for direct burial*	Rigid metal conduit and intermediate metal conduit
Over 600 V through 22 kV	30	18	6
Over 22 kV through 40 kV	36	24	6
Over 40 kV	42	30	6

Cover is defined as the shortest distance in inches measured between a point on the top surface of any direct-buried conductor, cable, conduit, or other raceway and the top surface of finished grade, concrete, or similar cover.

Note: For SI units, 1 in525.4 mm.

*Listed by a qualified testing agency as suitable for direct burial without encasement. All other nonmetallic systems shall require 2 in (50.8 mm) of concrete or equivalent above conduit in addition to above depth.

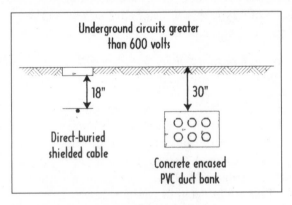

Figure 3.4 Minimum cover requirements, high voltage.

systems or in legacy power and lighting systems where transformer banks were installed in exterior banks. The presence of telecommunication cable and fiber optic will complicate safety operations, however. Comply with OSHA requirements for telecommunications personnel operating in electric power manholes. Telecommunication cables cannot share raceway with power cables and to the furthest extent that space and budgets allow, designers should avoid letting telecommunications and power raceway converge in the same manhole.

Q10. *Water is leaking from the underground conduit into the switchgear room. Does the NEC permit us to plug the raceway?*

A. Yes. Section 300-5(e) says that conduits or raceways through which moisture may contact energized live parts may be sealed or plugged in the manhole, in the room, or both. Where the leakage originates from direct burial cables, a bushing or terminal fitting with an integral bushed opening may be used at the end of a conduit or other raceway that terminates underground.

Q11. *Someone forgot to put the ground conductor in the new concrete-encased duct bank. Can we put it in a different trench?*

A. As long as it is in close proximity to the same trench. The **NEC** does not define close proximity, so use common sense. Section 300-5 requires that all conductors of the same circuit and, where used, the grounded conductor and all equipment grounding conductors must be installed in the same raceway. Conductors in parallel in raceways must be permitted, but each raceway must contain all conductors of the same circuit including grounding conductors. If you left room for future expansion in your duct bank, then you will have a spare empty duct to pull it through. Otherwise, you'll need to dig.

Q12. *How many cables can we put in a single underground raceway?*

A. The general rules for conduit fill apply. There are common situations in which you need to put one conductor in a raceway (Sec. 300-5, isolated phase raceway). Isolated phase installations are those in which there is only one phase per raceway. The construction method is common in the underground duct systems of supply stations where secondary conductors, on the order of 2000 kcmil, must be built one per duct. The spacing between isolated phase underground raceways should be as small as possible and the length of the run limited in order to avoid the increased circuit impedance and resulting increase in voltage drop. The **NEC** permits isolated phase installations in nonmetallic raceways in close proximity where conductors are paralleled as permitted in Sec. 310-4 and the conditions of Sec. 300-20 are met.

Q13. *Can we install medium-voltage cable directly in the ground?*

A. Yes, but it must be shielded if it is energized above 2000 V. Section 310-7 says that conductors used for direct burial applications must be of a type identified for such use. Nonshielded multiconductor cables rated 2001 to 5000 V are permitted if the cable has an overall metallic sheath or armor and is grounded through an effective grounding path meeting the requirements of Sec. 250-51.

Q14. *Explain what is new in the 1999 **NEC** regarding underground construction for circuits operating above 600 V.*

A. Two things: Former Art. 710 has vanished completely. Underground wiring methods for medium voltage now appears in Sec. 300-50. A new Art. 490 (Over 600 Volts, Nominal) has been added to the end of Chap. 4. This change makes sense because medium-voltage distribution is far more common now than ever, and it no longer needs to be regarded as a "special condition" relevant to Chap. 7. Despite the editorial change, there is practical similarity between earlier editions of the **NEC** and the 1999 **NEC** regarding medium-voltage circuit construction.

Q15. *Does the **NEC** have anything to say about where we locate manholes for medium-voltage distribution?*

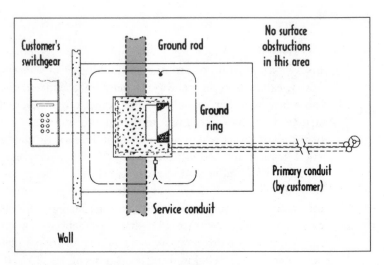

Figure 3.5 Typical customer pad mount installation—plan view.

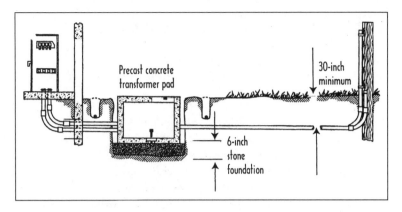

Figure 3.6 Typical customer pad mount installation—section view.

A.   Nothing explicit, only inferences that may be drawn from Art. 90
      requirements for good design. Manholes are typically placed
      where pulling points are required for either straight or angular
      conduit segments. In the case of 350-kcmil 15-kV class EPR
      pulled through 4-in PVC, for example, manholes need to be
      placed no greater than 500 ft apart because of sidewall pressure,
      among other variables. You should not ignore the horizontal
      dimension either because gravity can make it difficult for electri-
      cians to pull cable. A rigorous engineering treatment would apply
      calculation methods that many cable manufacturers are eager to
      demonstrate. In cases where manholes are not indicated, you
      should follow the **NEC** rules for conductor bending radius in Sec.
      300-34 and shown for underground cable in Fig. 3.7. The rule
      that appears in Sec. 300-34 applies to medium-voltage systems
      and will often have significant effect on an excavation budget.

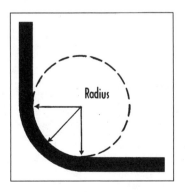

Figure 3.7
Radius of underground conduit bends.

OVERHEAD ELECTRIC POWER TRANSMISSION

Q16. *How high above the roof do the service conductors need to be?*

A. The general rule is to keep them at least 8 ft above the roof and at least 10 ft above grade. Much depends upon the voltage level and the type of adjacent buildings. The vertical clearances of all service-drop conductors must be based on conductor temperature of 60°F (15°C), no wind, with final unloaded sag in the wire, conductor, or cable. The important thing is to keep them out of reach, even to people who have no business on the roof. There are exceptions, as always. The vertical clearance above the roof level must be maintained for a distance not less than 3 ft in all directions from the edge of the roof.

Table 124-1 of the National Electric Safety Code (ANSI/IEEE C2) provides specifications for the guarding of live parts (see Table 3.3). Each column of this table must be used with reference to Fig. 3.8. Details about where guarding is required and the strength and types of guards should be gathered from the original source. See core reference in Appendix item A7. Refer the dimensions in Fig. 3.8 to the columns of Table 3.3.

TABLE 3.3 CLEARANCE FROM LIVE PARTS

Nominal voltage between phases	Vertical clearance of unguarded parts		Horizontal clearance of unguarded parts		Clearance guard to live parts	
	ft	in	ft	in	ft	in
151–600	8	8	3	4		2
2,400	8	9	3	4		3
7,200	8	10	3	4		4
13,800	9	0	3	6		6
23,000	9	3	3	9		9
34,500	9	6	4	0	1	0
46,000	9	10	4	4	1	4
69,000	10	5	4	11	1	11
115,000	11	7	6	1	3	1
138,000	12	2	6	8	3	8
161,000	12	10	7	4	4	4
230,000	14	10	9	4	6	4

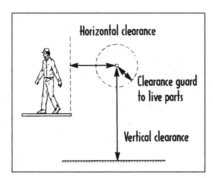

Figure 3.8
Clearance from live parts.

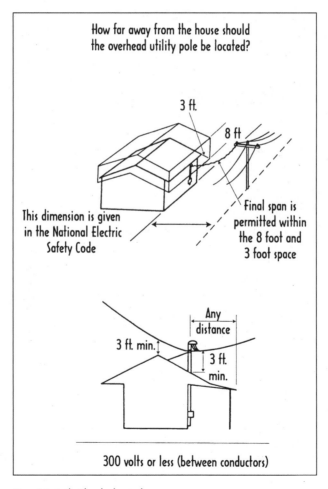

Figure 3.9 Residential overhead service drops.

On steep sloped roof (4-in rise, 12-in run) the vertical clearance of overhead conductors may be reduced from 8 ft to 3 ft. There are no restrictions on the length of the conductors over the roof or the horizontal distance over the roof. The NESC is silent on the matter of how far away the pole may be, only specifying that the conductor clearance comply with Table 234-1 (Appendix item A7).

Q17.   *What are the special conditions for vertical clearances?*

A.   There are always exceptions. Section 230-24(a) grants an exception for low voltage conductors over a roof with a rise of 4 in for every 12-in run. With any pitch steeper than this the greater hazard probably lies in falling off the roof. Thus, a reduction to 3 ft is permitted.

COMMUNICATION TRANSMISSION

Q18.   *What are the rules for underground communications wires and cables entering buildings?*

A.   Underground communications wires and cables in a raceway, handhole, or manhole containing electric light, power, class 1, or non-power-limited fire alarm circuit conductors must be in a section separated from such conductors by means of brick, concrete, or tile partitions. Where the entire street circuit is run underground and the circuit within the block is so placed as to be free from likelihood of accidental contact with electric light or power circuits of over 300 V to ground, the insulation requirements of Secs. 800-12(a) and 800-12(c) must not apply, insulating supports must not be required for the conductors, and bushings must not be required where the conductors enter the building.

Q19.   *Should communications wires and cables be installed with overhead power cables?*

A.   Section 800-10 states that when communications wires and cables and electric light or power conductors are supported by the same pole or run parallel to each other in-span, the following conditions must be met: (1) Where practicable, the communications wires and cables must be located below the electric light or power conductors. (2) Communications wires and cables must not be attached to a crossarm that carries electric light or power conductors. (3) The climbing space through communications wires and cables must comply with the requirements of Sec. 225-14(d). (4) Supply service drops of 0 to 750 V running above and parallel to communications service drops are permitted to have a minimum separation of 12 in at any point in the span, including the point of and at their attachment to the building, provided the nongrounded conductors are insulated and that a clearance of 40 in is maintained between the two services at the pole.

TABLE 3.4 Clearance between Supply and Communications Facilities in Joint-Use Manholes and Vaults

Phase-to-phase supply voltage	Surface to surface	
	in	mm
0–15,000	6	150
15,001–50,000	9	230
50,001–120,000	12	300
120,001 and above	24	600

Exception 1: These clearance do not apply to grounding conductors.
Exception 2: These clearances may be reduced by mutual agreement between the parties concerned when suitable barriers or guards are installed.

Q20. *Do I need a service mast for the communications wires?*

A. No. Section 800-10(b) says that communications wires and cables must have a vertical clearance of not less than 8 ft from all points of roofs above which they pass. Exceptions are made for very steep roofs and auxiliary buildings, such as garages and the like.

Q21. *Should power and telecommunications wiring be installed together?*

A. In general, no. Section 800-11 lays out the rules for underground circuits entering buildings. Underground communications wires and cables in a raceway, handhole, or manhole containing electric light, power, class 1, or non-power-limited fire alarm circuit conductors must be in a section separated from such conductors by means of brick, concrete, or tile partitions. Where the entire street circuit is run underground and the circuit within the block is so placed as to be free from likelihood of accidental contact with electric light or power circuits of over 300 V to ground, the insulation requirements of Secs. 800-12(a) and 800-12(c) do not apply, insulating supports are not required for the conductors, and bushings are not required where the conductors enter the building.

Q22. *What NEC requirements apply to overhead lines on the exterior of the building?*

A. Where communications wires and cables and electric light or power conductors run parallel to each other, the communications wires and cables must be located below the electric light or power conductors. The climbing space through communications wires and cables must be 24 to 30 in [or as otherwise required by Sec. 225-14(d)] so that lineworkers can climb over or through conductors to work safely with conductors on the pole. Supply service drops of 0 to 750 V running above and parallel to communications service drops must be permitted to have a minimum separation of 12 in at any point in the span, including the point of and at their attachment to the building, provided the nongrounded conductors are insulated and that a clearance of 40 in is maintained between the two services at the pole. Communication wires and cables cannot hang lower than 8 ft from all points of roofs above which they pass. There are exceptions, notably on auxiliary buildings, garages, and the like, as well as on roofs that are so steep that the presence of communication wires would be the least of your problems. The details appear in Art. 800-10.

Q23. *What are the NEC site requirements for electric service apparatus?*

A. The local utility generally has just as much to say about this as the NEC. Its requirements are designed to ensure safety without discouraging the construction of billable square footage. As a minimum the rules for Arts. 110 and 230 must be met with respect to the construction and operation of the building premises wiring. A typical utility will ask for 3 ft of clear, level, and unobstructed workspace provided in front of all service equipment, including pedestals and underground pull sections.

Many utilities will ask for one or more of the following: a concrete housekeeping pad, a pedestal, an exterior metering or an interior meter room with 24-h access. The construction standards of a national work group called the *Electric Utilities Service Equipment Requirements Committee* (EUSERC), which consists of about 80 utilities in 11 states, are gaining acceptance in the utility industry at large.

Q24.   *If an electric service is installed at a readily accessible location outside of a building, how far from the building may it be located?*

A.   In Sec. 230-70a no particular distance is shown, but Sec. 225-8 does require a disconnect at the building if the installation must comply with Secs. 230-71 and 230-72. In general, burial depths are based upon what it takes to keep the conductor safe from damage and from corrosion in proportion to the safety hazard it presents. Obviously high-voltage circuits are more dangerous than low-voltage circuits, so the rules are more rigorous. Supply circuits to landscape lighting and sprinkler systems that operate at less than 30 V require less. Wires in conduit is safer and stronger. Wires in conduit in concrete is even safer. The type of wire and the type of conduit makes a difference. It can be as simple as burying the low-voltage wire 30 in below grade, and as complicated as concrete around steel conduit when underground pipes box you in from below and street pavement boxes you in from above.

TRANSFORMER AND SWITCHGEAR SITING

Q25.   *Can we build medium-voltage circuits in the building?*

A.   Yes. Such circuits are common in high-rise buildings, hospitals, and on campuses with municipal class bulk distribution grids. Section 710-4 says that these circuits may be built within a building as long as they are installed in rigid metal conduit, in intermediate metal conduit, in rigid nonmetallic conduit, in cable trays, as busways, as cablebus, in other identified raceways, or as open runs of metal-clad cable suitable for the use and purpose. Open-frame switchyards with type MV cables, bare conductors, and bare busbars must also be permitted as long as places are accessible.

Q26. *Where do we locate the service transformer?*

A. Frequently, the answer is: where the utility company wants it, owner and architect-engineer notwithstanding. The most general rule appears in Sec. 230-70. The service disconnecting means must be installed at a readily accessible location either outside of a building or structure or inside the nearest point of entrance of the service conductors. Transformers and related switchgear must be readily accessible to qualified personnel for inspection and maintenance.

Q27. *Where do NEC rules for transformer location appear?*

A. Article 450 is divided into sections titled: (A) General Provisions, (B) Specific Provisions Applicable to Different Types of Transformers, and (C) Transformer Vaults. General provisions that are relevant to Division 02000 involve mechanical protection, transformer guarding, and ventilation. Special provisions that are relevant to Division 02000 involve the size, type, and voltage class of transformers. The subject of transformer vaults, whether they are interior or exterior, falls completely within Division 02000. Building planners should be aware that a service transformer must usually be installed with the appropriate primary and/or secondary switchgear. This switchgear may be at least equal in size to the transformer. There are situations when all the switchgear should be located on the building exterior, times when all the switchgear should be located in the building interior, and many situations in which the optimal transformer plus switchgear arrangement involves putting some of it inside and outside.

Q28. *Why does the transformer insulation system affect the dimensions of its footprint?*

A. Mother nature presents us with one of those immutable trade-offs between size and efficiency. An oil-filled transformer will have a smaller footprint than an equivalent kVA dry-type transformer, but special attention must be given to the insulating fluid. A more energy-efficient dry-type transformer (80° rise) will

TABLE 3.5 SEPARATION DISTANCE FOR LIQUID-INSULATED OUTDOOR TRANSFORMERS AND BUILDINGS

Liquid	FMRC-approved transformer	Liquid volume (gal)/(cu m)	Horizontal distance			Vertical distance (ft)/(m)
			Fire-resistant construction (ft)/(m)	Noncombustible construction (ft)/(m)	Combustible construction (ft)/(m)	
Less flammable (approved fluid)	Yes	N/A		3(0.9)		5(1.5)
	No	≤1000(3.8)	5(1.5)	5(1.5)	25(7.6)	25(7.6)
		>1000(3.8)	15(4.6)	15(4.6)	50(15.2)	50(15.2)
Mineral oil or unapproved fluid	N/A	<500(1.9)	5(1.5)	15(4.6)	25(7.6)	25(7.6)
		500–5000(1.9–19)	15(4.6)	25(7.6)	50(15.2)	50(15.2)
		>5000(19)	25(7.6)	50(15.2)	100(30.5)	100(30.5)

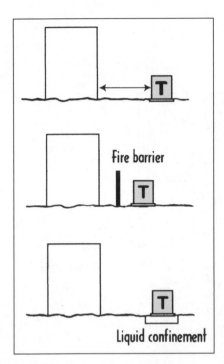

Figure 3.10
Liquid-filled transformer installation requirements, UL use restrictions for HMWH.

require more space because of the heat-transfer characteristics of the insulation. With a cheaper insulation you will have a smaller transformer, but you may have to spend money on ventilation. This may be a critical determinant in whether or not a transformer is located indoors or on the exterior.

To determine the proper space separation for a liquid-filled transformer, the amount of liquid and its particular heat-producing characteristics must be evaluated. Listing restrictions will vary with the listing agency and with the particular product listed.

Q29.   *If the transformer has a primary voltage of 40 kV, can the transformer be installed indoors?*

A.   Yes, but once you exceed 35 kV the transformer needs to go into a vault, regardless of the type of dielectric fluid.

Q30.   *Can a dry-type transformer be installed outdoors?*

A.   Yes, but it must have a weatherproof enclosure and it should be mounted appropriately. If you select a dry-type transformer in excess of 112^1/$_2$ kVA you must locate it at least 12 in away from combustible material of any building. An exception exists for 80° rise transformers (Sec. 450-22).

Q31. *What are the installation rules for liquid-filled transformers to be located outdoors?*

A. Section 450-23 says that transformers insulated with listed less flammable liquids having a fire point of not less than 300°C are permitted to be installed outdoors attached to, adjacent to, or on the roof of buildings, where installed in accordance with (1) or (2):

(1) For type I and type II buildings, the installation must comply with all restrictions provided for in the listing of the liquid. Here is a specific case in which NFPA 70 requires explicit information from NFPA 220. (2) Section 450-24 says that transformers insulated with a dielectric fluid identified as nonflammable are permitted to be installed indoors or outdoors.

Q32. *What types of transformer fluids are available?*

A. There are four in general use. Mineral oils are the most common and present the greatest fire hazard owing to the flammability of the oil. Askerels will burn only if subjected to extreme temperatures for long periods of time—thus, this type is treated as nonflammable but contains PCBs. Two liquids that can replace askerel are less flammable and nonflammable. Less flammable fluids have a fire point of at least 300°C. Nonflammable fluids have no practical fire point.

Q33. *What is a "less flammable" transformer liquid?*

A. It is listed by Factory Mutual Research Corp. and Underwriter's Laboratories, Inc., as a liquid with a fire point of at least 300°C. The FM listing is based on the use of the FM approved less flammable fluid in a transformer tank that meets certain criteria. Pressure-relief devices must be provided. FM also recommends the use of enhanced electrical protection.

Q34. *How do we handle nonflammable fluid-insulated transformers?*

A. Section 450-24 says that transformers insulated with a dielectric fluid identified as nonflammable are permitted to be installed indoors or outdoors. Such transformers installed indoors and rated over 35,000 V must be installed in a vault. Such transformers installed indoors must be furnished with a liquid confinement area and a pressure-relief vent. The transformers must be furnished with a means for absorbing any gases generated by arcing inside the tank, or the pressure-relief vent must be connected to a chimney or flue that will carry such gases to an environmentally safe area.

Q35.   *If oil is the most flammable of all transformer fluids, are the NEC rules as rigorous for outdoor installations?*

A.   Yes. Section 450-27 says that combustible material, combustible buildings, and parts of buildings, fire escapes, and door and window openings must be safeguarded from fires originating in oil-insulated transformers installed on roofs, attached to, or adjacent to a building or combustible material.

Space separations, fire-resistant barriers, automatic water spray systems, and enclosures that confine the oil of a ruptured transformer tank are recognized safeguards. One or more of these safeguards must be applied according to the degree of hazard involved in cases where the transformer installation presents a fire hazard.

Oil enclosures are permitted to consist of fire-resistant dikes, curbed areas or basins, or trenches filled with coarse crushed stone. Oil enclosures must be provided with trapped drains where the exposure and the quantity of oil involved are such that removal of oil is important.

Q36.   *What is the difference between a transformer and a transformer vault?*

A.   A vault as described by the **NEC** is an enclosure on the interior of a building with fire-resistive properties. A manhole is usually outside the building proper, although it may also have transformers in it. Section 450-41 says that vaults must be located where they can be ventilated to the outside air without using flues or ducts wherever such an arrangement is practicable.

The walls and roofs of vaults must be constructed of materials that have adequate structural strength for the conditions with a minimum fire resistance of 3 h. The floors of vaults in contact with the earth must be of concrete not less than 4 in thick, but where the vault is constructed with a vacant space or other stories below it, the floor must have adequate structural strength for the load imposed thereon and a minimum fire resistance of 3 h. Studs and wallboard construction are not acceptable.

Q37.   *Can the type of transformer fluid determine where we site the transformer?*

A.   Yes. See Sec. 450-23a2. Given an unlisted transformer which uses listed less flammable liquid for its fluid after having been converted from askerel, if the transformer is located indoors and doesn't meet the clearances called for in the listing of the liquid, Sec. 450-23a2 specifically allows a violation of the liquid-listing conditions (often unachievable in retrofits) as a trade-off for installing automatic fire suppression and a liquid-confinement area. Section 450-24 generally allows nonflammable fluid-insulated transformers indoors with a liquid-confinement area and vent.

SOIL PREPARATION

Q38.   *What rules does the NEC assert for backfill?*

A.   Section 300-5 says that backfill containing large rock, paving materials, cinders, large or sharply angular substance, or corrosive material must not be placed in an excavation where materials may damage raceways, cables, or other substructures or prevent adequate compaction of fill or contribute to corrosion of raceways, cables, or other substructures. Where necessary to prevent physical damage to the raceway or cable, protection must be provided in the form of granular or selected material, suitable running boards, suitable sleeves, or other approved means.

Q39.   *How shall we manage transformer drainage?*

A.   In Sec. 450-46 the **NEC** says that vaults containing a transformer 100 kVA or larger must be provided with a drain or other means that will carry off any accumulation of oil or water. The **NEC** is silent on just exactly where this runoff should go. There is deference to practicality and local conditions that may or may not make it legal for you to drain transformer oil into a storm sewer line. Make sure the drawings you send to the city building department for review show where you expect the drainage to go.

Q40.   *Can we reduce ground resistance with soil preparation?*

A.   Yes. Refer to Appendix Item A8, Resistivity of Soils and Resistance of Single Rods, which is adapted from IEEE Standard 142. When ground resistance is too high, water solutions may bring the resistance down. When structures are built on hills— radio transmitters, for example—groundwater is sparse and you may observe higher ground resistance. The opposite is the case with swampland.

LANDSCAPE ACCESSORIES

Q41.   *For purposes of safety and aesthetics, what can we do to hide electrical switchgear?*

A.   The **NEC** has no rules for aesthetics, only safety. Refer to Figs. 3.11 through 3.14, which describe the manner in which a leading U.S. utility company permits its customers to screen switchgear. All of these arrangements comply with the **NEC**.

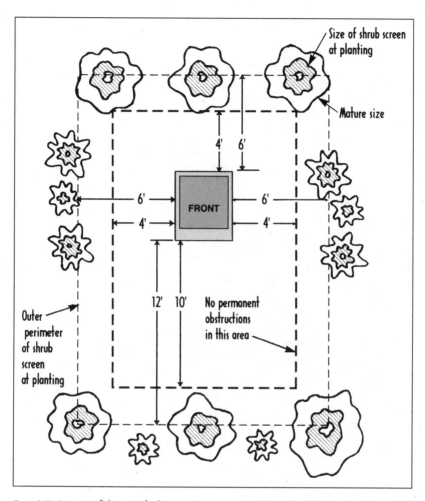

Figure 3.11 Customer installed screening of pad mount service apparatus—pictorial view. (Adapted from Pennsylvania Power & Light, *Customer Reference Specifications.*)

Figure 3.12 Customer installed screening of pad mount service apparatus—plan view with suggestions for shrubbery and plantings.

Figure 3.13 A fence or wall screening of pad mount service apparatus—pictorial.

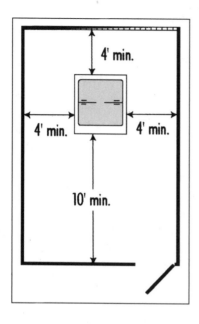

Figure 3.14
Fence or wall screening of pad mount service apparatus—
plan view.

4

Division 03000—Concrete

Remarks. The type of construction that applies
to underground work falls under the CSI sub-
heading of this division titled, "Cast-in-Place
Concrete," which indicates a general-use finish
with no appearance criteria. In many construction projects the cost of building an
underground concrete-encased conduit system
dominates the construction cost budget, often
to an extent that designers are forced to dream
ugly but NEC-compliant kludges to avoid
having to build one. (Many of the controver-
sies surrounding Art. 230 are rooted in the
desire to avoid building concrete-encased
conduit.) Per-unit costs drop dramatically in
direct proportion to the length and condition
of the right-of-way.

*In developed municipalities, underground dis-
tribution is the rule. In all likelihood the local
building department endorsed the practice of
the regional power and telecommunication
companies a long time ago. The designer and
contractor will find those construction standards very helpful. In "develop-ing" municipalities, underground distribution systems may be
optional. The leadership of fast-growing sub-
urban or ex-urban governmental bodies should
establish concrete-encased conduit systems as
a requirement before development makes it
even more expensive to build them. World
economies are becoming increasingly dependent
upon reliable electrotechnology, and it is the
power and telecommunication infrastructure
that is at greatest risk because so much of it
is either suspended in the air or too close to
the surface.*

Q1.   *What **NEC** requirements commonly appear in Division 03000?*

A.   Most of them have to do with concrete encasement of raceway or with providing a structural framework for transformer vaults. These requirements appear in various passages throughout Chaps. 3 and 4. Requirements for metal grounding electrodes encased in concrete appear in Sec. 250-50. Requirements for metal raceways and equipment encased in concrete appear in Sec. 300-6. The use of hollow cells of cellular concrete floor slabs as electrical raceways appears in Art. 358. Underground feeder and branch circuit wiring is covered in Arts. 339 and 343. Cellular concrete floor raceway is in Art. 358. New to the 1999 **NEC** is Part D to Art. 370, which merges **NEC** requirements with the NESC safety rules for the installation and maintenance of underground electric supply and communication lines.

Q2.   *Anything new in the 1999 **NEC** that concerns the concrete trades?*

A.   Yes. Section 364-6 deals with busways that run vertically through a building. Paragraph 2 is a new requirement for 4-in concrete curbing around vertical busway risers that penetrate two or more dry floors. The curb must be built within 12 in of the busway floor penetration. The idea is to keep liquids from entering vertical busways and causing faults. Electrical equipment must be located safely away from the retaining basin. The requirement is illustrated in Fig. 4.1.

Q3.   *What is the difference, if any, between underground "duct" and underground "conduit"?*

A.   The NESC explains the difference. While it is often the practice to use the terms duct and conduit interchangeably, the term duct (as used in the NESC, for example) is a single enclosed raceway for conductors and cable. The term "conduit" is a structure containing one or more ducts. The term "conduit system" is the combination of conduit, conduits, handholes, and/or vaults joined to form an integrated whole. The **NEC** does not define either "duct" or "conduit," though it does define raceway. Refer to Art. 100.

Q4.   *Does the **NEC** have any specific requirements for building a concrete-encased conduit system?*

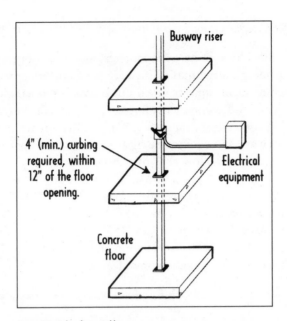

Figure 4.1 Curbing for vertical busway.

A.   None other than that the system should be designed adequate for the purpose. (And a good design always begins with a good land survey.) Although some related material appears in Sec. 300-6, the bulk of the material that appears here has to do with direct buried underground installations for low-voltage systems.

Conduit systems for medium-voltage distribution are built by utilities or organizations with large bulk distribution systems. Construction and performance standards for these systems have evolved generally along the line of NESC requirements. Substantial modifications have been made by individual utilities due to regional soil and climatic conditions.

The broad principles that apply are that ducts should be installed at a depth below the surface to avoid possible damage and, if possible, should be below the frost line to prevent dislocation from motion caused by severe changes in temperature. They should be graded between manholes so that water will not accumulate in them. Where there are bends or offsets in the line, ducts should be gradually bent to accommodate the minimum bending radii of the largest cables to be installed; this bending radius should be from 7 to 20 times the diameter of the cable, depending on its size, voltage classification, type of insulation, type of sheath, and other characteristics. The beds in the duct line should preferably be located near the manholes, as each bend will increase the pulling tension on the cable and thus reduce the maximum length or distance between manholes.

Conduit systems are most commonly built from type EB or DB Schedule 80 PVC. When the system is within 10 ft of the building, under a roadway or parking lot, the ducts are sometimes specified as rigid steel.

In unstable ground the conduit system and the manholes to which it is connected should move as a whole (Fig. 4.2). Make sure that adequate reinforcement is placed at the manhole-conduit intercept to protect against cables being sheared by earth movements. Reinforcing bars are not a substitute for ground wires.

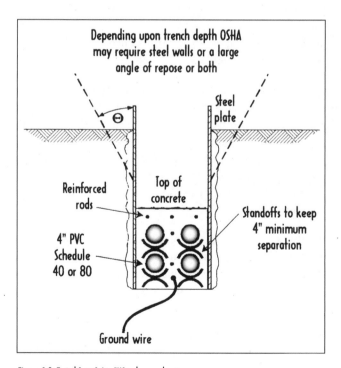

Figure 4.2 Typical 2 × 2 4-in PVC underground section.

Q5.   *What do I need to look for in Division 03000 specifications for cast-in-place concrete?*

A.   A good set of specifications should include information about formwork accessories, reinforcing materials, concrete admixtures, curing, and bonding compounds. In some variants of the CSI format, the construction of underground raceway is broken out separately. Something should be said about cold and hot weather concreting. Of particular interest to electrical people installing underground PVC for, say, medium-voltage cables, is the notion of "slump." A detailed explanation of concrete slump is beyond the scope of this book, but its practical effect will cause a newly installed conduit system to float to the top of the pour, thus exposing the raceway. If you want to pass inspection, make sure the concrete people have got the slump right.

Q6.   *What are the broad principles that apply to the design and installation of a conduit system—before the concrete has been poured?*

A.   It starts with a good survey. A good survey will include information about all existing underground infrastructure, not just the horizontal plan-view information but profile information about vertical invert elevations.

Conduit should be adequately sized for the maximum size cable to be pulled through them. They should not contain traps where water will accumulate; thus they should be sloped downward toward manholes and buildings; perhaps 6 in per 100 ft or more. They should be grounded with a 4/0 AWG bare stranded ground wire run through the bank. A marker strip, either red or yellow, should be fastened to the top of the duct bank before the trench has been refilled.

Finally, a good survey is completed by a follow-up survey indicating the as-built condition of the underground conduit system. Architects and owners should include the cost of the follow-up survey in the up-front professional services contract.

In a fashion, a manhole (Fig. 4.3) is just another box covered in Art. 370, Part D. Manholes are most commonly precast or cast-in-place, frequently using masonry in combination. The 1999 **NEC** merges requirements for conduit systems with Sec. 32 of the 1997 NESC.

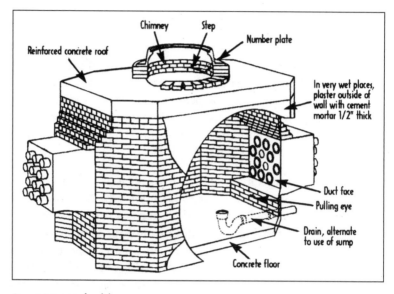

Figure 4.3 Isometric of manhole.

Q7.   *What are the broad principles that apply to the design, specification, and installation of electrical manholes?*

A.   With either concrete and masonry construction they must come with a full complement of hardware: cast-iron frame with cover, a hot dipped galvanized steel ladder and pulling eyes, and a soil pipe drain with grate. They should contain vertical concrete inserts in each wall, suitable for use with unistrut cable racks. The racks themselves should come with insulators for supporting cables. They should be grounded by driving 10-ft ground rods at each corner with connections to the duct bank ground wire.

Q8.   *Can underground feeder cable be embedded in concrete?*

A.   No. This is a cable type in common use on residential lots and installed by weekend handypersons. It is listed as type UF cable in sizes No. 14 copper or No. 12 aluminum or copper-clad aluminum through No. 4/0 with an overall covering that is moisture-, fungus-, and corrosion-resistant and suitable for direct burial in the earth. Article 339 says that type UF cable must *not* be used embedded in poured cement, concrete, or aggregate.

Q9.   *What are the general considerations for manhole loading?*

A.   Again, the NEC makes no specific requirements other than that the manhole be designed for the practical safeguarding of persons and property from hazards arising from the use of electricity. This much said, you need to consult other standards to meet this requirement, not the least of which are the publications produced by American Association of State Highway and Transportation Officials Guidelines (AASHTO).

A rigorous investigation involves looking into the allowable bearing pressures for soil. Unless organic clays or silts are encountered, a value of 1.5 tons/sq ft may be used as a conservative bearing value. If a manhole or vault is to be installed on clay, clayey solid, or organic material, careful valuation should be made of the potential for settlement. The use of a crushed stone base or piles may be required and soils boring may be necessary.

The loading on the several parts of the manhole depends on the maximum load imposed on the sidewalk or street surface. The live load on the surface affects the design of both the roof slab and the walls. Wheel loads of 21,000 lb and impacts of 50 percent are typical values for heavily traveled streets over which truck traffic may be concentrated. For small wheel areas concentrated load may be as high as 63,000 lb/sq ft. The type of pavement, the nature of the solid beneath the pavement, and the thickness of the solid above the roof of the manhole will serve to magnify or reduce the effect of the concentrated load. Live load requirements are established by the AASHTO. Electrical engineers should consult with a civil engineer with some practical experience in the application of these standards.

Q10.   *How do we build the transformer and switchgear pad?*

A.   There are no specific NEC requirements for grade-level equipment pads, only the core rules of Chap. 3 for raceway construction and Chap. 4 for transformer protection. If the owner will be buying power from the serving utility at a commercial rate, for example, then the service planning people will provide the architect-engineer with a switchgear pad detail. Figure 4.4 is a typical section view of one such layout in circulation at a major

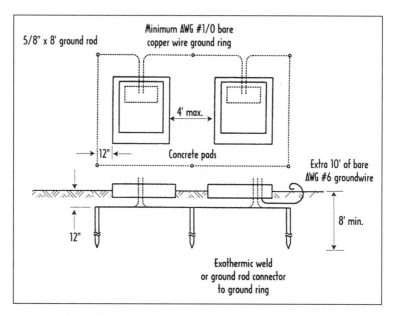

Figure 4.4 Switchgear pad detail. (*Adapted from Detroit Edison Customer Electric Service Installation Guide.*)

U.S. utility. There are hundreds of variants of it to suit local climates and soil conditions.

Q11. *What does the NEC require for transformer vaults built in concrete structures?*

A. Section 450-42 says that the walls and roofs of vaults must be constructed of materials that have adequate structural strength for the conditions with a minimum fire resistance of 3 h. The floors of vaults in contact with the earth must be of concrete not less than 4 in (102 mm) thick, but where the vault is constructed with a vacant space or other stories below it, the floor must have adequate structural strength for the load imposed thereon and a minimum fire resistance of 3 h. Six-inch-thick reinforced concrete is a typical 3-h construction.

Q12. *What is cellular concrete floor raceway?*

A. Article 358 covers cellular concrete floor raceways, the hollow spaces in floors constructed of precast cellular concrete slabs, together with suitable metal fittings designed to provide access to the floor cells. A "cell" is a single enclosed tubular space in a floor made of precast cellular concrete slabs (the direction of the cell being parallel to the direction of the floor member). A "header" is the transverse metal raceway for electric conductors.

An important operation and maintenance rule for discontinued outlets appears in this article. When an outlet is abandoned, discontinued, or removed, the sections of circuit conductors supplying the outlet must be removed from the raceway. No splices or reinsulated conductors, such as would be the case of abandoned outlets on loop wiring, must be allowed in raceways. Particulars regarding wiring method all appear here.

Q13.   *Can grounding electrodes be built into the concrete footings?*

A.   Yes. Section 250-50(c) contains a description of several permissible approaches for putting together a grounding electrode system. The specifications for a concrete-encased electrode system should include, as a minimum, (1) encasement by at least 2 in of concrete, (2) located within and near the bottom of a concrete foundation or footing that is in direct contact with the earth, (3) consisting of at least 20 ft of one or more bare or zinc galvanized or other electrically conductive coated steel reinforcing bars or rods of not less than $1/2$ in diameter, or consisting of at least 20 ft of bare copper conductor not smaller than AWG No. 4. Reinforcing bars may be bonded together mechanically or electrochemically to meet the 20-ft minimum required of an electrode.

Figure 4.5 is included in this chapter to demonstrate an ingenious method of using an underground concrete structure to vent on-site generator exhaust. Design eliminates vertical pipe runs and also muffles noise. Its use depends upon the availability of a few square meters of protected grade-level space at an appropriate distance from the building. It might be a reasonable alternative to be considered by owners and architects because it eliminates the need for rooftop exhaust stacks, which can be as large as 5 ft wide and 50 ft high. Exhaust flows through a PVC transit pipe into the rock-filled gravel pit contained by precast concrete. For more details refer to the May 1998 edition of *Consulting-Specifying Engineer,* "*Power Standing By,*" by Harry B. Zackrison, Jr.

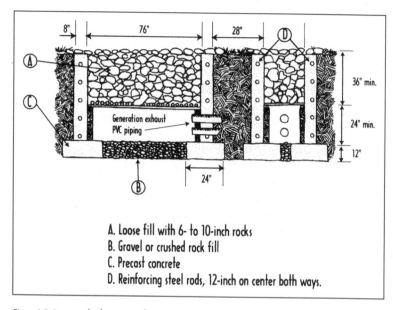

A. Loose fill with 6- to 10-inch rocks
B. Gravel or crushed rock fill
C. Precast concrete
D. Reinforcing steel rods, 12-inch on center both ways.

Figure 4.5 Concrete and rock generator exhaust pit.

Q14. *What is a Ufer ground?*

A.   A concrete-encased electrode. The concrete becomes a semiconducting medium because of its hygroscopic nature. The reinforcing steel is electrically connected to the earth through the moisture absorbed by the concrete and the effects of capacitive coupling (see Fig. 4.6).

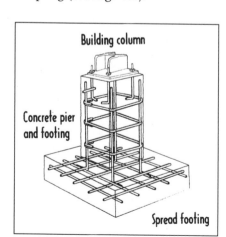

Figure 4.6
Ufer Ground. (*Adapted from Bierals, "Applying the 1993 National Electric Code."*)

Steel reinforcing bars in Fig. 4.6 are at least 20 ft and not less than $1/2$ in in diameter, or alternatively at least 20 ft of bare copper conductor not smaller than AWG No. 4 encased in at least 2 in of concrete and located near the bottom of a concrete foundation or footing.

Q15.   *What other standards apply to concrete-encased conduit systems?*

A.   Particularly on the subject of medium-voltage distribution, the reader may find the following publications useful: Edison Electric Institute, *Underground Systems Reference Book, American Association of State Highway and Transportation Officials Guidelines* (AASHTO), *Application & Design Manual of Consolidated Edison Company of New York, Underground Systems Design Guide of the Detroit Edison Company.*

5

Division 04000—Masonry

Remarks. *In the building trades this division is generally understood to include stone and bricking. From the standpoint of the electrical trades, there is not much difference in these materials. All of them are regarded as grounded surfaces, a fact that frequently has profound influence upon architectural design of interior electric service switchgear rooms.*

Q1.   *What NEC requirements commonly appear in Division 04000?*

A.   There are many passages throughout the first four chapters of
the Code which refer to how electrical materials are joined with
masonry. All of them should be fairly intuitive: electric equipment
should be firmly secured to the surface on which it is mounted,
where exposed to the weather; raceways enclosing service-entrance
conductors must be raintight and arranged to drain. Masonry
walls typically do not require exposure protection unless unprotected openings or windows are present. (In these cases the openings should be protected in accordance with exposure protection
recommendations for outdoor oil-filled transformers.) Insulated or
bare aluminum or copper-clad aluminum grounding conductors
must not be used where in direct contact with masonry. The conditions for application of several wiring methods depend upon
whether or not they will be applied in a wet location.

Q2.   *Is brick, stone, and concrete a grounded surface?*

A.   Yes. Section 110 says that when the dimensions of the working
space in the direction of access to live parts is likely to require
examination, adjustment, servicing, or maintenance while energized, concrete, brick, or tile walls are considered as grounded.
The basic idea is that it should be impossible for electricians to
touch an energized part and the adjacent masonry or concrete
wall simultaneously. If they can, then the wall needs to be moved
farther away or an insulating material (such as wood) needs to
cover the grounded surface.

Q3.   *What are the NEC rules for the installation of electric heating
cables in masonry?*

A.   Constant-wattage heating cables must not exceed $16^{1}/_{2}$ W per
lin ft of cable. The spacing between adjacent runs of cable must
not be less than 1 in on centers. Cables must be secured in place
by nonmetallic frames or spreaders or other approved means
while the concrete or other finish is applied. Leads must be
protected where they leave the floor by rigid metal conduit,
intermediate metal conduit, rigid nonmetallic conduit, electrical
metallic tubing, or other approved means. Bushings or approved
fittings must be used where the leads emerge within the floor
slab. Refer to Art. 424.

6

Division 05000—Metals

Remarks. This division covers structural steel, structural aluminum, metal wall framing, raceway deck systems, sheet metal, handrails, aluminum joists, gratings, ladders, metal stairs, castings, fasteners, and the like. Other than our natural affinity to copper and aluminum as an energy transport medium, issues involving metals that are of concern to the electrical trades generally have to do with raceway systems and grounding. These are core NEC topics that appear in nearly every article. They are covered in detail in Divisions 16100 and 16450, respectively.

Q1.   *What **NEC** requirements commonly appear in Division 05000?*

A.   The two issues involving metals that underlie all of the wiring methods described in the **NEC** is the matter of the use of dissimilar metals, the matter of electrical continuity, and the use of metal conduit for underground conduit systems.

Q2.   *Why the concern with the use of dissimilar metals?*

A.   Because of the effect that the different mechanical and thermal characteristics of copper and aluminum have upon electrical devices. Pressure terminal or pressure splicing connectors and soldering lugs, for example, should be identified for the material of the conductor and must be properly installed and used. Conductors of dissimilar metals must not be intermixed in a terminal or splicing connector where physical contact occurs between dissimilar conductors (such as copper and aluminum, copper and copper-clad aluminum, or aluminum and copper-clad aluminum), unless the device is identified for the purpose and conditions of use.

Q3.   *Do metal raceways carry current?*

A.   Yes, but only under fault conditions in a solidly grounded system. This fact usually surprises nonelectrical people, that the electrical conduit itself, not just the phase conductors, is the energy transport media. Metal raceways, cable armor, and other metal enclosures for conductors must be metallically joined together into a continuous electric conductor and must be so connected to all boxes, fittings, and cabinets as to provide effective electrical continuity. Raceways and cable assemblies must be mechanically secured to boxes, fittings, cabinets, and other enclosures.

Q4.   *Can the structural metal frame of a building be used as a grounding electrode conductor?*

A.   Yes. Part C of Sec. 250 covers the subject of grounding electrode systems and grounding electrode conductors. Section 250-50(b) permits the use of the metal frame of a building or structure, *where the frame itself is effectively grounded.*

Q5.   *Can the structural metal frame of a building be used as the equipment grounding conductor?*

A.   No. Part G of Sec. 250 describes various methods of equipment grounding. Section 250-136 says that equipment secured to grounded metal supports must be considered effectively grounded. The structural metal frame of a building, however, must not be used as the required equipment grounding conductor (Sec. 250-136).

Q6.   *What is exothermic welding?*

A.   Frequently called Cad-welding, it is a superior alternative to mechanical bonding of ground current conducting paths. Section 250-70 refers to various methods for making the connection between the grounding conductor and the electrode. The NEC does not permit solder connections for this purpose.

Q7.   *How do you run raceway through metal wall frames?*

A.   The same way you run it through wood. Section 300-4 says that where a cable- or raceway-type wiring method is installed parallel to framing members, such as joists, rafters, or studs, the cable or raceway must be installed and supported so that the nearest outside surface of the cable or raceway is not less than $1^{1}/4$ in from the nearest edge of the framing member where nails or screws are likely to penetrate. Where this distance cannot be maintained, the cable or raceway must be protected from penetration by nails or screws by a steel plate, sleeve, or equivalent at least $1/16$ in thick.

Table 6.1 summarizes Underwriter's Laboratories' guide card information for corrosion protection of various metallic conduit types.

TABLE 6.1 METAL CONDUIT CORROSION PROTECTION RECOMMENDATIONS

	Required	Optional
In concrete:		
Rigid steel		X
Intermediate steel		X
Aluminum rigid	X	
Steel EMT	Below grade may need	On or above grade

TABLE 6.1 METAL CONDUIT CORROSION PROTECTION RECOMMENDATIONS (*Continued*)

	Required	Optional
Aluminum EMT	X	
In soil:		
Rigid steel		X
Intermediate steel		X
Aluminum rigid	X	
Steel EMT	Severely corrosive soil	
Aluminum EMT	X	

Q8.   *How do you protect metal raceway from corrosion?*

A.   Section 300-6 says that metal raceways, cable armor, boxes, cable sheathing, cabinets, elbows, couplings, fittings, supports, and support hardware must be of materials suitable for the environment in which they are to be installed. Ferrous raceways, cable armor, boxes, cable sheathing, cabinets, metal elbows, couplings, fittings, supports, and support hardware must be suitably protected against corrosion inside and outside (except threads at joints) by a coating of approved corrosion-resistant material such as zinc, cadmium, or enamel. Table 6.1 summarizes the current thinking on corrosion protection for metal conduit. Figure 6.1 offers a few remedies. The reader seeking more depth should consult Soare's "Grounding," available from the IAEI.

Per NEC 300-6 ferrous metal raceways may be installed in concrete or in direct contact with the earth, or in areas subject to severe corrosive influences where made of material judged suitable for the condition, or where provided with corrosion protection approved for the condition. Asphalt paint will work.

Q9.   *What requirements does the code provide for the compensation of thermal expansion of metallic raceway?*

A.   Section 300-7 says that portions of an interior raceway system which are exposed to widely different temperatures must be

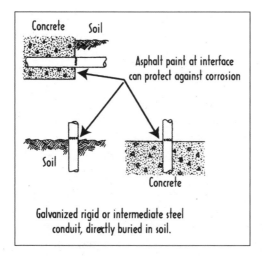

Figure 6.1 Asphalt paint to avoid corrosion. (*Adapted from McGraw-Hill's National Electric Code Handbook.*)

provided with expansion joints where necessary to compensate for thermal expansion and contraction. Table 347-9 provides the expansion information for PVC.

Q10. *What are the NEC requirements for cellular metal floor raceways?*

A. Article 356-1 says that no conductor larger than No. 1/0 must be installed, except by special permission. The combined cross-sectional area of all conductors or cables must not exceed 40 percent of the interior cross-sectional area of the cell or header. Splices and taps must be made only in header access units or junction boxes.

In general, cellular metal floor raceways must be so constructed that adequate electrical and mechanical continuity of the complete system will be secured. They must provide a complete enclosure for the conductors. The interior surfaces must be free from burrs and sharp edges, and surfaces over which conductors are drawn must be smooth. Suitable bushings or fittings having smooth rounded edges must be provided where conductors pass.

Q11. *How do I build a service mast?*

A.   Sometimes this work is claimed by other architectural or mechanical trades. No matter who does the work, the clearances need to be right. Section 230-27 says that the point of attachment of the service-drop conductors to a building or other structure must provide the minimum clearances as specified in Sec. 230-24. In no case can this point of attachment be less than 10 ft above finished grade. Where a service mast is used for the support of service-drop conductors, it must be of adequate strength or be supported by braces or guys to withstand safely the strain imposed by the service drop. Where raceway-type service masts are used, all raceway fittings must be identified for use with service masts. Only power service-drop conductors are permitted to be attached to a service mast. Service-drop conductors passing over a roof must be securely supported by substantial structures. Where practicable, such supports must be independent of the building.

7

Division 06000—Wood and Plastics

Remarks. There are frequent references in the
NEC to other NFPA standards that deal with
fire safety issues in the integration of electrical
systems with architectural materials. We cover
some of them here and in the following chapter.
The reader is referred to Chap. 3 of the NEC
Handbook, also published by the NFPA, which
contains more detailed information about the
fire rating of various architectural finishes, not
the least of which is a table listing the fire resistance ratings (in minutes) of common building
materials.

Q1. *What **NEC** requirements commonly appear in Division 06000?*

A.   Almost everything in the **NEC** regarding wood has to do with keeping the heat-generating capabilities of electrical systems as far away as possible from a combustible material such as wood (and some types of plastics). Signs and outline lighting systems, for example, should be constructed and installed so that adjacent combustible materials cannot be subjected to temperatures in excess of 90°C. The spacing between wood or other combustible materials and an incandescent or HID lamp or lamp holder must not be less than 2 in. Wood may be used to insulate a grounded surface in workspaces with electrical apparatus. Refer to Art. 110.

Q2. *How is the issue of fire-rated walls linked to the **NEC**?*

A.   The whole idea behind a firewall is to contain or reduce oxygen supply to a space. But raceway for power and communication cables has a way of piercing fire-rated walls and leaving leaks in them for the spread or feeding of fire. Sections 300-21 and 820-52(b) cover the spread of fire or products of combustion. It is the intent of this section that electrical equipment such as raceways and cables be installed in such a manner that they will not contribute to the spread of fire or the products of combustion through the specified component parts of a building. Fire-stopping materials and methods are applied to maintain the fire-resistance rating of the building component that is pierced.

Q3. *Can plastic electrical boxes be used in a 2-h fire-rated wall?*

A.   It depends upon the AHJ. Many jurisdictions will allow only metal boxes in this type of construction. When plugging the holes in plastic boxes, fire-stop paddy pad is recommended. Check the product for its listings and applications.

Q4. *What are the rules for installing raceway in (or on) wood walls?*

A.   Section 300-4 says that where subject to physical damage, conductors must be adequately protected. In both exposed and concealed locations, where a cable or raceway-type wiring method is installed through bored holes in joists, rafters, or wood members, holes should be bored so that the edge of the hole is not less than $1^1/4$ in from the nearest edge of the wood member. Where this distance cannot be maintained, the cable or raceway should

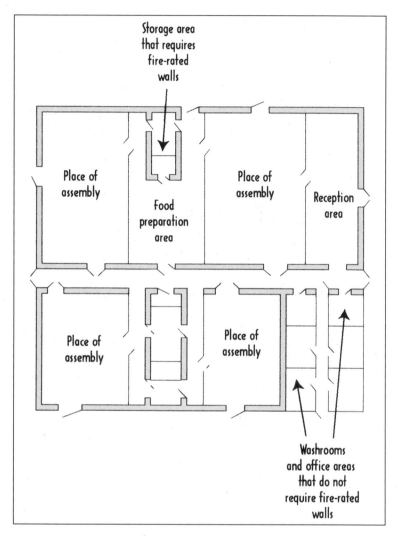

Figure 7.1 General principles for establishing fire walls.

be protected from penetration by screws or nails by a steel plate or bushing at least $1/16$ in thick and of appropriate length and width installed to cover the area of the wiring. Where there is no objection because of weakening the building structure, in both exposed and concealed locations, cables or raceways may be permitted to be laid in notches in wood studs, joists, rafters, or other wood members where the cable or raceway at those points is protected against nails or screws by a steel plate at least $1/16$ in thick installed before the building finish is applied. Some exceptions exist for other wire types and wiring methods described in Chap. 3.

In mixed-use facilities, fire-rated walls are needed in only those spaces that are regarded as a place of assembly (Fig. 7.1). Article 518 of the **NEC** defines a place of assembly as a building (or portion of a building) designed or intended for the assembly of 100 or more persons. The washrooms and the office area of this multipurpose facility, for example, would not be considered a place of assembly, and ordinary wiring methods can be applied. Single lines represent walls not required by the building code to be of fire-rated construction.

Division 07000—Thermal and Moisture Protection

Remarks. This chapter is a continuation of the preceding discussion on the integration of electrical systems with architectural materials. Here the discussion is of a more general nature.

Q1.   *What **NEC** requirements commonly appear in Division 07000?*

A.   This CSI division covers waterproofing, insulation, fireproofing, and fire stopping, manufactured roofing and siding, membrane roofing, roof maintenance and repairs, sealant and caulking, among others. The preoccupation with wiring methods in places of assembly is that a place of assembly is a hazardous location and some measures must go into keeping burning cable insulation from becoming toxic and killing more people than the fire would.

Q2.   *What is a finishing rating?*

A.   A finish rating is established for assemblies containing combustible (wood) supports. The finish rating is defined as the time at which the wood stud or wood joist reaches an average temperature rise of 121°C or an individual temperature rise of 163°C as measured on the plane of the wood nearest the fire. A finish rating is not intended to represent a rating for a membrane ceiling. Fire-rated construction is the fire-resistive classification used in building codes.

Q3.   *Do the rules for fire stopping apply to rough wood walls?*

A.   Yes. Section 300-21 makes no distinction between rough and finish carpentry. Electrical installations in hollow spaces, vertical shafts, and ventilation or air-handling ducts must be so made that the possible spread of fire or products of combustion will not be substantially increased. Openings around electrical penetrations through fire-resistant-rated walls, partitions, floors, or ceilings must be fire stopped using approved methods to maintain the fire-resistance rating. Directories of electrical construction materials published by qualified testing laboratories contain many listing installation restrictions necessary to maintain the fire-resistive rating of assemblies where penetrations or openings are made.

Q4.   *How is the issue of fire-rated walls related to Code requirements for places of assembly?*

A.   Because the rules for wiring a place of assembly—defined by the Code in Art. 518 as any occupancy intended for the assembly of

100 persons or more—have always accounted for the hazards that electrical wiring presents in such places. Places of assembly frequently require emergency wiring for systems intended to reduce panic hazards, and these requirements are stated in Sec. 700-9.

The fixture in Fig. 8.1 is suitable for use in insulated ceilings in direct contact with thermal insulation. Thermal protection is provided to deactivate the lamp should the fixture overheat.

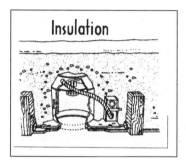

Insulation

Figure 8.1 A listed type IC recessed fixture. (*Adapted from the NEC Handbook, published by the NFPA.*)

Q5. *Do the references to walls and ceilings in the NEC also apply to drywall over wood, and are they considered combustible or noncombustible?*

A. Section 370-20 applies to framing or sheathing which is combustible. This section asserts the conditions under which the box may be set back from the surface of a wall or ceiling or must be flush with or extend from the surface. Drywall over wood studs is considered combustible construction. Drywall of different sizes and thicknesses has a combustible point at high temperature. Half-inch sheet rock with either metal studding or wood studding on both sides of the partition is rated for 3/4 h. Five-eighths of an inch on both sides of the partition is rated for 1 h, and using wood studding adds to the problem of combustion. When using metal studding with drywall, it does not mean that it is noncombustible construction.

Q6.   *What are the rules for keeping lights from igniting combustible materials?*

A.   Section 410-65 says that fixtures should be so installed that adjacent combustible material will not be subjected to temperatures in excess of 90°C. Where a fixture is recessed in fire-resistant material in a building of fire-resistant construction, a temperature higher than 90°C, but not higher than 150°C, may be considered acceptable if the fixture is plainly marked that it is listed for that service.

Q7.   *What are the rules for recessed incandescent fixtures?*

A.   Incandescent fixtures must have thermal protection and must be identified as thermally protected. Exceptions exist for recessed incandescent fixtures identified for use and installed in poured concrete. Listed recessed incandescent fixtures provide, by construction design, the equivalent temperature performance characteristics of thermally protected fixtures and are so identified.

Q8.   *What are the specific code requirements for keeping lighting fixtures from being damaged by moisture?*

A.   Section 410-4 says that fixtures installed in wet or damp locations must be so installed that water cannot enter or accumulate in wiring compartments, lampholders, or other electrical parts. All fixtures installed in wet locations must be marked "Suitable for Wet Locations." All fixtures installed in damp locations must be marked "Suitable for Wet Locations" or "Suitable for Damp Locations."

Q9.   *What kind of thermal protection is required for transformer vaults?*

A.   Section 450-42 says that the walls and roofs of vaults must be constructed of materials that have adequate structural strength for the conditions with a minimum fire resistance of 3 h. The floors of vaults in contact with the earth must be of concrete not less than 4 in thick, but where the vault is constructed with a vacant space or other stories below it, the floor must have adequate structural strength for the load imposed thereon and a minimum fire resistance of 3 h. For the purposes of this section, studs and wallboard construction is not acceptable. When transformers are protected with automatic sprinkler, water spray, carbon dioxide, or halon construction, the fire rating can be reduced to 1 h.

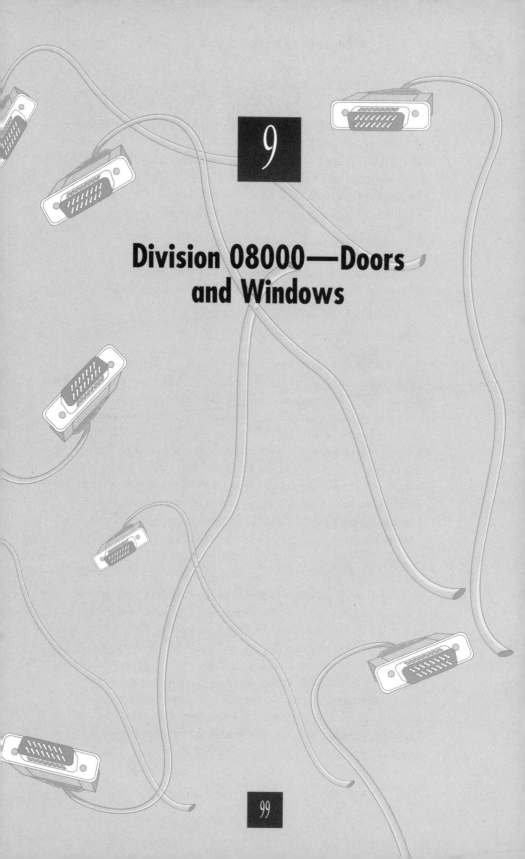

9

Division 08000—Doors and Windows

Q1.   *What NEC requirements commonly appear in Division 08000?*

A.   Per Art. 450, doors to transformer vaults must be kept locked at all times to prevent access of unqualified persons to the vault. This may complicate meter reading operations, and architects need to satisfy both requirements. The walls and roof for a transformer vault must have at least a 6-in thickness. The doorsills for a transformer vault must be at least 4 in high.

Q2.   *Should electrical room doors swing in or out?*

A.   Out. The path of egress should be from the hazardous enclosure or area to the one of lesser hazard; thus the doors should swing into an area of lesser hazard.

Q3.   *What should be the fire rating of the door?*

A.   Each doorway leading into a vault from the building interior must be provided with a tight-fitting door having a minimum fire rating of 3 h. The authority having jurisdiction may require a door for an exterior wall opening where conditions warrant. Doors must be equipped with locks, and doors must be kept locked, access being allowed only to qualified persons. Personnel doors must swing out and be equipped with panic bars, pressure plates, or other devices that are normally latched but open under simple pressure. A similar requirement appears in the National Electric Safety Code in the Sec. 11 passage regarding "Exit Doors."

Q4.   *What is the difference between door exit signs and exit markers?*

A.   Exit signs—fixtures that include the legend EXIT are listed by UL under the category of Exit Fixtures. The basic standard used to evaluate exit fixtures is the Standard for Emergency Lighting and Power Equipment, UL 924, and refer to NFPA 70 and 101 for installation requirements. Exit "markers," on the other hand, are products classified by UL as floor proximity egress path marking systems. These products are evaluated in accordance with UL 1994, the Standard for Low Level Path Marking and Lighting Systems, and are supplemental exit signs required by the Life Safety Code.

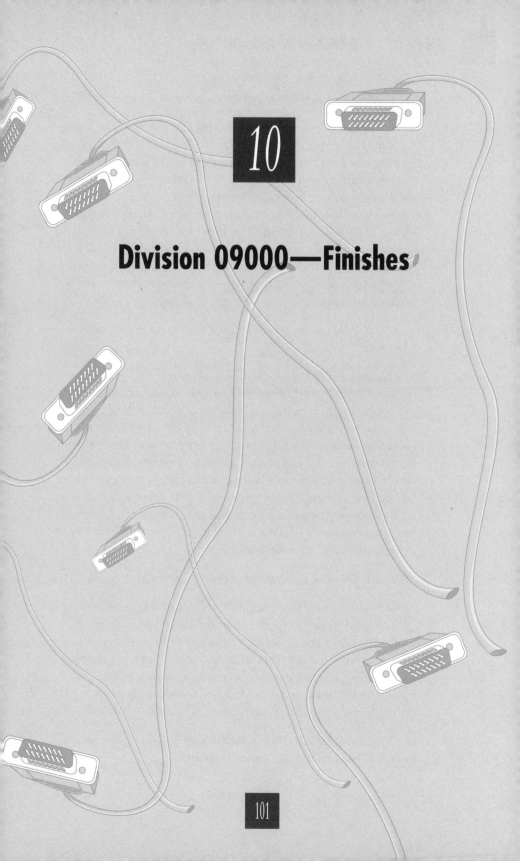

10

Division 09000—Finishes

Q1.   *What NEC requirements commonly appear in Division 09000?*

A.   Since suspended ceilings fall under this CSI division, the wiring methods that apply to them appear in Chap. 3. Cables and raceways are not permitted to be supported by ceiling grids. Wiring located above a fire-rated floor-ceiling or roof-ceiling assembly must not be secured to or supported by the ceiling assembly, including the ceiling support wires. An independent means of secure support must be provided. A similar rule applies to non-fire-rated floor and ceiling assemblies. Exceptions exist for branch circuit wiring and associated equipment where installed in accordance with the ceiling system manufacturer's instructions.

Q2.   *Can suspended ceiling panels be used for access to electrical equipment?*

A.   Yes. Section 300-23 says that cables, raceways, and equipment installed behind panels designed to allow access, including suspended *ceiling* panels, must be so arranged and secured as to allow the removal of panels and access to the equipment. Above suspended ceilings the suspended ceilings provide a thermal barrier of material that has at least a 15-min finish rating as identified in listings of fire-rated assemblies, except as permitted in Sec. 331-3. Fire barriers must be provided where fire walls, floors, or ceilings are penetrated. Access to equipment must not be denied by an accumulation of wires and cables that prevents removal of panels, including suspended ceiling panels.

Q3.   *What are the rules for suspended ceilings with light fixtures?*

A.   Section 410-16 says that framing members of suspended ceiling systems used to support fixtures must be securely fastened to each other and must be securely attached to the building structure at appropriate intervals. Fixtures must be securely fastened to the ceiling framing member by mechanical means, such as bolts, screws, or rivets. Clips identified for use with the type of ceiling framing member(s) and fixture(s) are also permitted.

Q4.   *Do NEC rules exist for the installation of heating cables on dry board, in plaster, and on concrete ceilings?*

A.   Yes. Article 424 says that heating cables must be applied only to gypsum board, plaster lath, or other fire-resistant material. With metal lath or other electrically conductive surfaces, a coat of plaster must be applied to completely separate the metal lath or conductive surface from the cable. The entire ceiling surface must have a finish of thermally noninsulating sand plaster having a nominal thickness of $1/2$ in, or other noninsulating material identified as suitable for this use and applied according to specified thickness and directions.

Q5.   *Does the NEC have anything to say about static electricity?*

A.   No explicit requirements, though much of the underlying thinking that goes into making hazardous locations safe has to do with managing static electricity (refer to Fig. 10.1). Serious voltages are involved—many charges rise to the level of 10,000 to 20,000 V. An average person may not notice a 3000-V spark, though this voltage level is sufficient to disrupt a computer circuit. A 20,000-V charge can create a spark an inch long and can be harmful to human beings.

Q6.   *How can static electricity be managed?*

A.   Humidity control to achieve a 50 percent relative humidity will greatly inhibit electrostatic problems. Where this is not possible, local application of specialty aerosol sprays (so-called Electrician-in-a-Can) may provide temporary relief. Static drain paths, using metal contacts to create charge flow from floor tiles to mats to the nearest grounded metal surface, may also work. Carpeting with low propensity to static electricity and suitable for computer installations is available in the form of mats, continuous material in customary widths, and squares which may be permanently attached to each of the removable floor tiles. Where underfloor access is needed, the continuous material is usually feasible. Some mats with heavy flexible backing and bound edges will lie flat without being permanently attached to the floor. They can easily be lifted for underfloor access, cleaning, or replacement.

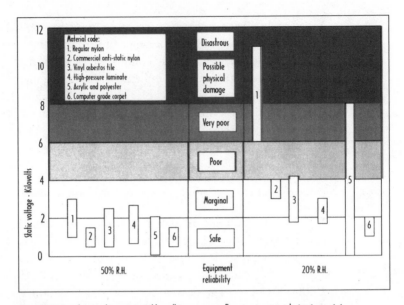

Figure 10.1 Typical static voltages generated by walking on common floor covering materials. (*Adapted from "Guideline on Electrical Power for ADP Installations," Federal Information Standards Publication 94, 1983.*)

Division 10000—Specialties

Remarks. The next four divisions of the CSI format, Specialties, Equipment, Furnishings, Special Construction, corresponding to Divisions 10000 through 13000, have titles that may make it difficult to distinguish content. Almost all of it involves electric end-user apparatus that requires some kind of supply-circuit design and construction.

Q1.   *What NEC requirements commonly appear in Division 10000?*

A.   The NEC does not cover the subassembly electrical systems for which this CSI division applies. In the main, if you specify listed equipment, you will comply with NEC requirements for wiring method, marking, and safety. That much said, the NEC does cover design and construction of the supply circuits to the utilization apparatus that is specified here: supply circuit wiring for signs and outline lighting, clocks, electronic detection and counting systems, and access flooring. Mechanical trades fire-protection specialties appear here, i.e., the extinguishing medium, fire hoses, and standpipe systems, etc. The actual electrics of smoke detection and fire alarms are covered in Division 16700.

Q2.   *What are the broad principles for complying with NEC requirements for electric signs?*

A.   They appear in Art. 600. Signs should be listed. Every commercial building must be provided with at least one outlet for sign or outline lighting system use. Each sign and outline lighting system must be controlled by an externally operable switch or circuit breaker. Signs and metal equipment of outline lighting systems must be grounded. Live parts other than lamps and neon tubing must be enclosed. A sign or outline lighting system equipment must be at least 14 ft above areas accessible to vehicles unless protected from physical damage. A working space at least 3 ft high, 3 ft wide, by 3 ft deep must be provided at each ballast, transformer, and electronic power supply or its enclosure where not installed in a sign. All pretty basic stuff.

Q3.   *What are the NEC requirements for wiring signs?*

A.   Circuits that supply signs and outline lighting systems containing incandescent, fluorescent, and high-intensity-discharge forms of illumination can be rated no more than 20 A. Circuits that only supply neon tubing installations cannot be rated in excess of 30 A. The load for the required branch circuit must be computed at a minimum of 1200 VA. The wiring method used to supply signs and outline lighting systems must terminate within a sign, an outline lighting system enclosure, suitable box, or conduit body.

Signs and transformer enclosures are permitted to be used as pull or junction boxes for conductors supplying other adjacent signs, outline lighting systems, or floodlights that are part of a sign, and are permitted to contain both branch and secondary circuit conductors. UL Standards UL 48, UL 1433, and UL 879 contain the recommended particulars for safe and reliable performance of electric signs.

Q4.   *What are the broad principles for complying with NEC requirements for access flooring?*

A.   They appear in Art. 645 and all of them are common sense. Power, communications, connecting and interconnecting cables, and receptacles associated with the data-processing equipment are permitted under a raised floor, provided the raised floor is of suitable construction and the area under the floor is accessible. Ventilation in the underfloor area must be for the data-processing equipment and data-processing area only (to minimize spread of fire). Openings in raised floors for cables protect cables against abrasions and minimize the entrance of debris beneath the floor. Cables must be listed as type DP cable having adequate fire-resistance characteristics suitable for use under raised floors of a computer room.

Q5.   *Can a metal access floor be used as a signal reference grid?*

A.   Yes. Experience has demonstrated that a grid of conductors on approximately 2-ft centers provides a satisfactory constant-potential reference network over a very broad range of frequencies from dc (0 Hz) to well above 30 MHz. These grids may be built on the subfloor, just underneath the metal access floor.

Refer to Fig. 11-1. Floor height for crawling access is 30 in. (Less than 18 in restricts airflow.) For large computer rooms, install firewall separation barriers to confine fire and halon extinguishing gas. Bolted struts assure low-electrical-resistance joints. Isolate from building steel except via computer system grounding conductors.

Questions on this subject also appear in Divisions 16450 and 16700.

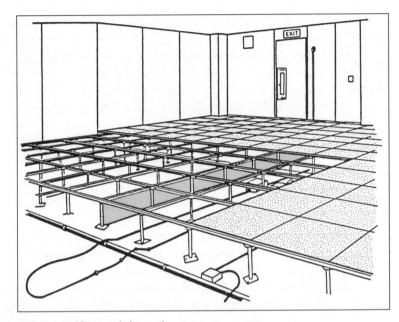

Figure 11.1 Raised floor as signal reference grid.

Q6.   *Is the area under a raised floor in an information technology equipment room considered an air-handling area?*

A.   No. Section 645-5(d) describes the permitted wiring methods under raised floors. Ventilation in the underfloor area is for the information technology room only. Wiring methods cited in Secs. 300-22(c) and (d) and Sec. 645-5(d) are permitted.

Division 11000—Equipment

Remarks. The "equipment" types covered in this CSI division involve utilization apparatus such as central vacuum cleaning systems, security and emergency systems, automatic teller machines, theater and stage, commercial laundry and dry cleaning, loading dock, packaged pumps, water supply and treatment, food service, navigation, operating room equipment, commercial food-preparation equipment, audiovisual systems, and industrial processes. Although noteworthy grand rewrites of NEC Chap. 6 articles have been undertaken, we must limit our coverage of the subject to a few of the most common types of specialty equipment here. The most radical restructuring has occurred in the articles pertaining to electric signs, electric welders, audio signal processing, amplification and reproduction equipment, and fire pumps.

At the very least, this chapter may help the reader gain appreciation for the NEC as a large body of core generic rules modified by a similar-sized body of exceptions. In every code cycle improvements are made in reducing the number of exceptions by casting requirements in positive language. Just as quickly, technological and market innovations create another exception.

Q1.   *What **NEC** requirements commonly appear in Division 11000?*

A.   The wiring rules for a great deal of specialty equipment appear in Chap. 6. Each section covers a different type of equipment. It is easy to follow **NEC** requirements because all the special exceptions to the code basics for workspace, disconnects, overcurrent protection, and grounding are covered in parallel. Table 12.1 provides an overview of equipment types covered in Chap. 6 that have modifying conditions placed upon the core generic **NEC** rules that appear in Chaps. 1 through 3.

Table 12.1 should give you some sense of how generic rules of **NEC** Chaps. 1 through 3 apply to the various specialty equipment classes. Column headings indicate **NEC** generic rules. Where a particular equipment type has an installation rule that differs from the **NEC** generic rule, an X appears. Refer to the applicable **NEC** article for specifics about the exception.

Food Service Equipment

Q2.   *What broad **NEC** principles apply to the branch circuits of commercial kitchen equipment?*

A.   Appliances, in general, are covered in Chap. 4. The bulk of the requirements for branch circuit wiring of them refer back to the basics of Chap. 2, Sec. 220-20 particularly. This section does not apply to conductors that form an integral part of an appliance. It asserts some very basic requirements for making sure branch circuits are capable of carrying appliance currents without overheating under the conditions specified. Ratings of an individual branch circuit cannot be less than the marked rating of the appliance or the marked rating of an appliance having combined loads as provided in Sec. 422-32.

Q3.   *Do permanently connected appliances require a disconnect?*

A.   No. Section 422-21 says that for permanently connected appliances rated at not over 300 VA or $1/8$ hp, the branch-circuit overcurrent device is permitted to serve as the disconnecting

	Location Workspace Guarding	Wiring Method	Supply Conductor Sizing	Overload Overcurrent Protection	GFCI	Disconnect	Grounding
600 -- ELECTRIC SIGNS	X	X			X	X	X
604 -- MANUFACTURED WIRE SYSTEMS		X					
605 -- OFFICE FURNISHINGS		X					
610 -- CRANES AND HOISTS		X	X	X		X	X
620 -- ELEVATORS, ETC	X	X	X	X		X	X
625 -- ELECTRIC VEHICLES, ETC.		X		X			X
630 -- ELECTRIC WELDERS		X	X	X		X	
640 -- SOUND EQUIPMENT, ETC.	X						
645 -- INFORMATION EQUIPMENT, ETC.	X	X	X			X	
650 -- PIPE ORGANS		X	X				
660 -- X-RAY EQUIPMENT			X	X		X	
665 -- INDUCTION HEATING EQUIPMENT, ETC.	X	X					
668 -- ELECTROLYTIC CELLS	X	X					
669 -- ELECTROPLATING		X	X	X		X	
670 -- INDUSTRIAL MACHINERY	X			X			
675 -- IRRIGATION MACHINES		X	X	X		X	
680 -- SWIMMING POOLS, FOUNTAINS, ETC.	X	X	X		X	X	X
690 -- SOLAR PHOTOVOLTAIC SYSTEMS	X	X	X	X	X	X	X
695 -- FIRE PUMPS	X	X	X	X		X	

TABLE 12.1 Overview of NEC Chap. 6—Core NEC Rules

means where the switch or circuit breaker is within sight from the appliance or is capable of being locked in the open position.

Electric Welders

Q4.   *What are the NEC basics for welding equipment?*

A.   Power-supply circuits for welding equipment are sized with knowledge of maximum peak current, duration of the current, and the duty cycle. This much said, it was somewhat tedious doing the basic branch-circuit design arithmetic in previous versions of the **NEC** because of the variety in the nature of electric welding equipment. The 1999 **NEC** has cleaned all this up by merging Art. 630, the rules for motor generator, ac transformer, and dc rectifier welders, into a single set of rules for arc welders.

Part B of this article deals with arc welding. Arc welding with capacitors has power factors on the order of 70 to 80 percent, but without them, power factor is on the order of 35 to 60 percent. Part C of this article deals with resistance welders. Resistance welders have very low power factors. They operate on the principle that the impedance of the joint between dissimilar metals is different for the flow of positive vs. negative current.

Thus the 1999 **NEC** contains an important change regarding the nominal ampere draw for electric welders. Rating plates of electric welders will have a new rating I_{eff}, the "effective rated primary current." A rigorous formula for the determination of I_{eff} is new to the 1999 **NEC**. Supply conductors must be sized accordingly.

Q5.   *How do you size supply conductors to a welding unit?*

A.   Section 630-11 says that the ampacity of the supply conductors must not be less than the I_{eff} value on the rating plate. If the I_{eff} is not given, the ampacity of the supply conductors cannot be less than the current value determined by multiplying the rated primary current in amperes given on the welder rating plate and the factors listed in Sec. 630-11. Current values are determined by multiplying the rated primary current in amperes given on the welder rating plate and the factor based upon the duty cycle or time rating of the welder. A duty cycle table is given in Sec. 630-11 with multipliers that are new in the 1999 NEC revision.

Q6.   *How do you size the supply conductors to a group of welders?*

A.   Section 630-11 says that the ampacity of conductors that supply a group of welders is permitted to be less than the sum of the currents. The conductor rating is determined in each case according to the welder loading based on the use to be made of each welder and the allowance permissible in the event that all the welders supplied by the conductors will not be in use at the same time. The load value used for each welder should take into account both the magnitude and the duration of the load while the welder is in use.

Q7.   *What are the rules for overcurrent protection of welders?*

A.   Section 630-12 says that each welder must have overcurrent protection rated or set at not more than 200 percent of the rated primary current of the welder. Conductors that supply one or more welders must be protected by an overcurrent device rated or set at not more than 200 percent of the conductor rating. Where the nearest standard rating of the overcurrent device used is under the value specified in this section, or where the rating or setting specified results in nuisance tripping, the next higher standard rating or setting is permitted.

Audiovisual Systems

Q8.   *The owner wants a great deal of specialty audiovisual equipment in the facility. How do I specify it according to NEC requirements?*

A.   This is another situation where standard architect-engineer practice in preparing bidding documents can cause difficulty for the electrical contractor. Because such equipment is complicated and continually changing, an up-to-date electrical designer will be hard to come by. Usually, you must depend upon a given equipment manufacturer for engineering information.

The blueprints will typically indicate general layout of the various items of equipment and their functional relationships. However, layout of the equipment, accessories, and conduit systems is diagrammatic unless specifically detailed and does not necessarily indicate every item required for a complete assembly. Quantities of all major installed and all portable equipment, including any add or delete alternates, will be indicated on the system and electrical drawings. Quantities of fittings and hardware related to the major equipment must be determined by examining the various functional diagrams, plans, and risers. Electrical drawings for receptacle backbox locations will appear on the blueprints.

Language such as *"supply and install all conduit and wireways to the extent not included in Division 16100 in order to provide a complete and operable system"* may put the electrical contractor at risk unless the form of proposal allows the cost of this system to be broken out of the overall contract.

Q9. *What broad principles apply to building audiovisual systems according to NEC requirements?*

A. This article has been changed so extensively in the 1999 NEC that it might as well be regarded as a completely new article. The old title to Art. 640, "Sound-Recording and Similar Equipment," was rooted in vacuum-tube technology. The title to Art. 640 has been updated to "Audio Signal Processing, Amplification, and Reproduction Equipment" and takes into account, among other things, the integration of various technologies, the combining of fire alarm and audio systems, for example.

New Art. 640 is broken up into three commonsense parts: Part A (general), Part B (permanent systems), Part C (temporary systems). Noteworthy exceptions and/or modifications to the NEC generics are as follows:

- 640-7. Where the wireway or auxiliary gutter does not contain power-supply wires, the equipment grounding conductors are not required to be larger than AWG No. 14. Where the wireway contains power-supply conductors, the equipment grounding conductor cannot be smaller than specified in Sec. 250-122.

- 640-9(d). The generic rules for power transformers that require that transformers operate within their rating have been extended to audio transformers and autotransformers. Now the AHJ may be able to refer rock concert roadies to the NEC when the volume of the music gets too loud. System grounding is not required for circuits on the load side of the equipment.

  - 640-9(a)(2) allows separately derived systems with 60 V to ground for the purpose of reducing objectionable noise in audiovisual equipment.

  - 640-9(c) provides that amplifiers with output circuits carrying audio program signals are permitted to use Class 1, 2, or 3 wiring where the amplifier is listed and marked for use with the specific class of wiring method. Wiring classes cannot be mixed in the same raceway.

  - 640-10. Audio system equipment supplied by branch circuit power cannot be placed within 5 ft of the inside wall of a pool, tub, or fountain.

  - 640-21(e). Now the NEC recognizes a "technical power system." The practical effect of this is that cords can be used to connected permanently installed equipment racks for isolating what audiovisual people call the "technical ground" from the building premises ground.

- 640-22 says that wiring equipment racks must be made of metal and grounded. Supply cords must terminate within the equipment rack in an identified connector assembly.

The reader should refer to core reference (*EC&M* Illustrated Changes in the 1999 National Electric Code and Analysis of the 1999 National Electric Code) for a comprehensive treatment.

Information Technology Equipment

Q10.   *What broad principles apply to the design and installation of information technology equipment?*

A.   Some requirements appear in Art. 645; others appear in Art. 830. Only listed information technology equipment is allowed, and the room can be occupied only by those personnel needed for the maintenance and functional operation of the installed information technology equipment. The room must be separated from other occupancies by fire-resistant-rated walls, floors, and ceilings with protected openings. The building construction, rooms, or areas and occupancy must comply with the applicable building code.

Q11.   *Does an information technology room require a separate HVAC system?*

A.   Yes. A separate heating, ventilating, and air-conditioning (HVAC) system is provided that is dedicated for information technology equipment use and is separated from other areas of occupancy. Any HVAC system that serves other occupancies is permitted to also serve the information technology equipment room if fire and smoke dampers are provided at the point of penetration of the room boundary. Such dampers must operate on activation of smoke detectors and also by operation of the disconnecting means required by Sec. 645-10.

X-Ray Equipment

Q12.   *What are the electrical operating characteristics of x-ray equipment?*

A.   To produce the x-ray (from the vacuum tube) the applied voltage is measured in terms of the peak voltage which may be anywhere within the range of 10 or 1000 kV. Thus, conductors leading to the x-ray tube must be heavily insulated. The current flowing in the high-voltage circuit may range from 5 to 1000 mA.

X-ray generating apparatus in health-care facilities typically produce momentary high-power factor loads on the order of 20–160 kVA on the equipment power supply circuit. Exposure times may range from less than 2 s up to about 7 s. Voltage regulation should be in the 3 to 10 percent range in order for the equipment to operate properly. Since x-ray generating apparatus needs to have a specific voltage and current available in order to properly excite the x-ray tube, the constraining factor for most radiographic equipment is more commonly the impedance of the power supply system rather than the current capacity.

Fluoroscopy is an x-ray technique that is used in industry. It is similar to radiography but operates at a much lower voltage (less than 250 kV) and, instead of producing a film, a shadow picture is projected on a screen such as that used for security checks of luggage at airport terminals. In either application, the high voltage is obtained by means of a transformer, usually operating at 230 to 240 V primary.

Q13.   *What broad principles apply to the design and construction of x-ray equipment?*

A.   The rules that appear in **NEC** Art. 660 cover all x-ray equipment operating at any frequency or voltage for industrial or other non-medical use. (The **NEC** requirements for x-ray equipment in health-care facilities appear in Art. 517, part E, and should not be confused with industrial x-ray requirements that appear in Art. 660.) Nothing in Art. 660 should be construed as specifying safeguards against the useful beam or stray x-ray radiation. That much said, we have the following code basics:

a. Unless approved for the location, x-ray and related equipment cannot be installed or operated in hazardous (classified) locations.

b. A disconnecting means of adequate capacity for at least 50 percent of the input required for the momentary rating or 100 percent of the input required for the long-time rating of the x-ray equipment, whichever is greater, must be provided in the supply circuit. The disconnecting means must be operable from a location readily accessible from the x-ray control.

c. The ampacity of supply branch-circuit conductors and the overcurrent protective devices cannot be less than 50 percent of the momentary rating or 100 percent of the long-time rating, whichever is the greater. The rated ampacity of conductors and overcurrent devices of a feeder for two or more branch circuits supplying x-ray units cannot be less than 100 percent of the momentary demand rating [as determined by (*a*)] of the two largest x-ray apparatus plus 20 percent of the momentary ratings of other x-ray apparatus.

d. All high-voltage parts—except the leads to the x-ray tube—must be in grounded metal enclosures unless the equipment is in a separate room or enclosure and the circuit to the primary of the transformer is automatically opened by unlocking the door to the enclosure.

Very few changes to this article occurred during the 1999 **NEC** revision cycle.

13

Division 12000—Furnishings

Remarks. The modular office furniture industry claims that about 80 percent of all office furniture "systems" are wired. In many ways, the popularity of these systems has simplified the work of electrical designers; pushing outward to specialty contractors, the job of wiring telecommunication, computer, power, and lighting branch circuits. It all must be brought back to a panelboard or to a local distribution frame, however. Designers should work closely with the furniture supplier to determine the configuration and capacity of branch circuiting.

Q1.   *What NEC requirements commonly appear in Division 12000?*

A.   The bulk of them appear in Art. 605, which covers modular office furniture, related lighting accessories, or similarly wired partitions used for subdividing office space. Other topics that appear in this division are casework, window treatments, furniture accessories, and stadium and arena seating. Rarely do any of these require special consideration beyond the core rules for branch-circuit wiring that appear in **NEC** Chaps. 2 and 3.

Q2.   *What are the general requirements for modular office furniture?*

A.   Section 605-2 says that such wiring systems must be identified as suitable for providing power for lighting accessories and appliances in wired partitions. In the interest of fire safety these partitions cannot extend from floor to ceiling unless given permission by the authority having jurisdiction. Wired partitions are permitted to extend to the ceiling but cannot penetrate the ceiling.

Q3.   *Are the wired partitions required to be permanently connected to the building electrical system?*

A.   According to Secs. 605-7 and 605-8, they are permitted to be but they are not required to be. When they are, they must be connected to the building electrical system by one of the wiring methods of Chap. 3. The rule applies to fixed and free-standing-type partitions. The free-standing power poles associated with modular furniture are regarded as a multioutlet assembly and must follow the core **NEC** rules for multioutlet assemblies.

Q4.   *Can the wired partitions be connected to the building electrical system by a flexible cord?*

A.   Yes. Individual partitions of the free-standing type, or groups of individual partitions that are electrically connected, mechanically contiguous, and do not exceed 30 ft when assembled, are permitted to be connected to the building electrical system by a single flexible cord and plug. A few conditions apply:

    a. The flexible power-supply cord must be of the extra-hard-usage type with No. 12 or larger conductors with an insulated grounding conductor and not exceeding 2 ft in length, *b.* The receptacle(s) supplying power must be on a separate circuit serving only panels and no other loads. It cannot be located more than 12 in from the partition to which it is connected.

c. Individual partitions or groups of interconnected individual partitions cannot contain more than *thirteen* 15-A, 125-V receptacle outlets. Individual partitions or groups of interconnected individual partitions cannot contain multiwire circuits.

Q5.   *Does the* **NEC** *permit the use of three-phase circuits in modular office partitions?*

A.   No. The requirement is plainly stated in Sec. 605-8. There is a reference to a fine print note in Sec. 210-4 which draws attention to the possibility of high harmonic currents in the neutral conductor when a neutral is shared. Office partitions typically have receptacles that are used for electronic office equipment that generates harmonic currents.

14

Division 13000—
Special Construction

Remarks. This CSI division bears close resemblance to Chap. 5 of the NEC: Special Occupancies. It is one of the more tedious articles of the code because it deals with processes and facilities that are extremely hazardous. The dangerous issues covered in Chap. 5 do not lend themselves easily to the broad, general treatment we give NEC topics in this book. The reader should consult one of the fine references in this book listed in the Appendix for the comprehensive treatment these topics deserve, especially the topic of health-care facilities. The NEC topics that we cover in this chapter are in more common use by the general public: high-rise commercial buildings, health-care facilities, hot tubs, swimming pools, and saunas. A more detailed discussion of engineering specifics appears in the applicable Division 16000 section in the second half of this book.

Q1. *What NEC requirements commonly appear in Division 13000?*

A. *Facilities* (as opposed to equipment) with special usage or with special equipment classes should have the modifying conditions detailed here. Typically those modifying conditions are conditions which require more rigorous engineering and construction practice than the generic NEC rules asserted in Chaps. 1 through 3. Arguably, however, all facility classes have special usage, and thus special electrical design and construction requirements. We will handle a few of the most common of these in this book, with special focus on particularly the requirements of high-rise and health-care facilities. Related questions on the subject of elevator recall appear in Chap. 14; smoke detection and control in Chap. 16, and fire alarms and related communications in Chap. 25.

Q2. *What broad principles of building life safety are reflected in the NEC?*

A. The NEC makes frequent reference to NFPA 101, the *Life Safety Code,* and the two should be applied together. Each standard ensures a reasonable level of life safety in building design and arrangement. The broad, common principles are as follows:

■ Adequate exits without dependence on any one single safeguard should be provided. The requirement for two doors in substation rooms is an example of this principle.

■ Any door in a means of egress must be capable of swinging from any direction to the full use of the opening. Doors must swing in the direction of egress when serving a room or area with 50 or more occupants.

■ Construction should be sufficient to provide structural integrity along the path of egress during a fire.

■ Exits should be designed to the size, shape, and nature of the occupancy. Again, electrical equipment and/or utility rooms are an example of a particular type of occupancy.

■ Exits should be clear, unobstructed, and *unlocked.* Intensified security staffing and technological advancements in electronic security systems may complicate the achievement of this objective.

■   Exits and routes of escape should be clearly marked so that there is no confusion in reaching an exit. Electrical exit signs, for example, should be readable and remain lit during a fire.

■   All available technologies should be applied to ensure early warning of a fire. Alarms should be both audible and visible.

High-rise facilities are an important class of building type and pose special, but not unmanageable, challenges in applying the foregoing principles. Refer to Table 14.1 for an overview of the manner in which the various U.S. building codes handle the high-rise problem.

Table 14.1 offers a broad overview of the requirements for high-rise facilities of four major U.S. building codes. Group A, R, and B buildings refer to assembly, residential, and business-type buildings.

Q3.   *How high does a building have to be in order for building codes to classify it as a high-rise building?*

A.   Any building at least seven stories high is defined as a high-rise, although such buildings may also be defined in terms of linear height rather than stories. The NFPA 101, Life Safety Code, defines a high-rise as a building measuring more than 75 ft from the lowest level of fire department vehicle access to the highest occupiable floor. Note grain elevators, storage, manufacturing, and industrial facilities may be as tall as high-rise buildings without having lots of floors. Four types of high rises are apartments, hotel and motel, facilities that care for the sick, and offices.

Q4.   *What are the general requirements of the Life Safety Code with respect to the four high-rise types?*

A.   The Life Safety Code has provisions for existing and new high-rises in each of the four property classes. Existing high-rise hotels must be protected throughout by an approved supervised automatic sprinkler system unless every guest room or suite has exterior access in accordance with NFPA 101. Existing high-rise health-care occupancies, as well as those three to six stories high, must be of fire-resistive construction.

Other construction types are permitted only when an automatic sprinkler protection is provided. Existing high-rise apartment

TABLE 14.1 High-Rise Code Comparison

Item	BOCA	UBC	SBCCI	MASS
Definition: Occupied floors >75 ft (23 m) above lowest vehicle access	X	X	X	
Occupied floors >70 ft (21.5 m) above lowest vehicle access				X
Automatic fire suppression	X	X		X
Automatic sprinkler system or areas of refuge for buildings less than 12 stories			X	
Group R buildings and group B buildings greater than 12 stories or 150 ft (46 m) shall have an automatic sprinkler system			X	
Secondary water supply may be required	X	X	X	
Smoke control on all floors		X		
Smokeproof enclosure for all exit stairs	X	X		
One smokeproof enclosure, other stairs to be pressurized				X
Group B buildings of greater than 15,000 sq ft (1400 sq m) per floor are exempt from smokeproof enclosure requirements when areas of refuge are provided		X		
Smoke detectors in utility rooms	X	X	X	X
Smoke detectors in shafts more than 2 stories high		X	X	X
Smoke detectors in main return and exhaust air plenums for each air-conditioning system		X		

TABLE 14.1 HIGH-RISE CODE COMPARISON (*Continued*)

Item	BOCA	UBC	SBCCI	MASS
Smoke detectors in main return and exhaust air plenums for each air-conditioning system serving more than one story			X	
Provide smoke detectors in ceilings of elevator lobbies		X	X	
Voice alarm systems	X	X	X	X
Public address systems	X	X	X	X
Fire department communication system	X	X	X	X
Fire department central control station	X	X	X	X
Emergency power	X	X		X
Stairway communication system	X	X	X	X
Elevators shall open into lobbies that are separated from other floor areas		X	X	
Elevators cannot be vented through machine rooms		X		
All rooms and enclosures shall have at least one means of exit that does not pass through an elevator lobby			X	
In seismic zones 2, 3, and 4, anchorage of mechanical/electrical systems for life safety shall be anchored to resist earthquake loads		X		

All numbers shown in brackets are metric equivalents.

buildings must have an approved supervised automatic sprinkler system installed throughout. They are exempted only if every living unit has an exterior exit access or if the building has an "engineered life safety system" that has been approved by the AHJ. The latter exception is also the only exception to the requirement for a complete approved automatic sprinkler system in existing high-rise office buildings.

Health-Care Facilities

Q5.   *Does the category "patient care areas" apply to a doctor's or dentist's office where located in one suite of a commercial building?*

A.   Yes. Refer to the definition of patient care areas in Art. 517. If the description fits, these offices may need to comply with Sec. 517-25 requiring limited lighting and power source during normal electrical service interruption. For example, in a doctor's office, where examinations only are being done, then the answer is No. In a dentist's office where root canals are being done, then the answer is Yes. You would want lighting and power for a suction pump so that a person would not choke.

The outlook for electrical power security and occupant safety is likely to become more complex as the health-care industry changes. There will be considerable mixing of facility classes to reckon with as the health-care industry continues to seek new ways to deliver its services at the lowest possible cost.

Q6.   *Do health-care facilities have special requirements for specifying smoke detectors?*

A.   Yes, the Life Safety Code is the primary national standard addressing corridor smoke detectors in health-care facilities. Section 713-10 of the 1997 Uniform Building Code, for example, requires duct detectors be provided within 5 ft of smoke dampers. This requirement can be eliminated when dampers are located in a corridor wall or ceiling where the corridor is protected by a smoke-detection system.

Q7. *How many different types of power systems are typically present in a hospital?*

A. Four: emergency, essential, life-safety, and critical, and they are all interconnected. You might say that the normal power source is the first of five different systems. Emergency power systems are required for all health-care facilities. (Refer to Fig. 14.1.) The basic requirements are defined in various local, state, and federal codes. Health-care industry associations such as the Joint Commission on Accreditation of Hospitals also develop consensus standards that should be followed.

Q8. *What is the emergency system?*

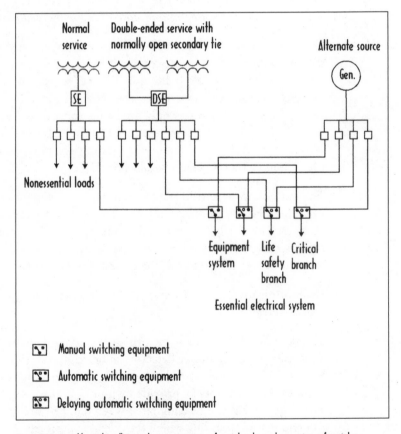

Figure 14.1 Typical hospital installation with a nonautomatic transfer switch and several automatic transfer switches.

A.   A system of feeders and branch circuits meeting the requirements of Art. 700, and intended to supply alternate power to a limited number of prescribed functions vital to the protection of life and patient safety, with automatic restoration of electrical power *within* 10 s of power interruption.

The emergency system is then further broken down. Those functions of patient care depending on lighting or appliances that are connected to the emergency system must be divided into two mandatory branches: the life safety branch and the critical . branch, described in Secs. 517-32 and 517-33.

The branches of the emergency system are installed and connected to the alternate power source so that all functions specified for the emergency system will be automatically restored to operation within 10 s after interruption of the normal source.

Q9.   *What is the essential electrical system?*

A.   A system comprised of alternate sources of power and all connected distribution systems and ancillary equipment, designed to ensure continuity of electrical power to designated areas and functions of a health-care facility during disruption of normal power sources and also designed to minimize disruption within the internal wiring system.

Per Sec. 517-25 the essential electrical system for these facilities must comprise a system capable of supplying a limited amount of lighting and power service, which is considered essential for life safety and orderly cessation of procedures during the time normal electrical service is interrupted for any reason. This includes clinics, medical and dental offices, outpatient facilities, nursing homes, limited-care facilities, hospitals, and other health-care facilities serving patients.

Q10.   *What is the Life Safety Branch?*

A.   A subsystem of the emergency system consisting of feeders and branch circuits, meeting the requirements of Art. 700 and intended to provide adequate power needs to ensure safety to patients and personnel, and which are automatically connected to alternate power sources during interruption of the normal power source.

Q11.   *What systems may be connected to the Life Safety Branch?*

A.   No function other than those listed below must be connected to the Life Safety Branch.

> *Illumination of means of egress.* Illumination of means of egress, such as lighting required for corridors, passageways, stairways, and landings at exit doors, and all necessary ways of approach to exits. Switching arrangements to transfer patient corridor lighting in hospitals from general illumination circuits to night illumination circuits must be permitted provided only one of two circuits can be selected and both circuits cannot be extinguished at the same time.

> *Exit signs.* Exit signs and exit directional signs.

> *Alarm and alerting systems.* Fire alarms and alarms required for systems used for the piping of nonflammable medical gases.

> *Communications systems.* Hospital communications systems, where used for issuing instructions during emergency conditions.

> *Generator set location.* Task illumination battery charger for emergency battery-powered lighting unit(s) and selected receptacles at the generator set location.

> *Elevators.* Elevator cab lighting, control, communications, and signal systems.

Q12.   *What is the critical branch?*

A.   A subsystem of the emergency system consisting of feeders and branch circuits supplying energy to task illumination, special power circuits, and selected receptacles serving areas and functions related to patient care, and which are connected to alternate power sources by one or more transfer switches during interruption of the normal power source. An important generic requirement that an insulated grounding conductor be used for feeder and/or branch circuit makeup in critical care areas appears in Sec. 517-13(a). The requirement applies to metal raceway and the conductor is sized according to Table 250-122.

Swimming Pools, Etc.

Q13.   *What are the broad principles of the* **NEC** *that apply to swimming pools, hot tubs, saunas, and other similar facilities?*

A.   The provisions of Art. 680-1 apply to the construction and installation of electric wiring for and equipment in or adjacent to all swimming, wading, and permanently installed therapeutic pools. The term "fountain" as used in Art. 680 includes fountains, ornamental pools, display pools, and reflection pools. The term is not intended to include drinking-water fountains.

Q14.   *What are the rules for wiring the accessories for such facilities?*

A.   Receptacles on the property must be located at least 10 ft from the inside walls of a pool or fountain. Where a permanently installed pool is installed at a dwelling unit(s), at least one 125-V receptacle must be located a minimum of 10 ft from and not more than 20 ft from the inside wall of the pool. This receptacle must be located not more than 6 ft 6 in above the floor, platform, or grade level serving the pool. All 125-V receptacles located within 20 ft of the inside walls of a pool or fountain must be protected by a ground-fault circuit interrupter. Disconnecting means must be accessible, located within sight from pool, spa, or hot-tub equipment, and must be located at least 5 ft horizontally from the inside walls of the pool, spa, or hot tub.

Q15.   *What are the requirements for electric pool water heaters?*

A.   Section 680-9 says that all electric pool water heaters must have the heating elements subdivided into loads not exceeding 48 A and protected at not more than 60 A. The ampacity of the branch-circuit conductors and the rating or setting of overcurrent protective devices must not be less than 125 percent of the total load of the nameplate rating.

Q16.   *What are the rules for underground wiring?*

A.   Section 680-10 says that underground wiring must not be permitted under the pool or within the area extending 5 ft horizontally from the inside wall of the pool. Some exceptions apply. Where space limitations prevent wiring from being routed 5 ft or more from the pool, such wiring is permitted where installed in rigid metal conduit, intermediate metal conduit, or a nonmetallic raceway system. All metal conduit must be corrosion-resistant and suitable for the location. Division 2 contains information about minimum burial depths.

Miscellaneous

Q17.   *What is the basis for the rigor in* **NEC** *requirements for wiring methods in places of assembly?*

A.   There are several reasons:

- Electrical fires at public events will injure more people.

- The by-products of combustion are usually more dangerous than the fire itself.

- Many places of assembly employ temporary wiring.

- Places of assembly present panic hazards.

Traditionally, nonmetallic wiring methods were not permitted as a wiring method in places of assembly. However, there has been some relaxation in the requirements in the *1999 NEC.* They appear in Sec. 518-4, which lists the permissible wiring methods.

Q18.   *What special considerations should there be for designing supply circuits to x-ray apparatus used in health-care facilities?*

A.   The **NEC** requirements appear in Part E of Art. 517. Designers need to size supply circuit elements on the basis of the largest operating rating per Sec. 517.73. What is not stated explicitly in the **NEC**, however, is the need for excellent voltage regulation.

It is relatively simple to design an x-ray supply circuit with excellent voltage regulation when the power source is an infinite bus. This is usually the case when the power supply originates from the serving utility. To help designers specify a supply circuit to an x-ray machine in a health-care facility, manufacturers will typically specify minimum feeder sizes for a given increment of supply circuit length. (An equivalent approach involves specifying minimum transformer sizes and/or minimum circuit impedance.)

Consideration should be given to the operation of selected x-ray machines if they are to be supplied power from an on-site generator in a power emergency. Because of the high real power requirements for x-ray machines, x-ray exposures can cause slowing of standby generators, reducing the frequency of the supply. Reduced frequency can be a problem for x-ray generators if the unit operates on a resonance principle.

15

Division 14000—Conveying Systems

Remarks. The mechanical trades claim some jurisdiction over these requirements, and a good set of contract documents will state where the mechanical trades end the electrical trades begin. ASME Standard A17.1, the Safety Code for Elevators and Escalators, first published in 1921, looms large in all aspects of elevator design. The bulk of it is devoted to mechanical specialty issues such as counterweights, guide rails, and general hoistway construction. Without it, high-rise buildings would not have been possible.

Questions in this chapter relate primarily to the unit-field interface and focus upon elevators— the most commonly applied conveying system. Owners and engineers should be mindful that physically handicapped persons are at great risk when power is lost to elevators or other personnel conveyors. While it may be unthinkable that some jurisdictions permit this condition to exist, electrical professionals should look for ways to provide an alternate source of power to elevators so that a safe means of egress is provided for the physically disabled.

Q1.   *What NEC requirements commonly appear in Division 14000?*

A.   The bulk of the requirements for conveying systems are covered in
Art. 620, with modifications for health-care facilities appearing
in Secs. 517-32 and 517-42. Elevators, cab lighting, control, communication, and signal systems are one of the eligible system types
that may be connected to the life safety branch of health-care facility power systems. All of the exposed, non-current-carrying metal
parts of an elevator must be grounded per Sec. 250-112. Two
detailed quantitative examples of elevator feeder ampacity design
appear in Chap. 9 of the NEC. These examples demonstrate the
use of Table 620-14, Feeder Demand Factors for Elevators. Where
construction elevators are present, follow Art. 305 rules.

Q2.   *How are elevator systems generally built?*

A.   Conveying systems are so well packaged by manufacturers that
electrical people need only connect them to power supply terminals, save for providing lighting and a few branch circuit receptacles for auxiliary space. Almost everything else is done by the
manufacturer's field installation people. Since the NEC effectively
ensures that manufacturers are part of the NEC process, there is
little likelihood that there will be conflict with respect to compliant voltage levels, conductor types, and controls.

Q3.   *What are the broad principles that govern the design of the elevator machine room?*

A.   Part H of Art. 620 lays out the general rules for all conveying
systems. Machinery spaces can include such equipment as secondary and deflector sheaves, governors, communication panels,
controls, and other auxiliary equipment. All related motor-
generator sets, motor controllers, and disconnecting means must
be installed in a room or enclosure set aside for that purpose.

Some special considerations are given to the controllers and the
driving machinery. Motor controllers are permitted outside
these spaces provided (*a*) that they are in enclosures with doors
or removable panels capable of being locked in the closed position and (*b*) that the disconnecting means is located adjacent to
or is an integral part of the motor controller. Motor controller
enclosures for escalator or moving walks are permitted in the

balustrade on the side located away from the moving steps or moving treadway. If the disconnecting means is an integral part of the motor controller, it must be operable without opening the enclosure. In all cases, the room or enclosure must be secured against unauthorized access.

Various sections of the code refer to lighting of the machine rooms and truss interiors. Basically, the designer needs to provide at least one 15-A, 120-V work receptacle and illumination levels of 10 footcandle (fc) at the floor level. The lighting control switch must be located within easy reach of the access to such rooms. Where practicable, the light control switch must be located on the lock jamb side of the access door.

Q4.   *Do elevator systems have more relaxed working clearance rules?*

A.   Yes, but only where qualified persons are doing the maintenance. Section 620-5 says that working clearance requirements of Sec. 110-26 must be waived as permitted as long as there is no exposure of live parts. The relaxed requirements apply to working space around controllers, disconnecting means, or other electrical equipment.

Q5.   *Why is regenerated power a concern for motor-generator-type elevator applications?*

A.   In some elevator applications, the motor is used as a brake when the elevator is descending and generates electricity. If the source is commercial utility power, it can easily be absorbed. If the power source is an engine-driven generator, the regenerated power can cause the generating set and the elevator to overspeed. The ASME standard says that "When a power source is used which, in itself, is incapable of absorbing the energy generated by an overhauling load, means for absorbing sufficient energy to prevent the elevator from attaining governor tripping speed or a speed in excess of 125% of rated speed, whichever is lesser, must be provided on the load side of each elevator power supply line disconnecting means."

Q6.   *How is the overspeed problem solved?*

A.   To prevent overspeeding of the elevator, the maximum amount of power that can be pumped back into the generating set must be known. The permissible amount of absorption is about 20 percent of the generating set's rating in kilowatts. If the amount pumped back is greater than 20 percent, other loads must be connected to the generating set, such as emergency lights or dummy load resistances. Emergency lighting should be permanently connected to the generating set for maximum safety. A dummy load can also be automatically switched on the line whenever the elevator is operating from the engine-driven elevator.

Q7.   *What are the general rules for wiring elevators?*

A.   Cables must not twist during their rise and fall. All conductors must have the necessary strength and durability for the conditions under which they will be exposed. Thus supports for cables or raceways in a hoistway or in an escalator, etc., must be securely fastened to the guide rails, escalator or moving walk truss, or to the hoistway, wellway, or runway construction. Section 620-21 says that conductors and optical fibers located in hoistways, in escalator and moving walk wellways, in wheelchair lifts, stairway chair lift runways, and machinery spaces, in or on cars, and in machine and control rooms, not including the traveling cables connecting the car or counterweight and hoistway wiring, must be installed in rigid metal conduit, intermediate metal conduit, electrical metallic tubing, rigid nonmetallic conduit, or wireways, or shall be type MC, MI, or AC cable. Some exceptions apply.

Q8.   *How do I select the motor controller?*

A.   Section 620-15 says that motor controllers must comply with Sec. 430-83. The rating is permitted to be less than the nominal rating of the elevator motor, when the controller inherently limits the available power to the motor and is marked as power-limited. Elevators having polyphase ac power supply must also be provided with means to prevent the starting of the elevator motor if the phase rotation is in the wrong direction or if there is a failure of any phase. This protection is considered to be provided if a reversal of phase of the incoming polyphase ac power will not cause the elevator driving machine to operate in the wrong direction.

The power-supply conductors to an adjustable-speed SCR drive (Fig. 15.1) shall be sized from the calculation that produces the greater number of amperes. Compute the amperes available at the optional power transformer and compare it with the drive input amperes. Use the larger number.

Q9. *Where are the rules for wiring service receptacles for the machinery rooms?*

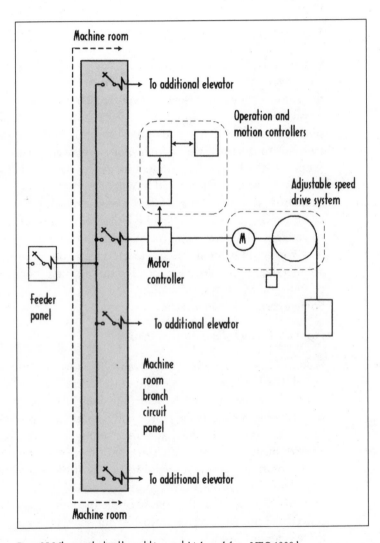

Figure 15.1 Elevator with adjustable-speed drive control. (*Adapted from* NEC 1999.)

A.   Section 620-23 says that (*a*) a separate branch circuit must supply the machine room and machinery space lighting and receptacle(s). Required lighting must not be connected to the load-side terminals of a ground-fault circuit-interrupter receptacle(s). (*b*) The machine room lighting switch must be located at the point of entry to such machine rooms and machinery spaces, and (*c*) at least one 125-V, single-phase, duplex receptacle must be provided in each machine room and machinery space. Similar rules apply to the hoistways, per Sec. 620-24. Signs indicating the location of supply-side overcurrent devices will assist elevator mechanics in troubleshooting a power loss.

Q10.   *Do the convenience circuits get disconnects?*

A.   Yes, Sec. 620-53 says that elevators must have a single means for disconnecting all ungrounded car light, receptacle(s), and ventilation power-supply conductors for that elevator car. The disconnecting means must be capable of being locked in the open position and must be located in the machine room for that elevator car. Where there is equipment for more than one elevator car in the machine room, the disconnecting means must be numbered to correspond to the identifying number of the elevator car whose light source they control.

Q11.   *How do I size the feeder conductors?*

A.   First, work through either of the two examples of App. D. Example D-9 involves generator field control; Example D-10 involves an adjustable-speed drive. The general rules for supply conductor sizing appear in Sec. 620-13. These correlate with the requirements for general motor circuit wiring that appear in Art. 430. Special attention must be given to the coordination of overcurrent devices, especially in grouped elevators. Conductors supplying power to an adjustable-speed SCR dc drive must be sized from the calculation that produces the greater number of amperes. Conductors supplying more than one motor to operate elevators may be sized using the demand factors that appear in Art. 620 and are based upon service classification.

    With generator field control, the conductor ampacity must be based on the nameplate current rating of the driving motor of the motor-generator set that supplies power to the elevator motor. A solved problem is provided at the end of this chapter to illustrate, in step-by-step fashion, the broad design principles. For a more detailed example using SCR drives, see Stallcup's "Electrical Calculations Simplified," one of the core references indicated in the Appendix.

Q12.   *How do I locate and size conductors for the disconnect?*

A.   All motor circuits get disconnects within sight of the motor. Section 620-51 says that a single means for disconnecting all ungrounded main power-supply conductors for each unit must be provided so that no pole can be operated independently. Where multiple driving machines are connected to a single elevator, escalator, moving walks, or pumping units, there must be one disconnecting means to disconnect the motor(s) and control valve operating magnets. The disconnecting means for the main power-supply conductors must not disconnect the branch circuit required in Secs. 620-22, 620-23, and 620-24.

Q13.   *Can other wiring be installed in the hoistway or machine room?*

A.   No. Section 620-37 (Wiring in Hoistways and Machine Rooms) says that only such electric wiring, raceways, and cables used directly in connection with the elevator including wiring for signals, for communication with the car, for lighting, heating, air conditioning, and ventilating the elevator car, for fire-detecting systems, for pit sump pumps, and for heating, lighting, and ventilating the hoistway are permitted inside the hoistway and the machine room.

Q14.   *Where does the disconnecting means need to be located?*

A.   Section 620-53 says that the disconnecting means must be capable of being locked in the open position and must be located in the machine room for that elevator car. Where no machine room exists, the disconnecting means must be located in the machinery space for the given car. Where more than one driving machine disconnecting means is supplied by a single feeder, the overcurrent protective devices in each disconnecting means must be selectively coordinated with any other supply-side overcurrent-protective devices.

Q15. *What are the broad principles that apply to protecting elevator systems from damage by lightning?*

A.   Elevator equipment is typically located on the roof of the building and thus subject to lightning hazards. Equipment and wiring not associated with the elevator must not be installed in elevator machine rooms and hoistways. Where a lightning-protection system is provided and if the system grounding "down" conductors located outside the hoistway is within a critical horizontal distance of the elevator rails, bonding of the rails to the lightning-protection system grounding down conductors is required by NFPA 780 to prevent a dangerous side flash between the lightning system grounding down conductors and the elevator rails. A lightning strike on the building air terminal will be conducted through the lightning-protection system (LPS), and if the elevator rails are not at the same potential as the lightning-protection system, a side flash may occur. Down conductors are installed vertically near the perimeter of the structure; so this requirement may have application "outside" elevators. A more detailed discussion appears in Division 16600.

Q16. *Can lightning-protection wiring be installed in the hoistway?*

A.   Yes. Section 620-37, Wiring in Hoistways and Machine Rooms, says that bonding of elevator rails (car and/or counterweight) to lightning-protection system grounding down conductor(s), are permitted. The lightning-protection system grounding down conductor(s) must not be located within the hoistway. Elevator rails or other hoistway equipment must not be used as the grounding down conductor for lightning-protection systems.

Q17. *Under what circumstances can the elevator main feeder wiring be installed in the hoistway?*

A.   Section 620-37 says that main feeders for supplying power to elevators and dumbwaiters must be installed *outside* the hoistway unless (1) by special permission, feeders for elevators are permitted within an *existing* hoistway if no conductors are spliced within the hoistway, and (2) if the driving-machine motors are located in the hoistway or on the car or counterweight.

Q18.   *Are there any special requirements for heating, cooling, or lighting elevators or the hoistways?*

A.   The **NEC** asserts none, but as a matter of comfort for patrons, many parking lots have heated (or cooled) elevators. For the most part, elevator manufacturers include these conveniences as part of their package. Section 620-22 says that when these conveniences are provided, they must have dedicated branch circuits which do not exceed 600 V. The branch-circuit overcurrent protection device for these conveniences must be located in the machine room or space for the associated elevator.

Q19.   *What requirements exist for elevator car lighting?*

A.   Elevator cars must be provided with not less than two lamps that will shed a minimum illumination at the car threshold of not less than 5 fc for passenger elevators and 2.5 fc for freight elevators. The **NEC** requires that all elevators shall have a means for disconnecting all ungrounded car lighting conductors and this means must be located in the respective machine room. Passenger elevators must be provided with an emergency lighting power source that is automatically switched on instantly after normal lighting fails in the car. The emergency lighting power source is required to be on the car; thus the standby generator cannot be used. This requirement is typically met in the elevator package installed in the field by the elevator manufacturer and contractor.

Q20.   *Do elevator auxiliary branch circuits get GFCI protection?*

A.   Yes, Sec. 620-85 says that each 125-V, single-phase, 15- and 20-A receptacle installed in machinery spaces, pits, elevator car tops, and in escalator and moving walk wellways must be of the ground-fault circuit-interrupter type. All 125-V, single-phase, 15- and 20-A receptacles installed in machine rooms must have ground-fault circuit-interrupter protection for personnel. A single receptacle supplying a permanently installed sump pump does not require ground-fault circuit-interrupter protection.

Q21.   *Do elevators need to be connected to a standby generator?*

A.   Stipulations for the use of standby power are asserted in the ASME standard, paragraph 210.10. Rule 509.1 asserts requirements for a telephone connected to a central telephone exchange be installed in the car and an emergency signaling device operable from inside the car and audible outside the hoistway be provided. Article 620-52 of the **NEC** permits but does not require emergency or standby power. Codes vary from state to state, however, and you need to consider the implications of this upon maintenance budgets. Some states are considering the adoption of rules for testing of elevators which will add substantially to costs. It also figures heavily into how on-site generators are sized.

Q22. *How do I size the emergency generator to the elevator?*

A. The sizing of the generator depends upon the number of cars to be used, their rated load, intended running speed, and method of operation. Provisions should be made to dissipate the power regenerated when elevators are under negative or overhauling loads. This regenerated energy may cause overspeeding of the elevators unless a sufficient amount of power is dissipated by the emergency system, by the electric losses in the elevator equipment, or by some other positive load on the standby supply.

Where two or more elevators are in use in buildings three or more stories high, the elevators or banks of elevators should be connected to separate sources of power. There are situations in which increased standby power is required for all elevators within 15 s. Equipment savings may be made by first supply power during outages to one-half of all the elevators, provided that the traffic can be safely rerouted and the capacity of the operating elevators is adequate to handle the extra traffic. Power should then be transferred to the second bank of elevators shortly thereafter to clear stalled elevators. Power may be left on this bank until power returns to normal. When elevator service is critical, it is recommended that there be fully automatic generating plant starting and power transfer.

When SCR direct-drive elevators are used, the designer should use caution. This type of elevator may have isolation transformers applied which, even if only one elevator is going to run, should all be energized. The standby power system should be sized to supply this inrush current.

Q23. *When do I connect an elevator to an emergency power system?*

A. Part K of Art. 620 covers emergency and standby power systems, which merely states that elevator(s) are permitted to be powered by an emergency or standby power system. The matter is left to the AHJ. For elevator systems that regenerate power back into the power source, which is unable to absorb the regenerative power under overhauling elevator load conditions, a means must be provided to absorb this power. Other building loads such as power and lighting are permitted as the energy absorption means required provided that such loads are automatically connected to the emergency or standby power system operating the elevators and are large enough to absorb the elevator regenerative power. The disconnecting means required by Sec. 620-51 must disconnect the elevator from both the emergency or standby power system and the normal power system. Where an additional power source is connected to the load side of the disconnecting means, which allows movement of the car to permit evacuation of passengers, the disconnecting means required in Sec. 620-51 must include an auxiliary contact. This contact shall cause the additional power source to be disconnected from its load when the disconnecting means is in the open position.

Q24. *What is the NEC rule for selective coordination of overcurrent devices?*

A. It appears in Sec. 620-62 of the **NEC** and in Sec. 45.1 of NFPA 110, Emergency and Standby Power Systems. It is plainly stated: Where more than one driving machine disconnecting means is supplied by a single feeder, the overcurrent protective devices in each disconnecting means must be selectively coordinated. In some circuits full selectivity is very expensive, difficult, and in some cases impossible to achieve. Refer to core reference (Anthony, *Electrical Power System Protection and Coordination*) for comprehensive discussion of selective coordination.

Local codes may require one or more elevators to be operated when the main power supply fails, especially in health-care facilities. Standby generators need to be sized accordingly. Figure 15.2 shows how the on-site generator and logic in the controls of the entire four-elevator system will permit one elevator at a time in each bank to descend to the main floor. This system consists of an automatic transfer switch for each elevator, a sensing and control panel, and a remote selector station. When normal power fails, a preselected elevator is supplied power from the standby power system. The limit as to how many elevators operate at one time is established by the designer in cooperation with local authorities. Elevator control can automatically supervise the

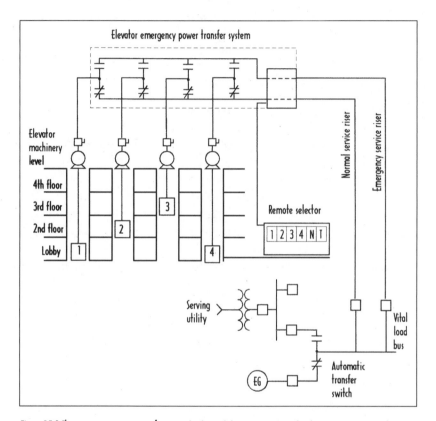

Figure 15.2 Elevator emergency power-transfer system. (*Adapted from IEEE Standard 446, 1987, p. 58.*)

return of each car and also switch cars to automatic service while providing interlocking protection and backup flexibility.

Q25.   *Does the NEC require that elevators work during a fire; or should everyone use the stairway?*

A.   The **NEC** is silent on the matter of whether the elevators need to work, thus leaving the matter to the local authorities. The modern practice is to allow electricity to remain on during a fire.

Elevator operation may be handled in two phases. Phase I requires that all automatic-operation elevators that have to travel 25 ft or more must have a switch at the main floor and smoke sensors in each elevator lobby that, when activated, shall automatically return all elevators nonstop to the "main" or "designated" landing. Phase II requires that a switch be provided in each elevator that, when activated, allows emergency personnel to have control of the elevator. Refer to Fig. 15.2. For a detailed description of elevator operation during a fire refer to ASME A17.1.

Q26. *Is a hospital elevator an emergency system?*

A. Yes. Article 517 defines an emergency system as a system of feeders and branch circuit meeting the requirements of Art. 700, except as amended by Art. 517, and intended to supply alternate power to a limited number of prescribed functions vital to the protection of life and patient safety, with automatic restoration of electrical power within 10 s of power interruption. Thus the rules of Art. 517 take precedence over rules in Art. 700. Emergency systems required in health-care facilities are to be installed according to the provisions of Art. 700. However, where there is a difference in rules between Arts. 517 and 700, those in Art. 517 are to be followed for health-care facilities. This change emerged from concern over designers being required to size standby generators in health-care facilities based on language in Art. 700.

Q27. *Does the NEC require that power to elevators be disconnected before the sprinklers turn on?*

A. Yes. The ASME standard requires that if sprinklers are installed on hoistways, machine rooms, or machinery spaces, a means must be provided to automatically disconnect the main-line power supply to the affected equipment prior to the application of water. Water on elevator electrical equipment can result in hazards such as uncontrolled car movement and movement of elevator with open doors. Other sprinklers in the building should not affect elevator power. A sprinkler system is not a fire-alarm system; it is a fire-suppression system.

Q28. *What other codes apply to conveying systems?*

A.   ASME A17.1, described at the head of this chapter, is the grand old standard for this equipment class. Electrical requirements appear in various places of this document and assert conformance to the wiring requirements of the National Electric Code and the certification requirements of CSA B44.1, also Elevator and Escalator Electrical Equipment Certification Standard, ASME/ANSI A17.5-1991 (CSA B44.1-M91), ASME A17.3, Safety Code for Existing Elevators and Escalators. Changes in the 1999 **NEC** have brought some U.S. elevator requirements in line with the Canadian Electric Code.

Solved Problem—Division 14000

Situation.

A 30-hp, 15-min-intermittent-duty, 460-V motor is applied in a packaged elevator system. The system also comes with a heating and air-conditioning feature which requires 12 A of three-phase power at continuous duty.

Requirements.

Select feeder conductor, disconnect, and fuses for the supply circuit.

Solution

1.   *Find total ampere load.*
       a.   From Table 430-152 a 30-hp motor requires 40 A.
       b.   Intermittent-duty-cycle percentage factor = 85 percent (from Table 430-22(a).
       c.   Heating and cooling requires 12 A (see Sec. 620-22).
       d.   40 × 85% + 12 A = 46 A.
2.   *Select supply conductors*
       a.   From Table 310-16, 46 A requires AWG No. 8.

   b. 90°C THHN conductors are good for up to 55 A in this environment.

3. *Select time-delay fuses.* Time-delay fuses must be sized so that they do not open during motor starting. Table 430-152 gives the maximum rating of the motor branch-circuit protective device at 175 percent; thus 46 A × 175% → 80 A.

4. *Select disconnect.* Based upon time-delay fuse of 80 A, select 100 A disconnect.

REMARKS.

A similar procedure may be followed for designing the supply circuit for a system of several elevators. Conductors supplying more than one motor to operate elevators may be sized using a demand factor which is based upon service classification. These factors appear in Art. 620.

16

Division 15000—Mechanical

Remarks. There is a fair amount of overlap in the mechanical and electrical trades, so much so that mechanical and electrical engineers will renumber passages in Divisions 15000 and 16000 as expertise and budgets allow. When mechanical engineers dominate building infrastructure design, electrical engineers will design the power backbone for the HVAC systems and concentrate on lighting and communications.

Here we cover broad principles, with a few quantitative example calculations that mechanical engineers often perform. Greater detail on motors appears in Division 16480. We have included a few handy tables for design reference purposes. As in other chapters of this book, the Code has a great deal of repetitious *information about the general character of listed equipment, and it will be a continuing assumption that requirements that are normally met by listed equipment manufacturers are omitted from the discussion. We will use the term variable-speed drive (VSD) to mean what the NEC refers to as adjustable-speed drive (ASD).*

Q1.   *What **NEC** requirements commonly appear in Division 15000?*

A.   The bulk of them are concentrated in Chap. 4, the articles
devoted to motors that drive air-handling equipment. The general rules appear in Art. 430, with more rigorous rules for hermetic rules appearing in Art. 440. Article 440 must be applied
in conjunction with Art. 430, and its requirements are more
strict than those of Art. 430. As always, the first principles of
Chaps. 1 through 3 apply unless otherwise modified by requirements that appear in Chap. 4.

Requirements for the heating and cooling of interior electrical
substation rooms appear in Art. 110. Wiring methods appear in
Chap. 3, Fixed Electric Space-Heating Equipment in Art. 424,
and covered here in Division 16800. The **NEC** covers industrial
process heating systems in Art. 665. Section 550-15 covers outside heating equipment requirements for mobile and manufactured homes. Heating appliances are covered in Art. 422. Many
HVAC systems have control wiring, which is covered in Art. 725.

Electrical demand for most building classes to which the **NEC**
applies is concentrated in rotating machinery associated with appliances and HVAC systems. In a typical household, for example,
there are 25 motors (Fig. 16.1). Basement: sump pump, washer-
dryer, furnace blowers and fans, air and water conditioner, air
dehumidifier, air intake louvers. Kitchen: refrigerator compressor,
ice maker, range exhaust fans, garbage disposal, dishwasher, can
opener, microwave oven fans, blenders. Living areas: clocks, ceiling
fans, vacuum cleaners, computer drives and cooling fans, fax
machines and printers, sewing machines, videocassette recorders.
Bathroom: exhaust fan, hair dryers, window operators, backdraft
dampers. Elsewhere: garage door openers, trash compactors, drills,
saws, and various tools.

Q2.   *Are there any other codes that I must apply with respect to the
building mechanical systems?*

A.   Yes, there are many, not the least of which are several shelf-feet
of ASHRAE requirements, the local building code in general and
the Life Safety Code (NFPA 101) in particular. In HVAC systems
there are two broad categories of Code concern: health and safety
considerations and energy provisions. Of the two, health and

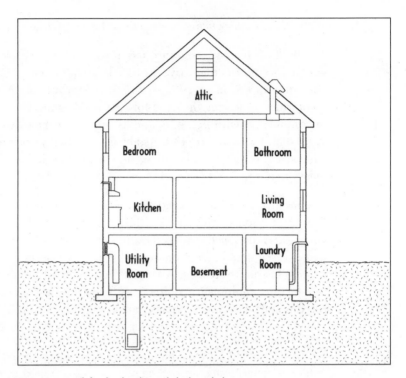

Figure 16.1 Pictorial of residential ventilation and related motor loads.

safety considerations have priority, although in fact they rarely conflict with energy concerns. The general life safety considerations are fire rating and fire separation.

Fire separation: Occupied spaces typically have a 1-h rating separating them from the corridor, and corridors have a 2-h separation from the building egress stair. Similarly, in mixed-use buildings, a retail floor would be separated from the upper office levels with a 2-h fire-rated floor. In an apartment building each tenant would be separated from the next by a fire separation of 1 h.

Buildings that are sprinklered are often permitted to reduce the fire ratings of fire separations, typically by 1 h. Thus for many jurisdictions if a building is sprinklered exitway access corridors and tenant spaces that normally have a 1-h fire-separation requirement are not required to be fire-separated. There are other fire-rating reductions commonly permitted for sprinklered buildings such as for shafts, elevators, and doors.

Vertical openings:. Any opening in a floor of more than 1 sq ft must be protected to prevent smoke and fire migration. *Means of egress:* Most codes require that buildings have at least two means of egress on each story for occupants to exit a building in a emergency. *Plenums:* Exit and access corridors cannot be used for supply or return air plenums. Ceilings above corridors may be used if they are fire-separated from the corridor. Plenums are generally limited to one fire area. Special insulation is required on exposed wiring systems.

Q3.   *Does the **NEC** permit water pipes in substation rooms?*

A.   Yes, but only if the sprinkler heads and pipes are located properly. This problem comes up frequently in renovation projects that involve major overhaul to mechanical and electrical infrastructure. Section 450-47, 384-4a1 says that any pipe or duct system *foreign to the electrical installation* must not enter or pass through a transformer vault. Piping or other facilities provided for vault fire protection or for transformer cooling must not be considered foreign to the electrical installation.

Q4.   *Does the **NEC** require substation rooms to have sprinklers?*

A.   No. Sprinkler systems are optional. Experience has shown the effect of a sprinkler discharge onto a fire scene in an electrical room is far less severe than the fire itself and preferable to a lack of protection. If you opt to install a sprinkler system, then, according to Sec. 110-26(1)(c), the sprinkler piping must comply with the dedicated space restrictions above electrical equipment. Thus sprinkler piping must be routed to avoid the projected footprint of the equipment. According to Sec. 450-47, any piping foreign to the electrical installation must not enter or pass through a transformer vault. Piping or other facilities provided for fire protection or for transformer cooling must not be considered foreign to the electrical installation. There is no need to install switchgear suitable for wet locations, for example.

Q5.   *How much transformer ventilation should we provide?*

A.   Section 450-9 says that ventilation must be adequate to dispose of the transformer full-load losses without creating temperature rise that is in excess of the transformer rating. Transformers with ventilating openings must be installed so that the ventilating openings are not blocked by walls or other obstructions. The required clearances must be clearly marked on the transformer.

Q6.   *What are NEC requirements for ventilating interior electric service apparatus?*

A.   The **NEC** does not require any for primary switches and secondary distribution equipment beyond what the manufacturers recommend. Transformers, however, are a different story, since they inherently generate heat. The basic rule is indicated in **NEC** 450-45.

Q7.   *How do I ventilate a transformer vault?*

A.   The **NEC** cites specific requirements for natural air circulation, and the IEEE makes a general recommendation for forced-air ventilation. The **NEC** says that transformers with ventilating openings must be installed so that the ventilating openings are not blocked by walls or other obstructions. (Required clearances must be clearly marked on the transformer.) One rule of thumb is to allow 3 cu ft/min per kVA of transformer capacity. A more rigorous calculation would involve obtaining information about load and no-load losses and ventilating efficiency.

Article 450-41 says a vault must be located where it can be ventilated to the outside air without using flues or ducts wherever such an arrangement is practicable. Ventilation openings must be located as far away as possible from doors, windows, fire escapes, and combustible material. A vault ventilated by natural circulation of air must be permitted to have roughly half of the total area of openings required for ventilation in one or more openings near the floor and the remainder in one or more openings in the roof or in the sidewalls near the roof, or all of the area required for ventilation must be permitted in one or more openings in or near the roof. For a vault ventilated by natural circulation of air to an outdoor area, the combined net area of all ventilating openings, after deducting the area occupied by screens, gratings, or louvers, must not be less than 3 sq in per kVA of transformer capacity in service, and in no case must the net area be less than 1 sq ft for any capacity under 50 kVA. Ventilation openings must be covered with durable gratings, screens, or louvers, according to the treatment required in order to avoid unsafe conditions. All ventilation openings to the indoors must be provided with automatic closing fire dampers that operate in response to a vault fire. Such dampers must possess a standard fire rating of not less than 1$^{1}/_{2}$ h. Ventilating ducts must be constructed of fire-resistant material.

Q8. *What rules does the NEC assert for wiring in ducts?*

A. Section 300-22 excludes from use in all air-handling spaces any wiring that is not metal-jacketed or metal-enclosed, to minimize the creation of toxic fumes due to burning plastic under fire conditions. The Fine Print Note to Sec. 300-22(c) makes it very clear that the area above a drop ceiling used for the return air is not a plenum. *Be mindful of the difference between a space, a plenum, and a duct.* The basic condition that must be satisfied is that the wiring materials and other construction of the equipment must be suitable for the expected ambient temperature to which they will be subjected. The designer and/or installer must check carefully with equipment manufacturers and with inspection agencies to determine what is acceptable in air-handling space above a suspended ceiling.

Q9. *Where does the NEC absolutely forbid wiring of any kind?*

A. In ducts used for dust, loose stock, or vapor removal. Section 300-22 asserts that no wiring systems *of any type* may be installed in such ducts or for ventilation of commercial-type cooking equipment. The intent of this section is to limit the use of materials that would contribute smoke and products of combustion during a fire and to provide an effective barrier against the spread of products of combustion into the ducts or plenums.

Q10. *Can we put wiring in ducts or plenums used for environmental air?*

A. Yes, under certain conditions. Section 300-22b says that only wiring methods consisting of type MI cable, type MC cable employing a smooth or corrugated impervious metal sheath without an overall nonmetallic covering, electrical metallic tubing, flexible metallic tubing, intermediate metal conduit, or rigid metal conduit may be installed in ducts or plenums specifically fabricated to transport environmental air. Flexible metal conduit and liquidtight flexible metal conduit are permitted, in lengths not to exceed 4 ft, to connect physically adjustable equipment and devices permitted to be in these ducts and plenum chambers. The connectors used with flexible metal conduit must effectively close any openings in the connection. Equipment and devices are permitted within such ducts or plenum chambers only if necessary for their direct action upon, or sensing of, the contained air. Where equipment or devices are installed and illumination is necessary to facilitate maintenance and repair, enclosed gasketed-type fixtures are permitted.

This section applies to ducts and plenums such as sheet-metal ducts specifically constructed to transport environmental air. Equipment and devices such as lighting fixtures and motors are not normally permitted in ducts or plenums.

Q11.   *Can we put wiring in space used for environmental air?*

A.   Electric equipment with metal enclosures is allowed within spaces used for environmental air; however, nonmetallic enclosures must be listed for their use. The space over a hung ceiling used for environmental air-handling purposes is an example of the type of other space to which Sec. 300-22(c) applies.

This part permits use of totally enclosed, nonventilated busway in air-handling ceiling space provided it is a non-plug-in-type busway that cannot accommodate plug-in switches or breakers. The one specific busway wiring method was added for hung-ceiling space used for environmental air. Surface metal raceway or wireway with metal covers or solid bottom metal cable with solid metal covers may be used in air-handling ceiling space provided that the raceway is accessible, such as above lift-out panels. This section recognizes the installation of motors and control equipment in air-handling ducts where such equipment has been specifically approved for the purpose. Equipment of this type is listed by UL and may be found in the Electrical Appliance and Utilization Equipment List under the heading Heating and Ventilating Equipment.

Q12.   *Does the Code allow transformers to be installed in space used for "environmental air"?*

A. The Code does not specifically prohibit transformers which have metal enclosures with ventilation openings from being in such space used for environmental air movement. The relevant passage of the Code appears in Sec. 300-22(c)(2), which allows electrical equipment with a metal enclosure or nonmetallic enclosure listed for the use and having adequate fire-resistant and low-smoke-producing characteristics within such space. An exception in Sec. 450-13 states that dry-type transformers not exceeding 600 V nominal and 50 kVA is permitted in fire-resistant hollow spaces of buildings not permanently closed in by structure and provided they meet the ventilation requirements of Sec. 450-9. A transformer under faulty conditions and ventilated as required would potentially allow smoke or other overheated products into a space used for environmental air and could be passed on through the system. If the installation in question does not meet the rating, the transformer should not be allowed and is prohibited if the 1-h fire-resistant rating cannot be verified.

Q13. *What is the correct nominal voltage to use, 440, 460, or 480 V?*

A. 460. That is the nominal value that appears in Table 430-150, one of the most frequently applied tables in the **NEC**. In the old days, when services were rather distant from the motor load, there was a great deal of voltage drop. You will still see motor starters with nominal voltages of 440 V, though the voltage-drop problem has long since been resolved. Prior to the late 1960s, low-voltage three-phase motors were rated 200 V for use on both 208-V and 240-V systems; 440 for use on 480-V system and 550 V for use on 600 V systems. The reason for this was that most three-phase motors were used in large industrial plants where relatively long circuits resulted in voltage considerably below nominal at the ends of the circuits. Also, utility supply systems had limited capacity and low voltages were common during heavy load periods. As a result, the average voltage applied to three-phase motors approximated 220, 440, and 550 V nameplate ratings. One of the first remedies to this problem was to separate lighting from "power" (or motor) circuits. In temporary wiring systems this is still required by the **NEC**. Another remedy came in the development of dry-type transformers and materials that made it easier to put substations closer to the load, typically inside the building rather than in a faraway underground vault.

Table 16.1 is based on a building with the insulation necessary to provide proper comfort and operating economy. With a known heat loss, the electric load in kilowatts can be obtained by dividing the estimated heat loss (Btu/h) by 3413, since there are 3413 Btu in 1 kWh of electricity. Usually it is necessary to use a demand factor of 100 percent for electric heating loads. Loads larger than a few hundred watts should be connected to the power panels in order to prevent excessive voltage drop on the lighting circuits. Installed heating loads should not be supplied from lighting panelboards.

TABLE 16.1 All-Weather Comfort Standard Recommended Heat-Loss Values

| | Design heat loss per sq ft of floor area | |
Degree days	Btu/h	Watts
Over 8000	40	11.7
7001–8000	38	11.3
6001–7000	35	10.3
5001–6000	32	9.4
3001–5000	30	8.8
Under 3001	28	8.2

Table 16.2 gives the approximate air-conditioning load that might occur in the average commercial building. Loads include compressors and all auxiliary equipment involved in the HVAC system. Actual ac loads are dependent on the internal heat load, which can vary considerably with building design and use. This information may be helpful in establishing the size of the service feeder. The **NEC** does not assert load requirements. Article 440 covers the construction requirements for ac units that have hermetic motors; all other ac motors are covered in Article 430. Branch-circuit construction requirements for new restaurants appear in Art. 220-36. Table 220-36 recognizes the effect of load diversity typical of restaurant occupancies; it also recognizes the amount of continuous load as a percentage of the total connected load. The data for this change was based on load studies of 262 restaurants. (See **NEC Handbook**.) These studies have shown that the demand factors decline as connected loads

increase. Elevators and show window lighting also figure into commercial facility loading.

TABLE 16.2 TOTAL CONNECTED ELECTRICAL LOAD FOR AIR CONDITIONING ONLY

Type of building	Conditioned area, VA/sq ft
Bank	7
Department store	3–5
Hotel	6
Office building	6
Telephone equipment building	7–8
Small store (shoe, dress, etc.)	4–12
Restaurant (not including kitchen)	8

For initial estimates, before actual loads are known, the factors shown in Table 16.3 may be used to establish the major elements of the electrical system serving HVAC&R systems.

TABLE 16.3 FACTORS USED TO ESTABLISH MAJOR ELEMENTS OF THE ELECTRICAL SYSTEM SERVING HVAC&R SYSTEMS IN HOSPITALS

Item	Unit
Refrigeration machines:	kVA/ton of chiller capacity
Absorption	0.10
Centrifugal/reciprocating	1.00
Auxiliary pumps and fans:	
Chilled water pumps	0.08
Condenser water pumps	
Absorption	0.15
Centrifugal/reciprocating	0.07

Table 16.3 Factors Used to Establish Major Elements of the Electrical System Serving HVAC&R Systems in Hospitals (*Continued*)

Item	Unit
Cooling tower fans	
Absorption	0.10
Centrifugal/reciprocating	0.07
Boilers:	kVA/boiler hp
Natural gas/fuel oil	0.07
Boiler auxiliary pumps:	kVA/boiler hp
Deaerator	0.10
Auxiliary equipment:	kVA/bed
Clinical vacuum pumps	0.18
Clinical air compressors	0.10

Q14.   *What are the basics for sizing conductors to a chiller motor?*

A.   To properly size field electrical wiring you must know the minimum circuit ampacity of the chiller. The **NEC** 440-33 defines the method of calculating the minimum ampacity. It is defined as the sum of three amperages: rated load current ratings (RLA), plus the full load amperes (FLA) all remaining loads, plus 25 percent of the highest motor or motor compressor in the group.

As an example, suppose we wanted to calculate the minimum circuit ampacity of a machine with a design RLA of 350 A and is to be operated on a 460-V power supply. Assume that the remaining load on the starter consists of a 3-kVA control power transformer that supplies power to the control, the oil pump motor, the oil sump heater, and the purge unit motor. The remaining load FLA equals 3000 divided by the unit design voltage.

$$\text{Min circuit ampacity} = (125\% \times 350\text{ A}) + (3000\text{ VA}/460)$$
$$= 437.5\text{ A} + 6.5\text{ A} = 444\text{ A}$$

The selection of conductors is based on a number of jobsite conditions, i.e., type of conductor, number of conductors, length, ambient, and temperature rating of conductors. Circuit breakers and fused disconnects should be sized by the electrical engineer in accordance with Sec. 440-21. This protection should be for motor-type loads and should not be less than 150 percent of the compressor motor rated load amps.

Q15. *How do I select motor overload protection?*

A. Refer to Fig. 16.2 and **NEC** Sec. 430-32. The procedure for sizing the overload protection for motors rated more than 1 hp is found by multiplying the nameplate FLA of the motor nameplate by 115 or 125 percent. If the overload protection is not sufficient to start the motor or carry the load, the multiplier may be increased to 130 to 140 percent according to **NEC** Section 430-34. The percentage factor is selected based upon service factor or temperature rise of the motor.

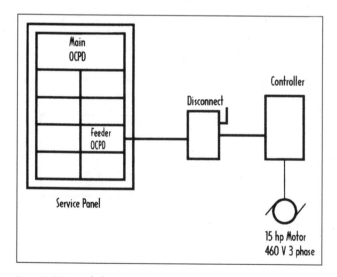

Figure 16.2 Motor overload protection.

If this motor has a service factor of 1.15, a temperature rise of 40°C, and a nameplate current of 18 A, you would use a multiplier of 125 percent for the selection of the time-delay fuses in the disconnect. Thus 18 A × 125% = 22.5 A. According to Sec. 240-6(a), you would select the 20-A fuse. The overload relay would be selected on the basis of the actual nameplate current using a manufacturer's table similar to Appendix item A-26.

Q16.   *How do I specify circuit elements to an air-conditioning unit?*

A.   Refer to Fig. 16.3. For units with a branch-circuit selection current, the circuit conductors are selected either (*a*) by the nameplate of the listed assembly or (*b*) based upon the nameplate values of the individual motors. Article 440-33 restates the branch-circuit conductor sizing rule for a group of motors. Conductors to one or more motor-compressor(s) must have an ampacity not less than the sum of the rated-load or branch-circuit selection current ratings, whichever is larger, of all the motor-compressors plus the full-load currents of the other motors, plus 25 percent of the highest motor or motor-compressor rating in the group.

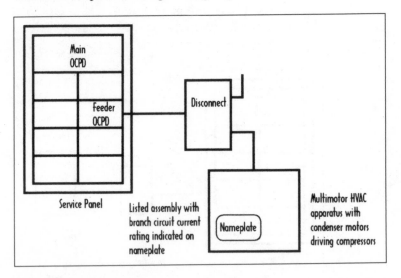

Figure 16.3 Feeder circuit to an ac unit.

For listed assemblies the testing laboratory standard includes the 25 percent increase for the largest motor-compressor in the group plus the ampere draw of any other nonrotating load (such as a heater or a control transformer). The actual nameplate full-load amperes for the complete assembly can be used to size the branch-circuit conductors. AC unit manufacturers are required by Art. 440-4(c) to provide a nameplate that indicates the rated-load current for branch-circuit makeup.

Q17. *How shall I specify supply-circuit elements to two ac units?*

A.   For units with a branch-circuit selection current, the circuit conductors are selected either by (*a*) the nameplate of the listed
assembly or (*b*) based upon the nameplate values of the individual
motors. Article 440-33 restates the branch-circuit conductor-sizing
rule for a group of motors that appears in Sec. 430-24. Refer to
Fig. 16.5. Conductors to one or more motor-compressor(s) must
have an ampacity not less than the sum of the rated-load or
branch-circuit selection current ratings, whichever is larger, of all
the motor-compressors plus the full-load currents of the other
motors, plus 25 percent of the highest motor or motor-compressor
rating in the group. For listed assemblies the testing laboratory
standard includes the 25 percent increase for the largest motor-
compressor in the group plus the ampere draw of any other nonrotating load (such as a heater or a control transformer). The
actual nameplate full-load amperes for the complete assembly can
be used to size the branch-circuit conductors. AC unit manufacturers are required by Art. 440-4(c) to provide a nameplate that
indicates the rated-load current for branch-circuit makeup.

The 240-V, 30-A circuit breaker is typically not within sight of
the exterior pad-mount ac unit (Figs. 16.4). Thus Art. 440-14
requires separate disconnect adjacent to the motor-controller or a

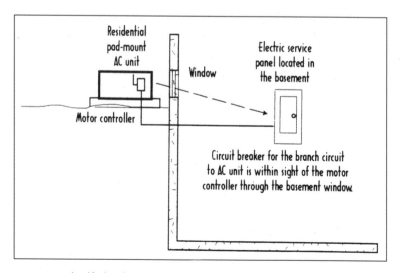

Figure 16.4 Residential fixed-wired ac unit.

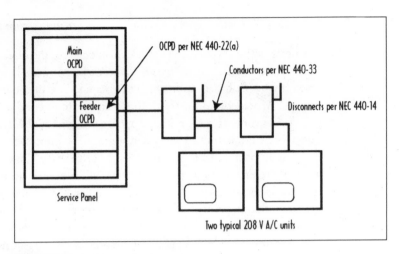

Figure 16.5 Feeder circuit to two ac units.

combination starter. Whichever disconnect apparatus is within sight it must also be readily accessible.

Q18.   *What part of the NEC applies to motors that are part of appliances such as dishwashers, disposals, coolers, and hydromassage tubs?*

A.   Article 430 applies to motor-driven appliances except where amended by Art. 422, Appliances. Section 422-4 specifies that the branch circuit be selected based on the marked rating of the appliance. If not marked, Art. 430 applies to motor-operated appliances. Hermetic motors are covered in Art. 440, which also amends provisions of Art. 430. Air handlers, which are part of fixed electric space heating equipment, are covered in Art. 424. Properly listed hydromassage bathtubs will have the motor attached and the piping fixed in place. To field install or remotely locate the pump is a violation of the listing.

Q19.   *What are premium-efficiency motors?*

A.   Because of motor manufacturing restrictions imposed by the EPACT 1997, most motors will be of the "energy-efficient" design only. NEMA MG-1 uses the term "energy-efficient," but the rest of the industry still uses "high efficiency." The law applies to any NEMA design motor that is a general-purpose,

T-frame, single-speed, foot-mounted, polyphase induction motor designated as Designs A and B, continuous rated, operating at 230/460, 60 Hz as defined in NEMA Standard MG-1 1987. Motors affected are rated 1 to 200 hp, dripproof, and totally enclosed. More discussion appears in Division 16480.

Variable-Speed Drives

*Q20.   Does the **NEC** require VSDs?*

A. No. This is a good example of what is meant by the **NEC** being a "permissive" code. The selection of a VSD does not necessarily pose a safety hazard and is almost always done to improve operating economics. A safety hazard only exists when the VSD installation does not meet **NEC** requirements.

Refer to Fig. 16.6. Good specification practice would ensure that the VSD and all accessories be provided by one supplier. This may be a requirement for listing. Some plant engineers, for example, do not install a VSD on a bus containing power-factor correction capacitors or harmonics-sensitive equipment.

If power-conversion equipment is part of an adjustable-speed drive system, the circuit conductors, the OCPD, and other related elements are sized on a basis of rated input current of the unit. The disconnecting means must have a rating not less than 115 percent of the rated input current of the conversion unit.

Q21.   What article of the Code has jurisdiction on the variable-frequency drive unit?

A.   If a variable-frequency drive is being used to supply a motor, then Art. 430 is applicable. One of the 1996 changes to this section has been to use the phase "solid-state motor-controller" in place of "adjustable-speed drive." This was done to include solid-state controllers that do not provide speed adjustment in the rule. You should know whether the drive you are working with is an adjustable *voltage* drive because the rules for sizing conductors, circuit breakers, and fuses are different from the rules for adjustable-speed drives.

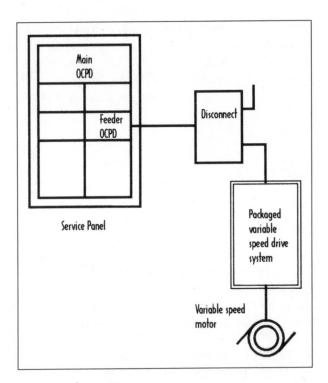

Figure 16.6 Supply circuit to VSD.

Q22.   *What kind of applications would suit a VSD retrofit?*

A.   When you use a motor for an application where torque requirements remain constant or reduced over the total speed range being applied, a VSD will be a good means to achieve speed control, provided that the motor is able to handle the distorted electrical power being delivered to it by the VSD. Applications where torque requirements remain constant or reduced over a motor's total speed range include fans, pumps, and conveyor belts.

Q23.   *What kind of applications would not suit a VSD retrofit?*

A.   Centrifugal pumps and fans, where as speed decreases, torque will usually decrease as the square of the speed, and horsepower will decrease with the cube of the speed. Thus, if the horsepower is established at the low end of the speed requirement (say 50 percent rated speed at 10 hp), the horsepower requirement at full speed will be eight times as much, or 80 hp. As you can see in this situation, the deciding factor for horsepower requirement must be based at full-rated load, and a new motor may be necessary.

Q24.   *How do I size the conductors to a variable-speed drive?*

A.   See Sec. 430-2. If power-conversion equipment is part of an adjustable-speed drive system, the circuit conductors, overcurrent protective devices, and other related elements are sized on the basis of *rated input current* of the unit. The incoming branch circuit or feeder to power-conversion equipment included as a part of an adjustable-speed drive system must be based on the rated input to the power-conversion equipment. Where the power-conversion equipment is marked to indicate that overload protection is included, additional overload protection is not required.

The disconnecting means is permitted to be in the incoming line to the conversion equipment and must have a rating not less than 115 percent of the rated input current of the conversion unit. Be mindful of electrical resonance, which can result from the interaction of the nonsinusoidal currents induced by this type of load when applied with power-factor-correction capacitors. Refer to Fig. 16.6.

Q25.   *Can I connect a new VSD to an existing general-purpose motor?*

A.   Maybe; maybe not. The existing motor was designed for at least 60-Hz excitation; 50/60 Hz at best. You must determine whether the variable-frequency change will affect the motor performance. Ultimately you will need to know whether the new VSD will cause greater vibration, heat rise, or even audible noise. Pulse-width modulation causes a high rate of voltage rise of the carrier frequency, can cause insulation to break down on the end turns of the motor windings as well as feeder insulation. The carrier frequency, a by-product of obtaining current at a variable fundamental frequency, is the cause for having additional watts in the motor. With frequencies other than the fundamental, a motor runs at very high slip and thus inefficiently. Because of this and other conditions you may need to derate an existing motor when it is connected to a VSD. A compromise may need to be struck between the motor's capability and the actual horsepower output. In many cases it may be better just to purchase a new motor having the requirements you need.

Q26.   *How do you size motor conductors to a VSD?*

A.   See Sec. 430-6, Ampacity and Motor Rating Determination. The size of conductors supplying equipment covered by this article must be selected from Tables 310-16 through 310-19 or must be calculated in accordance with Sec. 310-15(b). For motors used in alternating-current, adjustable-voltage, variable-torque drive systems, the ampacity of conductors, or ampere ratings of switches, branch-circuit short-circuit and ground-fault protection, etc., must be based on the maximum operating current marked on the motor or control nameplate, or both. If the maximum operating current does not appear on the nameplate, the ampacity determination must be based on 150 percent of the values given in Tables 430-149 and 430-150.

Q27.   *What section of the NEC requires overcurrent protection on VSDs?*

A.   The requirements for overcurrent protection of motor circuits are contained in Part D of Art. 430-51. Part D specifies devices intended to protect motor branch-circuit conductors, the motor control apparatus, and the motors against overcurrent due to short circuits or grounds. You should be mindful of the distinction between overcurrent and short circuit, which are fundamentally different electrical phenomena. Article 450-52 gives the general overcurrent rules. Table 450-152 establishes the rules for selecting fuses and circuit breakers that protect VSDs.

Q28.   *Do I need to provide overload protection to a VSD even if the manufacturer has indicated that I don't?*

A.   Section 430-2 asserts that if the equipment is so marked, additional overload protection is unnecessary.

Q29.   *What are the rules for grounding VSDs?*

A.   AC system grounding is not required for separately derived systems operating at 50 to 1000 V when they are used exclusively for rectifiers supplying only adjustable-speed industrial drives (Sec. 250-5b, ex 2). Harmonics are sometimes singled out as the culprit in VSD misoperation. Faulty drive operation may originate in grounding. Check the wiring of the shielded isolation transformer. Proper wiring requires the incoming ground, shield, core, and secondary neutral and ground all to bond at one point. This is the single-point ground, and it prevents unwanted ground-current loops by providing only one conducting path between it and all external grounds. When the secondary neutral is not bonded to the single-point ground and the transformer is not locally grounded to the building ground system you may have problems with the VSD nuisance shutdown. Check the neutral-to-case ground voltage. It should be zero.

Q30. *Does the* **NEC** *require any particular operating characteristics for VSDs?*

A. No, none that exceed that which the manufacturers must provide in order to secure a UL listing. A good specification might include some of the following operating features:

- Speed control that is stepless throughout the speed range under variable load on a continuous basis.

- Control of the pulse-width modulation type and with power factor of not less than 0.9 at any speed from 50 to 100 percent.

- The system with an incoming circuit breaker with manual bypass having safeties and DDC control active in both the VSD and bypass modes.

- Automatic restart on loss of power but limited to three restart attempts.

- A complete self-diagnostics capability which is automatic but can be run manually.

- Capable of restarting a coasting load.

- Manufacturer to furnish reactors to limit voltage THD to 0 percent.

- Motor noise to be limited to 3 dB above the across-the-line noise.

- The controls to include a VSD bypass selector switch for choosing VSD operation or manual bypass.

■ The VSD (including isolation transformer, reactors, or filters) to have a minimum efficiency of 85 percent at any motor speed from 50 to 100 percent.

■ The VSD to have a minimum power factor of 0.9 at any motor speed from 50 to 100 percent.

■ The VSD to limit the total harmonic distortion reflected back into the power system to values tolerable for the entire building distribution system at any motor speed from 50 to 100 percent.

Variable-frequency drives are listed under UL508C, "Power Conversion Equipment." Typically, VSD manufacturers are eager to assist you in putting together a specification once you have selected the appropriate system for your application.

Q31. *Are there special requirements for the conductors between the VSD and the motor itself?*

A.   None but the general rules for specifying conductors that do not exceed the voltage, current, and temperature limits. Section 310-13(FPN) indicates that if the conductors are installed in a wet location electroendomosis (damage to dc cables, due in part to cumulative absorption of water) can take place if thermoplastic insulation is used. Because a portion of a VSD's voltage cycle is dc, the use of thermoset or rubber insulation is recommended.

Q32. *Does only the motor determine the branch size?*

A.   No. A recent change that appears in the 1996 **NEC** requires that the branch circuit be based upon 125 percent of the rated input to the power-conversion equipment. Most VSD manufacturers supply this information. If there is any other circuitry (auxiliaries such as control transformers, oil pumps, factory automation apparatus, sensing devices, etc.) on the branch circuit, it must be added to the calculated load. If the motor is connected to a VSD, then Sec. 430-6(c) has further requirements, notably a 150 percent (rather than a 125 percent) ampacity multiplier for sizing conductors and protective devices.

Q33.   Can VSDs be installed in a plenum?

A.   No. Under Sec. 300-22(b) equipment and devices must be permitted within such ducts or plenum chambers only if necessary for their direct action upon, or sensing of, the contained air. This allows an HVAC blower motor in a plenum because it is acting directly on air. But the VSD itself is not permitted in the area. For a motor to be permitted, the UL investigation would include checks to make sure that the VSD would not cause the motor to overheat within the plenum. If, on the other hand, a room is being used in its entirety as part of a return-air system, the VSD is permitted within the room.

Q34.   Is a disconnect required adjacent to the VSD?

Yes. Section 430-102(a) requires a disconnect ahead of and in sight of motor controllers. Article 100 defines "in sight" as being visible and within 50 ft.

Fire Pumps

Q35.   How do I know when a fire pump is necessary?

A.   When local water pressure needs to be supplemented, typically to a high-rise facility. The architect will have consulted with local authorities. Then the BOCA requirements would have been assessed. Hopefully as a mechanical engineer you would have already been part of the process, assessing water pressure. Perhaps the insurance company would have been part of the discussion. Somehow having the architect, the local fire marshal, the insurance company, and the mechanical and electrical engineers should result in a decision.

Q36.   Where do NEC requirements for fire pumps appear?

A.   See the following core articles: 695, 665, 422, and 424, and App. A, which indicates all references to NFPA 20. For connection at services, refer to Secs. 320-72a, 230-82, and 230-94. For emergency power supply, refer to Art. 700; for remote control circuits Art. 430-72, and for service entrance overcurrent Art. 230-90. Article 695, with its genesis in NFPA 20, has been extensively rewritten in the 1999 update to the **NEC**. More discussion in this book appears in Division 16480.

Unlike most other equipment, fire pump controllers are required to be purchased under a unit contract along with the pump and driving means. This establishes single-source responsibility for the satisfactory operation of the entire fire pump system. Other NFPA rules define the requirements for fire pump motors, controllers, installation, and testing and refer to **NEC** sections that describe the design generics of feeder makeup, overcurrent protection, and grounding.

Q37.   *Where do we put the fire pump?*

A.   Section 695-7 says that electric motor-driven fire pump controllers and power-transfer switches must be located as close as practicable to the motors they control and must be within sight of the motors. The engine-drive controllers must be located as close as practicable to the engines they control and must be within sight of the engines. Storage batteries for diesel engine drives must be rack supported above the floor, secured against displacement, and located where they will not be subjected to excessive temperature, vibration, mechanical injury, or flooding with water. All energized equipment parts must be located at least 12 in above the floor level. Fire pump controllers and power-transfer switches must be located or protected so that they will not be damaged by water escaping from pumps or pump connections. Thus, all fire pump control equipment must be mounted in a substantial manner on noncombustible supporting structures.

Q38.   *How must we provide power to the fire pump?*

A.   The **NEC** mentions several in Art. 695, among them:

- *Electric utility service connection:* Where power is supplied by a service, it must be located and arranged to minimize the possibility of damage by fire from within the premises and exposing hazards.

■   *On-site power production facility:* Where power is supplied by on-site generation, the generation facility must be located and protected to minimize the possibility of damage by fire.

■   *Feeder sources:* Where reliable power cannot be obtained from either of the foregoing, the AHJ may permit a combination of one or more sources of power to the fire pump.

The largest load for fire protection will usually be a fire pump, which is required to maintain system pressure beyond the capacity of the city water system (Table 16.4). The fire pump is one of the few loads permitted to be connected to the power source ahead of the service disconnect. The reason for this is that when there is a fire, the fire-fighting contingent wants the option to turn the power to the building off, while still keeping the fire pump energized.

TABLE 16.4 TYPICAL POWER REQUIREMENT (KW) FOR FIRE PUMPS IN COMMERCIAL BUILDINGS (LIGHT HAZARD)*

Area/floor, sq ft	Number of stories			
	5	10	25	50
5,000	40	65	150	250
10,000	60	100	200	400
25,000	75	150	275	550
50,000	120	200	400	800

*Based on zero pressure at floor 1.

Generally, for a commercial building, the loads of plumbing and sanitation equipment are not large. Typical loads for water-pressure boosting systems appear in Table 16.5. Sump and sewage pumps are usually small, often applied in pairs with an electrical or mechanical alternator control so that allowing for several 2-hp duplex units is a satisfactory allowance for the basement of most buildings.

Q39.   *Does the* **NEC** *require a specific level of reliability for fire pumps?*

TABLE 16.5 TYPICAL POWER REQUIREMENT (kW) FOR HIGH-RISE BUILDING WATER PRESSURE BOOSTING SYSTEMS

Building type	Unit quantity	Number of stories			
		5	10	25	50
Apartments	10 apt./floor	—	15	90	350
Hospitals	30 patients/floor	10	45	250	—
Hotels/motels	40 rooms/floor	7	35	175	450
Offices	10,000 ft²/floor	—	15	75	250

A.   Section 695-3 covers the topic of multiple power sources to electric motor-driven fire pumps. Where reliable power cannot be obtained from a source described in Sec. 695-3(a), it must be from two or more of either of the above in combination, or one or more of the above in combination with an on-site generator, all as approved by the authority having jurisdiction. The power sources must be arranged so that a fire at one source will not cause an interruption at the other source. Supply conductors must directly connect the power sources either to a listed combination fire pump controller and power-transfer switch or to a disconnecting means and overcurrent protective device(s) meeting the requirements of Sec. 695-3(c). Where one of the alternate power sources is an on-site generator, the disconnecting means and overcurrent protective device(s) for these supply conductors must be selected or set to allow instantaneous pickup and running of the full pump-room load.

Q40.   *If the utility cannot supply power to the fire pump at 208 V, can we use a transformer?*

A.   Yes. Section 695-5 says that where a transformer is dedicated to supplying a fire-pump installation, it must be rated at a minimum of 125 percent of the sum of:

1. The rated full load of the fire-pump motor(s)

2. The rated full load of the pressure-maintenance pump motor(s) when connected to this power supply

3. The full load of any associated fire-pump accessory equipment when connected to this power supply.

Q41.   *What consideration should be given to the fire-pump control wiring?*

A.   Section 695-14 says that external control circuits must be arranged so that failure of any external circuit (open or short circuit) must not prevent the operation of the pump(s) from all other internal or external means. Breakage, disconnecting, shorting of the wires, or loss of power to these circuits may cause continuous running of the fire pump but must not prevent the controller(s) from starting the fire pump(s) due to causes other than these external control circuits.

Q42.   *How are fire-pump circuits protected, if at all?*

A.   One of the most important rules is that ground-fault protection must not be installed on fire-pump feeder circuits even though they may exceed 1000 A 480 V. Overload protection is not required for fire pumps.

Heat-Generation Equipment

Q43.   *Where do the NEC rules for electric space heating appear?*

A.   The bulk of them appear in Art. 424, which is focused upon fixed electric heating units—as opposed to the movable single-phase units that plug into a 15-A branch-circuit receptacle commonly used by consumers. Another discriminator is whether the electric heater contains a blower motor. Small single-phase residential units typically do not; commercial and industrial heaters do. Even though electric heating is thermostatically controlled and is a cycling load, it must be taken as a continuous load for sizing the elements of the branch circuit. The simplicity of these rules should not minimize the special attention that must be given to high-current devices. Heating equipment employing heating elements rated more than 48 A must have the heating elements subdivided, with each subdivided circuit loaded to not more than 48 A and protected at not more than 60 A. The subdivision is usually made by the manufacturer in the heat enclosure or housing.

Section 424-3(b) must be used to calculate the size of conductors to supply power to fixed electric space-heating units that are equipped with motor-operated equipment (Fig. 16.7).

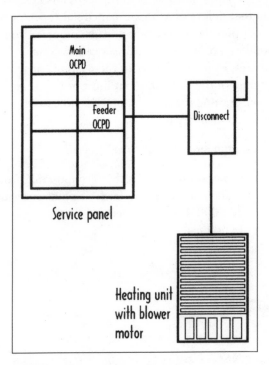

Figure 16.7 Small commercial electric space heater.

Q44.   *What are the basics?*

A.   First, a distinction between fixed and nonfixed (or flexible cord) connected equipment, most commonly, the difference between a single-phase electric heater that is used in residences and moved from room to room, and the three-phase commercial units that are installed in a loading dock, for example. Second, sometimes the heaters have motor loads that must be taken into consideration. The bulk of the rules appear in Art. 424, Fixed-Electric Space Heating Equipment.

Q45.   *Can I connect my home furnace to a branch circuit for the lighting?*

A.   No. Section 422-12 asserts that central heating equipment other than fixed electric space-heating equipment must be supplied by

Table 16.6 Typical Power Requirement (kW) for Electric Hot Water Heating System

Building type	Unit quantity	Load
Apartments/Condominiums	20 apt/condo	30
Dormitories	100 residents	75
Elementary schools	100 students	6
High schools	100 students	12
Restaurant (full service)	100 servings/h	30
Restaurant (fast service)	100 servings/h	15
Nursing homes	100 residents	60
Hospitals	100 patient beds	200
Office buildings	10,000 ft²	5

an individual branch circuit. Auxiliary equipment such as pump, valve, humidifier, or an electrostatic air cleaner directly associated with the heating equipment is permitted to be connected to the same branch circuit. This exception permits the electric motors, ignition systems, etc., natural gas or oil central heating equipment to be connected to the same individual branch circuit.

Q46. *Can I install receptacles on the same circuit as a fixed electric space heater?*

A. Only if the circuit is 15, 20, or 30 A. If the particular branch circuit you plan to use for heating has other outlets on it, then the branch circuit may only be a 15-, 20-, or 30-A branch circuit. The idea is to make sure that you don't plug in a high-power hair dryer when the electric heater is running full tilt.

Q47. *How can wall receptacles be provided in a room to satisfy the 6-ft requirement if electric baseboard heaters are in the way?*

A. You can't if you are to comply with the baseboard manufacturer's listing instructions. You must apply receptacle accessories supplied by baseboard heater manufacturers. Article 424-9 permits the use of these accessories on the condition that the receptacle outlets not be connected to the heater branch circuit. See Underwriter's

Laboratories Standard 1042, Electric Baseboard Heating Equipment, which warns that a heater must not be located below an electric convenience receptacle.

Q48. *Is a disconnect required for a small "window shaker" (residential air-conditioning unit)?*

A.   No. Section 440-13, Cord-Connected Equipment, for cord-connected equipment such as room air conditioners, household refrigerators and freezers, drinking water coolers, and beverage dispensers, a separable connector or an attachment plug and receptacle must be permitted to serve as the disconnecting means. See also Sec. 440-63.

Refrigeration Equipment

Q49. *Why does the NEC treat hermetic motor compressors differently than it treats standard electric induction motors?*

A.   Some of the basic differences are that it typically has no shaft, it combines motor and compressor in one unit, and the motor operates in the refrigerant. As the compressor builds pressure, the current increases, causing the windings to get hot. But at the same time, the refrigerant gets colder, passes over, and cools the motor windings. Thus, its current draw changes with the pressure developed by the compressor. The requirements for hermetic motor compressors are asserted in Art. 440.

A hermetic refrigerant motor-compressor is a compressor-motor combination which is enclosed in the same housing with no external shaft or seals in which the motor actually operates in the refrigerant fluid itself. Hermetically sealed equipment commonly uses halogenated hydrocarbon refrigerants, though variants are applied in industry and commercial systems.

Q50. *How do I determine the design amperes for the branch circuit?*

A.   The value in amperes is to be used instead of the rated load current. It is used to determine the ratings of motor branch-circuit conductors, disconnecting means, controllers, and branch-circuit ground fault and short-circuit protection whenever the running overload protection device permits a sustained current greater than the specified percentage of the rated load current. The value of the branch-circuit selection current will always be greater than the marked rated-load current. Larger units are usually equipped with inherent protectors, and thus manufacturers may not require additional external overcurrent devices. But they may be required by the AHJ, with fuse protection cited specifically.

Q51.   *Are the design rules any different for hermetic motors?*

A.   Section 440-6, Ampacity and Rating, says for a hermetic refrigerant motor-compressor, the rated-load current marked on the nameplate of the equipment in which the motor-compressor is employed must be used in determining the rating or ampacity of the disconnecting means, the branch-circuit conductors, the controller, the branch-circuit short-circuit and ground-fault protection, and the separate motor overload protection. Where no rated-load current is shown on the equipment nameplate, the rated-load current shown on the compressor nameplate must be used. For disconnecting means and controllers, see also Secs. 440-12 and 440-41.

Q52.   *How do I design a power circuit to a group of hermetic motors?*

A.   See Sec. 440-7. In determining compliance with this article and with Secs. 430-24, 430-53(b) and (c), and 430-62(a), the highest-rated (largest) motor must be considered to be the motor that has the highest rated-load current. Where two or more motors have the same highest rated-load current, only one of them must be considered as the highest-rated (largest) motor. For other than hermetic refrigerant motor-compressors and fan or blower motors as covered in Sec. 440-6(b), the full-load current used to determine the highest-rated motor must be the equivalent value corresponding to the motor horsepower rating selected from Tables 430-148, 430-149, or 430-150. There is one exception that comes up rather frequently: where so marked, the branch-circuit selection current must be used instead of the rated-load current in determining the highest-rated (largest) motor-compressor.

Q53.   *Where do I put the disconnect?*

A.   The same rules for general purpose motors apply to hermetic motors (see Sec. 440-14, Location). Disconnecting means must be located within sight from and readily accessible from the air-conditioning or refrigerating equipment. The disconnecting means must be installed on or within the air-conditioning or refrigerating equipment. Where the disconnecting means provided in accordance with Sec. 430-102(a) is capable of being locked in the open position, and the refrigerating or air-conditioning equipment is essential to an industrial process in a facility where the conditions of maintenance and the supervision ensure that only qualified persons will service the equipment, a disconnecting means within sight from the equipment is not required.

Miscellaneous

Q54.   *How do VSDs work?*

A.   Motor speed is adjusted by controlling the output voltage and frequency of the drive. This is accomplished by rectifying the incoming ac supply voltage to dc. The dc voltage is then inverted by a three-phase inverter section to an adjustable-frequency output whose voltage is adjusted proportionately to the frequency to provide constant volts per hertz excitation to the motor terminals. In this way, energy-efficient speed control is obtained by exciting the motor in the range of 2 to about 133 Hz—tolerable for most conventional low-voltage ac induction motors.

Q55.   *Do VSDs always contribute to harmonic distortion?*

A.   While VSDs are an undeniable source of harmonics, they should not be regarded as automatically blameworthy. The actual effect of the harmonics they generate is principally defined by the characteristics of the system itself: the overall system capacity, the system impedance, and the types of apparatus residing throughout the system. IEEE Standard 519 sets recommended safe levels of harmonics *for a given system,* but there is no standard which sets requirements for any individual types of equipment.

While harmonic currents and voltages will vary with differing speed cycles, the distortion may originate in intermittent wiring or grounding problems that mimic harmonic distortion. System construction problems may inadvertently create pathways for circulating electrical noise and may be mistakenly assigned to harmonic problems caused by a VSD. By taking measurements you will be able to distinguish between a power-related anomaly and a problem originating with the VSD. Most power quality problems with VSDs are manageable with appropriate application of circuit elements with harmonic trapping or dampening effects. Wiring problems that look like harmonic problems are often difficult to find though relatively inexpensive to repair.

Q56.   *Does the* **NEC** *allow a motor to restart automatically?*

A.   Only if it does not endanger people, which, in the case of small HVAC motors is often the case. The relevant section is 430-43, where language has been clarified with regard to motors restarting after an overload trip. A motor overload that can restart a motor automatically after overload tripping cannot be installed if automatic restarting of the motor can result in an injury to persons. Automatically restarting a motor after a power outage due to something other than a motor overload is not prohibited by this section. Other permissive conditions for automatic restarting appear in NFPA 79, Electrical Standard for Industrial Machinery.

Q57.   *What is a jockey pump?*

A.   Jockey pumps make up for minor leaks in the water lines. Their controllers must maintain specific pressures within fire lines and should conform to NFPA 20. The same standard prohibits fire pumps from being used to maintain water pressure within other parts of the facility.

Q58.   *What are* **NEC** *requirements for the control of fire alarm systems?*

A.   The application of **NEC** Sec. 300-21 is a very broad and expanding controversy in all contemporary electrical work. The general principle underlying this straightforward statement is that electrical installations be made to substantially protect the integrity of fire-rated walls, fire-resistant or fire-stopped walls, partitions, ceilings, plenums, and floors. Electrical installations must be so made that the possible spread of fire through hollow spaces, vertical shafts, and ventilating air-handling ducts will be reduced to the minimum.

This much said, the closed-circuit alarm system should be arranged to operate when the fire-protection system is tripped; the system should be interlocked to stop the cooling-tower fan. A key-operated bypass switch should be provided to take the fans off the circuit for testing the system. An accessible, manually operated fan shutoff switch should be provided for emergencies. Fire pumps can then be sequenced in the manner described in NFPA 20 and in Art. 695.

Q59.   *What does the BCSC engraving mean on a motor-compressor nameplate?*

A.   The manufacturer has designed better cooling and better heat dissipation into the unit. Thus, the motor-compressor can be continuously "worked" harder than equipment not so designed. This is safe insofar as the motor-compressor is concerned, but the conductors, disconnect switches, and other associated equipment must be capable of safely carrying this higher current draw. Sections 440-2 and 4404(c) assert that the BCSC current value must be used in the selection of disconnecting means instead of the locked-rotor current.

Q60.   *What are the most significant changes in the 1999* **NEC** *code cycle that affect mechanical system design?*

A.   A few of the most significant are as follows:

■   A relaxation in the generic rule for keeping a motor disconnect in sight of the motor appears in an exception to Sec. 430-102(b). If you can lock open a motor supply circuit at the motor control center, then there is no requirement for additional disconnect at the motor location. This will save dollars and wall space in many facilities.

■   The overhaul of Art. 695 resulted in clarification that fire pumps may be supplied power from an on-site power production *facility* per Sec. 695-3(a)(2) (as opposed to an on-site generator). This may affect the service supply design strategy for those organizations, such as process industries and large universities, with behind-the-fence generation. It also makes it easier to use some commonsense combination of feeder sources to increase reliable power to the fire pump as long as you can get the AHJ to agree with you.

■   Designers and contractors involved in pool, spa, fountain, and similar installations will find a fair amount of change in Art. 680 where new rules for placement of receptacles, disconnects, and emergency shutoff switches appear. Other changes include lighting, grounding, wiring methods, and GFCI.

Solved Problem —Division 15000

SITUATION.

A VSD drives a 460-V HVAC motor of unknown horsepower. It is not known whether the drive has the capability to open the circuit under locked-rotor conditions. Refer to Fig. 16.6.

REQUIREMENTS.

Design the power-supply circuit.

SOLUTION

1. Find the rated input current from the VSD nameplate. It is found to be 150 A.

2. Whichever OCPD we select must have a continuous rating of 150 A. Since so little is known about the VSD, select a fused disconnect. Section 430-52(b) says that we must select a fuse that does not open under starting conditions. Section 240-3(b)

allows us to select the next higher standard fuse as long as it is less than 800 A. Select a 150-A time-delay fuse.

3. Section 430-2 allows us to select a disconnect rated at least 115 percent of the rated input current of the conversion unit. $150 \times 115\% \rightarrow 172.5$. Select a 200-A, 600-V class general-duty switch.

4. Size conductors based upon Table 310-16. A No. 1/0 AWG THHN will carry up to 170 A up to 90°C.

5. From Table 250-122 select an equipment grounding conductor based upon the continuous rating of the fuse. Select No. 6 AWG Cu.

6. From App. C, Table C.1, find how many No. 1/0 AWG conductors you can put in a trade-size conduit. Assume that it is more economical to pull four of the same conductors into the raceway. Select the $1^{1}/_{2}$-in conduit to carry all four No. 1/0 AWG conductors.

REMARKS.

Many design engineers are in the habit of sizing all motor circuit components at 125 percent. This example shows application of minimum requirements.

Solved Problem 2—Division 15000

SITUATION.

A 50-hp, 480-V continuous-duty motor-line induction motor consumes its nameplate current of 60 A at 80 percent power factor during normal operation. The owner wants to improve the power factor by putting a capacitor at the motor.

REQUIREMENTS.

How do you select the motor running overload protection?

SOLUTION.

Per Sec. 430-32, the running overload protection, the nameplate current is used (not the 65 A value given in **NEC** Table 430-150). The actual running current *after* power factor correction is 60 A $\times$ 80 percent = 48 A. The value for overload protection given in Sec. 430-32 for continuous-duty motors is 125 percent. Thus 1.25 $\times$ 48 A = 60 A. This is the rating of the overload device you need to specify.

REMARKS.

The reason for applying the capacitor is to reduce input current and thereby save energy. Unity power factor results in the least amount of current needed to drive the motor, and this must be taken into account when selecting the overload protection. Motors have operating dynamics that make it nearly impossible to use the same device for both overloads and short circuits. The **NEC** asserts that overcurrent protection strategies involve the use of two or more devices applied together to protect motors and branch circuits from the effects of phase overloads and ground faults.

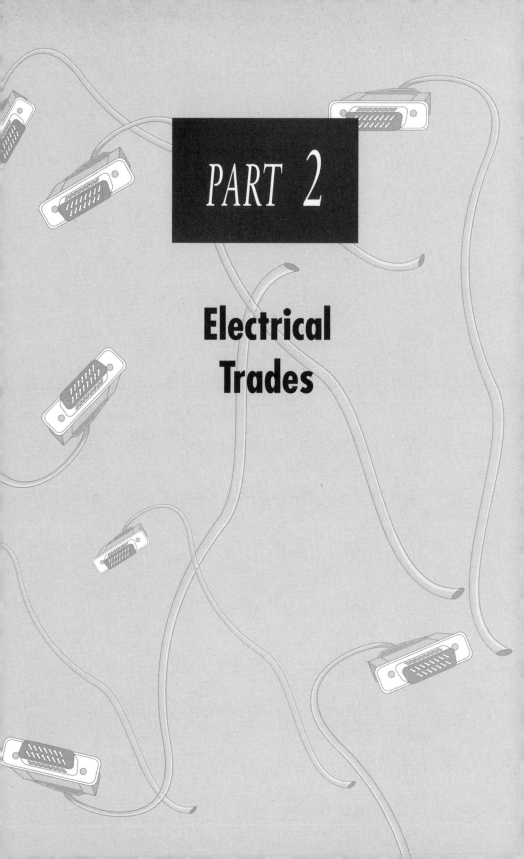

PART 2

Electrical Trades

Division 16000—General Information for Electrical Trades

Remarks. We begin the second half of this book with a chapter that should be read by architects and owners because it covers the sensitive topics of facility supervision, workspace, and accessibility. Selecting a location and designing service, feeder, and branch circuits is fundamentally an optimization problem in which the designer must strike the right compromise among the competing requirements for operability, safety, and economy. Legitimate code requirements should square neatly with good practice and designer preference.

Much of the difficulty in understanding the NEC lies in interpreting its well-intentioned effort to be as general as possible. In order to *help the class of building professionals for whom the meaning of the term "ungrounded conductor" simply means "hot wire," our discussion throughout Part 2 will be limited to mainstream applications in residential, commercial, industrial, and institutional facilities. We will assume 60-Hz ac systems using copper conductors installed in nonhazardous locations. We will assume dead front switchgear (exposed live metal equipment is typically in legacy industrial settings and is being progressively replaced). Secondary voltages will be conventional 480-, 240-, and 208-V three-phase systems and 120- and 277-V single phase branch circuits connected in parallel. We will assume the function of a neutral conductor is understood even though the NEC does not define it.*

Q1. *What NEC requirements commonly appear in Division 16000?*

A. Consensus standards specific to the electrical trades are typically listed here: NFPA, UL, ANSI/IEEE, NECA, NEMA, NETA, BICSI, USTA, and the like. Some variants of the CSI Masterspec© format include site electrical and temporary power requirements. (**NEC Answers** covers them in Chaps. 1 and 3, respectively.) Information about basic electrical requirements, basic materials and methods, and electrical identification appears here, along with procedural rules for submittals and coordination with other trades. Many engineering firms have specific coordination stipulations placed in this division that are intended to prevent conflicts between mechanical and electrical systems.

Special attention should be paid to the "cast of characters" who will have influence on the electrical construction. Among them are the host power and telecommunication utility representatives and the electrical inspector. Product quality assurance rules may be stated here with a list of qualified manufacturers, so the equipment suppliers need to be in the loop. The assumption is that these suppliers are approved by the owner and will meet **NEC** listing requirements. There may be a requirement that subcontractors be able to show experience in the field installation of specialty equipment. Specific requirements for licensing of professional engineers and electricians may appear here.

You should become acquainted with the OSHA representative, especially if the job is a renovation and will involve switching. OSHA may require that existing building one-line diagrams be brought up-to-date so that electricians can safely do switching. Owners should have up-to-date one-line diagrams for their own sake, not because OSHA mandates it. There is nothing more fundamental to electrical safety than having an up-to-date road map of the electrical system.

The concept of "separately derived service," so central to understanding the National Electric Code, can be better understood in terms borrowed from the Kirchhoff laws which in themselves are derived from the laws of energy conservation. All currents in a normally operating circuit which are supplied by a given transformer must eventually return to that transformer. Ampere return paths through the ground are established in each voltage level (Fig. 17.1).

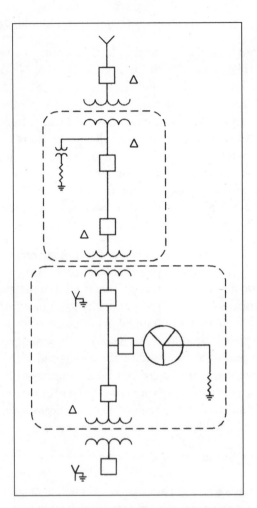

Figure 17.1 Transformers create the boundary of a closed electrical system.

Q2.   *What is utilization equipment?*

A.   The **NEC** defines utilization equipment as electrical equipment that converts electric power into some other form of energy, such as light, heat, or mechanical motion. It is the thing we use electricity *for*—as distinguished from the system through which the electric energy is merely distributed.

Every item of utilization equipment should have a nameplate listing, which includes, among other things, the rated voltage for which the equipment is designed. With one major exception most utilization equipment carries a nameplate rating that is the same as the voltage rated 120 V; for 208-V systems, 600 V; and so on. The major exception is motors and equipment containing motors. Table 17.1 will help proper selection of the motor nameplate voltage that is compatible with the specific available nominal system voltage. Motors are also about the only utilization equipment used on systems over 600 V.

Voltage

Q3. *Why does the word "nominal" appear so frequently in the NEC?*

A. It helps identify a generally agreed-upon midpoint of the voltage range at which practical circuits operate. The voltages of all real circuits will vary depending upon how far the utilization equipment is from the source and depending upon the voltage "draw" of the utilization equipment itself. For example, even though the nameplate on the back of your computer may indicate 115 V (the nominal value), the actual voltage at which the computer may effectively operate may be as low as 106 V and as high as 127 V. Nominal system voltages represent the desired voltage of the power source and are in multiples of 120 V. Rated voltages represent the ideal operating voltages of equipment. Motors are rated in multiples of 115 V. Ultimately, voltage class represents the level to which wiring and equipment are insulated.

Q4. *What voltage should we select for internal distribution, 480 or 208 V?*

A. It depends upon the load and the voltage level of the serving utility. The **NEC** makes no requirements except as they pertain to voltage drop. For renovations the decision may be very difficult, with either voltage operating as well as the other. The basic trade-off is between space for step-down transformers (480/208 transformer rooms for receptacles) and space for larger conduit (you may need more 208-V circuits and they will all be larger than their 480-V equivalent in terms of energy delivery).

Table 17.1 Standard Nominal System Voltages and Voltage Ranges*

Voltage class	Nominal system voltage			Nominal utilization voltage	Voltage range A			Voltage range A		
					Maximum	Minimum		Maximum	Minimum	
	Two-wire	Three-wire	Four-wire	Two-wire Three-wire Four-wire	Utilization and service voltage	Service voltage	Utilization voltage	Utilization and service voltage	Service voltage	Utilization voltage
Single-phase systems										
Low voltage	120			115	126	114	110	127	110	106
		120/240		115/230	126/252	114/228	110/220	127/254	110/220	106/212
Three-phase systems										
			208Y/120	200	218Y/126	197Y/114	191Y/110	220Y/127	191Y/110	184Y/106
			240/120	230/115	252/126	228/114	220/110	254/127	220/110	212/106
		240		230	252	228	220	254	220	212
			480Y/277	460	504Y/291	456Y/263	440Y/254	508Y/293	440Y/254	424Y/245
		480		460	504	456	440	508	440	424
		600		575	630	570	550	635	550	530
Medium voltage		2 400			2 520	2 340	2 160	2 540	2 280	2 080
			4 160Y/2 400		4 370/2 520	4 050/2 340	3 740Y/2 160	4 400 Y/2 540	3 950Y/2 280	3 600/2 080
		4 160			4 370	4 050	3 740	4 400	3 950	3 600
		4 800			5 040	4 680	4 320	5 080	4 560	4 160

200

	6 900	7 240	6 730	6 210	7 260	6 560	5 940
6 900	8 320Y/4 800	8 730Y/5 040	8 110Y/4 680		8 800Y/5 080	7 900Y/4 560	
	12 000Y/6 930	12 600Y/7 270	11 700Y/6 760		12 700Y/7 330	11 400Y/6 580	
	12 470Y/7 200	13 090Y/7 560	12 160Y/7 020		13 200Y/7 620	11 850Y/6 840	
	13 200Y/7 620	13 860Y/8 000	12 870Y/7 430		13 970Y/8 070	12 504Y/7 240	
	13 800Y/7 970	14 490Y/8 370	13 460Y/7 770		14 520Y/8 380	13 110Y/7 570	
13 800		14 490	13 460	12 420	14 520	13 110	11 880
	20 780Y/12 000	21 820Y/12 600	20 260Y/11 700		22 000Y/12 700	19 740Y/11 400	
	22 860Y/13 200	24 000Y/13 860	22 290Y/12 870		24 200Y/13 970	21 720Y/12 540	
23 000		24 150	22 430		24 340	21 850	
	24 940Y/14 400	26 190Y/15 120	24 320Y/14 040		26 400Y/15 240	23 690Y/13 680	
	34 500Y/19 920	36 230Y/20 920	33 640Y/19 420		36 510Y/21 080	32 780Y/18 930	
34 500		36 230	33 640		36 510	32 780	
46 000							
69 000							

From IEEE Std. 241, used with permission.

Almost all circuits within buildings are connected as in Fig. 17.2. These connection standards evolved over the 100-odd years of the power industry and were largely motivated by the need to push the maximum amount of power through the least amount of copper.

Q5.   *What is the difference between line-to-line and line-to-neutral voltages?*

A.   One of the difficulties that nonelectrical professionals have with understanding the **NEC** is that the space requirements are stated

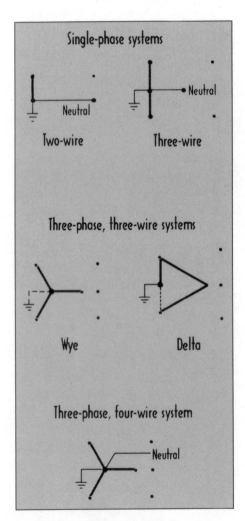

Figure 17.2 Principal transformer connections.

in terms of "nominal voltages to ground" or "nominal voltage between phases." In the vast majority of residential, commercial, industrial, and institutional facilities there are only two choices. For a nominal 208-V system, the line-to-line voltage (or nominal voltage between phases) is 208 V. The line-to-neutral voltage is 120 V. For a nominal 480-V system, the line-to-line voltage (or nominal voltage between phases) is 480 V. The line-to-neutral voltage is 277 V. Understanding the phase and neutral voltages differ by a factor of $\sqrt{3}$ will help understand some of the most important tables in the book.

Q6. *Does the NEC tell us where to locate the electric service switchgear?*

A. Not explicitly, though the practical effect of meeting **NEC** requirements for grounding and voltage drop will effectively limit your options. Frequently building aesthetics or space availability requires switchgear to be installed in a corner of a building without regard to what this adds to the cost of the building wiring to keep the voltage drop within limits. The best location for an electrical supply—from the standpoint of keeping voltage drop to a minimum—is in the center of the building. This is rarely practical even if all the **NEC** requirements for concrete encasement of service conductors can be met. Another practical reason for locating the switchgear near the building perimeter may be that the serving utility—and the building operation and maintenance crews—may want exterior access to the switchgear for operational purposes.

Figure 17.3 indicates that if a power supply is located in the center of a horizontal floor area, the area then can be supplied from circuits run radially from the locus of a circle. However, conduit systems are run in rectangular coordinates, so with this restriction, the area supplied is reduced to the interior squares, since the limits of the square are not parallel to the conduit. Thus, to fit the conduit system into a square building with walls parallel to the conduit system, the area must be reduced to the inner square. The reader should consult IEEE Standard 141 for more detail on this subject.

Q7. *What broad principles apply to the mounting and cooling of equipment?*

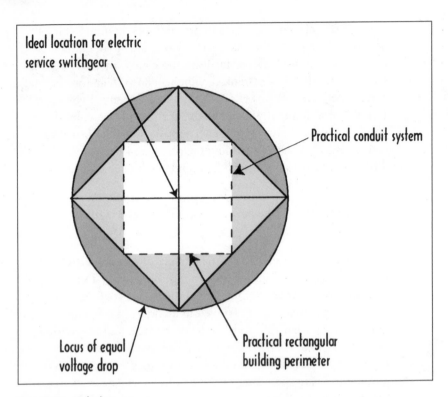

Ideal location for electric service switchgear

Practical conduit system

Locus of equal voltage drop

Practical rectangular building perimeter

Figure 17.3 Best site for electric service entrance.

A.   Section 110-13 says that electric equipment must be firmly
      secured to the surface on which it is mounted. Wooden plugs
      driven into holes in masonry, concrete, plaster, or similar materi-
      als cannot be used. Electrical equipment that depends upon the
      natural circulation of air and convection principles for cooling
      of exposed surfaces must be installed so that room airflow over
      such surfaces is not prevented by walls or by adjacent installed
      equipment. For equipment designed for floor mounting, clearance
      between top surfaces and adjacent surfaces must be provided to
      dissipate rising warm air. Electrical equipment provided with
      ventilating openings must be installed so that walls or other
      obstructions do not prevent the free circulation of air through
      the equipment. Secondary distribution supply should be located
      as close as possible to the center of the load area.

Q8.   *Explain voltage drop.*

A.   It is not unlike the phenomenon of fluid friction that progressively reduces pressure in a water pipe. Ampacities are just one factor in proper conductor sizing. Just as important are the **NEC** Chap. 2 rules for voltage drop along feeder and branch circuits. Section 215-2(b) in a fine-print note says that conductors for feeders as defined in Art. 100, sized to prevent a voltage drop exceeding 3 percent at the farthest outlet of power, heating, and lighting loads, or combinations of such loads, and where the maximum total voltage drop on both feeders and branch circuits to the farthest outlet does not exceed 5 percent, will provide reasonable efficiency of operation.

As a fine-print note, it is only a recommendation. Although the **NEC** does not contain any requirements that ungrounded or grounded conductors be sized to accommodate voltage drop, it is good practice. (Figure 17.4 offers an overview of the IEEE approach.) Voltage drop affects the efficiency of the equipment but is not a safety issue. The **NEC** recommends [FPNs to 210-19(a), 215-2(b), and 310-15] a maximum of 3 percent voltage drop for branch circuits, a maximum of 3 percent voltage drop for feeders, but a maximum of 5 percent voltage drop overall for branch circuits and feeders combined.

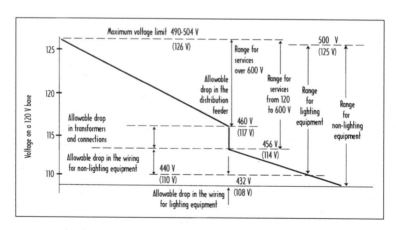

Figure 17.4 Voltage drops per IEEE/ANSI.

Q9.   Are there any rules of thumb for the determination of whether a motor will cause nuisance flicker?

A.   Yes. A motor requires about 1 kVA for each motor horsepower in normal operation, so the starting current of the average motor will be about 5 kVA for each motor horsepower. When the motor rating in horsepower approaches 5 percent of the secondary unit substation transformer capacity in kilovolt-amperes, the motor starting apparent power approaches 25 percent of the transformer capacity which, with a transformer impedance voltage of 5.75 percent, will result in a voltage sag on the order of 1 percent. A voltage change of 0.25 to 0.5 percent, for example, will cause a noticeable reduction in the light output of an incandescent lamp and a less noticeable reduction in the light output of HID lighting equipment.

Access

Q10.   *Does the NEC permit removing the covers of electrical equipment while it is energized?*

A.   Yes. Access to live parts is a normal and necessary aspect of maintenance and operation. The type of equipment that occasionally needs to be opened while energized is panelboards, switches, controllers, and controls for HVAC apparatus. Electricians frequently need to read a nameplate, assess the condition of bus or cable insulation, or get a replacement part number for a fuse without turning out the lights for building occupants. OSHA allows it as long as electricians are protected, and the NEC establishes a few minimums to do it safely. Part B of Art. 110 states the requirements for equipment operating up to 600 V. Part C of Art. 110 states the requirements for equipment operating above 600 V.

Q11.   *Can electrical equipment be installed against a wall?*

A.   Yes, and it frequently is, even though it plainly removes access capability while live. Equipment that does not require rear access for examination, adjustment, or service is not required to have any working clearances in the rear of the equipment. However, sufficient access and working space is required in the front according to Sec. 110-26. Junction boxes and compact, articulated unit substations ("packaged subs") are examples of equipment that are commonly installed against a wall. Manufacturers label metal-enclosed switchgear with instructions for proper installation of equipment in this regard.

Q12.   Is a hallway panelboard with a lock considered accessible?

A.   Yes. Enclosures housing electrical apparatus that are controlled by lock and key must be considered accessible to qualified persons. Key-controlled enclosures guard live parts from unqualified persons. A distinction needs to be made between panelboards where branch circuits are added under supervision, and those that are not. There are two parts in any electrical installation that must be protected: the circuit conductors and the elements of the equipment supplied. The **NEC** requires the OCPD to be of a rating small enough in size to protect both the conductors and equipment, or a second stage of protection must be provided (such as the overloads in a motor circuit).

Q13.   Can a mechanical maintenance person enter a high-voltage electrical room?

A.   Consult with the AHJ to whom the **NEC** assigns the authority to determine who is qualified The rule, which is stated in **NEC** 110-31, says that a room with a lock and key effectively limits access to the room. The AHJ will determine who gets a key.

It happens rather frequently that the electric service switchgear is located in an interior fenced-in area of the boiler room. The same approach will apply to electric meter readers who must gain access to high-voltage equipment in order to read a meter located on a substation. Many utilities are subcontracting their meter-reading operations to specialty consultants. These specialty consultants may or may not provide hazard training for their meter readers.

Low-Voltage Workspace

Q14.   What are the broad principles that apply to the determination of working space around electrical equipment?

A.   If you will design the workspace at least $6^{1}/2$ ft high and leave a minimum of 4 ft around all sides of any type of low-voltage equipment, you will have met just about every **NEC** rule for workspace without needing to read any further. That much said, however, building owners typically prefer not to surrender any more square footage to the electrical trades than they must. The **NEC** technical committees, ever mindful that electrical systems exist for the users of the building (and not the other way around), seek the right balance between safety and economy. In renovation projects, a "game of inches" ensues.

It is important to understand that the rules for workspace have evolved from engineering research into the physics of flash, i.e., the hazards to human beings when air ionizes in an electric field. The broad principles take into consideration voltage class, ampere rating, access, and the nature of the adjacent materials. Thus:

- "Flash" (or arcing) hazards are greater at higher voltages; thus step-back distances from the face of electrical equipment should be proportional to the voltage rating.

- Thermal hazards are greater at higher ampere load.

- Concrete walls have an effective path to ground (by way of metal structural reinforcement) and, as such, are grounded.

- It should be impossible for an electrician to simultaneously touch an energized surface and a path to ground. This can be accomplished by increasing distance between high potential and zero potential (ground).

- The working-space clearance requirements for the rear of equipment are generally the same as for the front of the equipment.

- It is sometimes necessary to open energized electrical equipment (remove the metal cladding for examination, adjustment, servicing, or maintenance).

- Distances between switchgear and anything else are measured from the exterior metal cladding (not from the interior live parts).

- An electrician needs at least 30 in of "shoulder" space in front of electrical equipment. The 30-in shoulder space does not necessarily have to be directly centered on the equipment.

■ An electrician ought to have at least 6.5 ft of headroom in which to work.

■ Equipment doors should swing at least 90°.

■ Equipment that cannot be serviced from all directions does not require as much space as other, more serviceable, equipment and thus should be granted some reduction in the minimum-space requirement.

■ Electrical renovation projects frequently result in a safer overall installation, and thus some relief from the game of inches in the workspace rules is appropriate in existing installations.

The **NEC** establishes three conditions for the determination of these distances, conditions 1, 2, and 3 in Tables 110-26 and 110-34. The conditions correlate with the surrounding environment and materials. It should go without saying that the surrounding environment should not be used for storage of any type.

Q15.   *Have the general rules for working space changed in the 1999 NEC cycle?*

A.   A few clarifications but, in the main, nothing that will impact an architectural space budget any more than before. Users of the **NEC** who had become familiar with Sec. 110-16 will find that it has been essentially renumbered as Sec. 110-26. With respect to the "game of inches" around electrical equipment, the clarifications are as follows:

■ Electrical equipment located above or below related equipment is permitted to extend not more than 6 in beyond the front of such equipment. For example, a 6-in-deep panelboard can be located above or below a 12-in wireway or auxiliary gutter.

■ Dedicated equipment space rules that formerly appeared in the **NEC** sections dealing with panelboards and medium-voltage switchgear have been put in one place, Sec. 110-25(f)(1)(a). The designated equipment space shall be an exclusive space for the width and depth of the equipment from the floor to a height of 6 ft or the structural ceiling, whichever is lower.

Nonelectrical professionals may need help determining which row of Table 17.2 applies (see Fig. 17.5). For 208/120 four-wire systems and 240/120 three-wire systems, row 1 applies. For 480/277 four-wire and 480/240 three-wire systems, row 2 applies. Where an ungrounded system is under consideration, the "voltage to ground" will be the greatest voltage between the given conductor and any other conductor of the circuit.

TABLE 17.2 WORKING SPACES [NEC TABLE 110-26(A)]

| Nominal voltage to ground | Minimum clear distance, ft | | |
	Condition 1	Condition 2	Condition 3
0–150	3	3	3
151–600	3	3½	4

Notes:
1. For SI units, 1 ft = 0.3048 m.
2. Where the conditions are as follows:

Condition 1—Exposed live parts one side and no live or grounded parts on the other side of the working space, or exposed live parts on both sides effectively guarded by suitable wood or other insulating materials. Insulated wire or insulated busbars operating at not over 300 V.

Condition 2—Exposed live parts on one side to ground shall not be considered as live parts. and grounded parts on the other side. Concrete, brick, or tile walls shall be considered as grounded.

Condition 3—Exposed live parts on both sides of the workspace (not guarded as provided in condition 1) with the operator between.

Q16.   *Where do I get started in designing the workspace?*

A.   Given the square footage and the use of the space, the electrical engineer will be able to determine the ampere load. The ampere load will then determine the size of the equipment. Panelboards

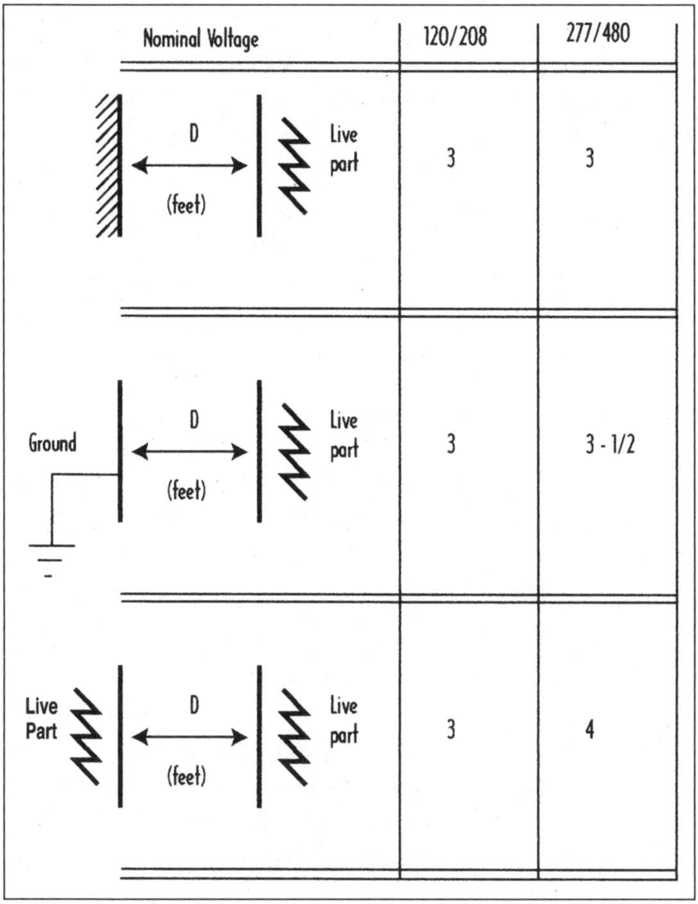

Nominal Voltage	120/208	277/480
D (feet) — Live part	3	3
Ground — D (feet) — Live part	3	3 - 1/2
Live Part — D (feet) — Live part	3	4

Figure 17.5 Pictorial representation of NEC Table 110-26 (Table 17.2).

and motor controllers, for example, may not require a lot of space from floor to ceiling. A free-standing switchboard or a motor-control center obviously will. In general, you will need 3 ft of space and a door at least 78 in high and 24 in wide as shown in Fig. 17.6.

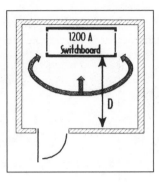

Figure 17.6 The basic workspace rule for low-voltage electric equipment. At least one entrance is required to provide access to the working space around electric equipment less than 6 ft wide. The equipment will typically contain overcurrent, switching, or control devices which may require servicing while energized. Dimension *D* depends upon whether the voltage level is 480 or 208 V. Refer to **NEC** Table 110-26(a) for *D* (Table 17.2). Distances are measured from the enclosure front or opening.

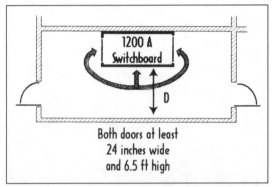

Figure 17.7 The basic workspace rule for low-voltage electric equipment greater than 1200 A and over 6 ft wide. Two entrances at each end are required to provide access to the working space around electric equipment. Each entrance must be greater than 24 in and 6.5 ft high. Dimension *D* depends upon the voltage level. Refer to **NEC** Table 110-26(a) for *D*.

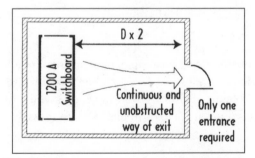

Figure 17.8 Another workspace rule for low-voltage electric equipment greater than 1200 A and over 6 ft wide. Where the workspace required by Sec. 110-16a is doubled, only one entrance is required. The equipment location must permit a continuous and unobstructed way of exit travel. Dimension *D* depends upon the voltage level. Refer to **NEC** table 110-26(a) (Table 17.2) for *D*.

For the case of a free-standing equipment greater than 1200 A (Fig. 17.7), note that you need two doors. At 480 V, a 1200-A switchboard will be delivering about a 1-MW load, a fairly serious load. Figure 17.8 shows another way to do it if you cannot get a second door to work. Simply double the space in front of the switchgear and you will have made it possible for an electrician to avoid a hazard from destructive arcing and/or violent burndown.

A comprehensive treatment of the "game of inches" with regard to workspace requirements is beyond the scope of this book. We have shown the basics here. Architects and owners are advised to exceed **NEC** minimums. The reader is encouraged to obtain copies of any of the core references for more detailed information than we have space for here. These core references contain many helpful pictorial diagrams of the fine points of Art. 110.

Q17.   *How much space is required in front of the switchgear?*

A.   Figure 17.9 cites the space requirement *for clear working space* in front of switchgear operating at voltages less than 600 V. It needs to be read carefully because the use of it assumes a grasp of line-to-neutral voltages (as distinguished from line-to-line voltages) and nominal voltage equipment class. Before laying out the walls surrounding electric service switchgear, nonelectrical professionals are advised to consult with an electrical professional in the use of Table 17.3 [**NEC** Table 110-34(a)].

Once the voltage to ground is determined, you need to consider each of the three conditions that apply. The conditions are classifications of the adjacent surfaces of the switchgear. The underlying principle is that you need to do all you can architecturally to prevent a human being from becoming the conducting path to ground. You can do this in either of two ways: (*a*) make it impossible for a human being to touch an energized part with one hand and a concrete wall with another (a concrete wall is a grounded surface) or (*b*) make the adjacent materials out of a nonconducting material (such as wood). It is always better to add the extra foot of working space. Refer to Fig. 17.9 for a pictorial representation. It shows the two workspace "zones" around interior switchboards. The details appear in **NEC** Sec. 384-4, which is intended to require the space outlined by the footprint of the panelboard or switchboard to be reserved for the installation of electrical equipment or for the installation of conduit, cable trays, etc., entering or exiting the equipment. The working clearance zone is the outlined area in front of the switchboard. Sprinkler protection is permitted as long as the piping does not enter the dedicated space.

Q18. *What are the broad principles that apply to the location of service switchboards with live parts?*

A. Section 384-5 says that switchboards that have any exposed live parts must be located in permanently dry locations and then only under competent supervision and accessible only to qualified

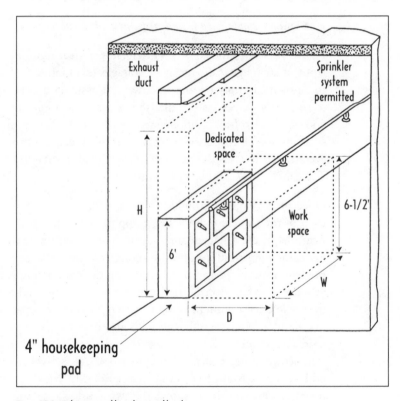

Figure 17.9 Workspace around low-voltage switchboards.

persons. Switchboards must be so located that the probability of damage from equipment or processes is reduced to a minimum. For other than a totally enclosed switchboard, a space not less than 3 ft (914 mm) must be provided between the top of the switchboard and any combustible ceiling, unless a noncombustible shield is provided between the switchboard and the ceiling. Clearances around switchboards must comply with the provisions of Sec. 110-26.

Q19. *How high should the ceiling be?*

A. As high as the structural ceiling. (A dropped or suspended ceiling that does not add strength to the building is not considered a structural ceiling.) See Fig. 17.9 and **NEC** Sec. 384-4, which says that the space should equal the width and depth of the equipment and extend from the floor to a height of 25 ft to the structural ceiling, *whichever is lower.* The intention of this requirement is not to make all switch, panel, and distribution board rooms 25 ft high but rather to make sure that when the structural ceiling does extend that high, the space above it is reserved for secondary conduit and other types of raceway associated with them.

The width and depth and height zone above the board must be clear of HVAC piping, duct, or other equipment foreign to the installation. (Sprinkler protection is, however, permitted outside the projected zone.) The work zone in front of the equipment is the dedicated space required by Sec. 110-26.

Q20. *Can several panels be mounted side by side with the required 30-in-wide working space overlapping for each panel?*

A. Yes, overlapping clearance of side-by-side electrical equipment is acceptable. The requirement of Sec. 110-26 is based on clear working space, not the size or configuration of equipment. The 30-in-wide clear working space is required where live parts, when exposed, are examined, adjusted, serviced, or maintained while energized

Medium-Voltage Workspace

Q21. *What are the broad principles that apply to workspace for medium-voltage equipment?*

A. Typically, this is equipment at the electric service entrance. The owner has a load over 1000 kVA with a demand profile that makes it economical to pay off the greater expense of medium-voltage equipment with savings from buying electricity at a higher voltage. Higher-voltage electricity is almost always cheaper per unit and is more reliable as long as you are buying large quantities of it.

In some cases the medium-voltage parts of the service equipment can be on the building exterior, leaving only the low-voltage parts of the service switchgear assembly on the inside. In other cases, this is not possible. In either case, the **NEC** rules for exterior and interior medium-voltage equipment appear in Secs. 110-30 through 110-34. Related material appears in an article that is new to the 1999 **NEC**—Article 490—Equipment, Over 600 Volts, Nominal.

Table 17.3 is the table you will need to refer to the most in determining workspace dimensions. For a majority of buildings where medium-voltage switchgear is present, you will use the first two rows and the first two columns. This is because the nominal voltage to ground for 2.4-, 4.8-, 4.16-, 12.47-, and 13.8-kV medium-voltage equipment falls within the ranges of the first two rows [conditions 1 and 2 (wood and concrete walls, respectively)]. The remaining rows and columns of Table 17.3 cover workspace dimensions for unguarded live equipment operating at 23 kV and above—something you just do not see very often anywhere but in an old manufacturing plant.

Q22. *How **high** does the workspace have to be?*

A. A related table to be used to determine how high a switchgear room has to be is Table 110-34(e), which appears in the 1999 **NEC** but is not shown here. It states the rules for the elevation of unguarded live parts above working space. In most commercial

TABLE 17.3 MINIMUM DEPTH OF CLEAR WORKING SPACE AT ELECTRICAL EQUIPMENT

	Minimum clear distance, ft		
Nominal voltage to ground	Condition 1	Condition 2	Condition 3
601–2500 V	3	4	5
2501–9000 V	4	5	6
9001–25,000 V	5	6	9
24,001–75 kV	6	8	10
Above 75 kV	8	10	12

Notes:
1. For SI units, 1 ft = 0.3048 m.
2. Where the conditions are as follows:

Condition 1—Exposed live parts on one side and no live or grounded parts on the other side of the working space, or exposed live parts on both sides effectively guarded by suitable wood or other insulating materials. Insulated wire or insulated busbars operating at not over 300 V shall not be considered live parts.

Condition 2—Exposed live parts on one side and grounded parts on the other side. Concrete, brick, or tile walls will be considered as grounded surfaces.

Condition 3—Exposed live parts on both sides of the workspace (not guarded as provided in conditions 1) with the operator between.

facilities, for example, unguarded live parts is simply not an option. While there are many legacy systems with single-phase banks with exposed live bus structurally built above the transformers, few utilities install this type of equipment any more. Where these legacy systems exist, they typically are in fenced, exterior enclosures or locked rooms, accessible from the exterior only by utility personnel. Some industrial settings, however, will have unguarded bus for cranes or hoists. Table 110-34(e) tells you how high above the floor the exposed live parts must be. Refer to Fig. 17.10.

Q23.   *How much space is required around switchgear batteries?*

A.   Medium-voltage switchgear usually requires a large dc battery to drive breakers and related control equipment. Storage battery systems also provide an alternate source of power for many information technology equipment systems that allow communication with medium-voltage equipment controls.

The generic rules of Art. 110 are supplemented by rules that appear in Art. 480. Manufacturers should be consulted on the parameters for proper ventilation. The vapors generated from storage battery systems are very corrosive and will most certainly damage cable insulation if the workspace is not adequately ventilated (see Sec. 310-9).

Working clearances should comply with Sec. 110-26 and should be measured from the edge of the battery rack. Typical voltages range from 2 vdc for a single telecommunications cell to 125–250 vdc for a rack of cells that drive a switchgear lineup. Common telecommunications practice allows at least 30 in of working clearance. The reader is referred to Bellcore GR-3, "Generic Requirements for Network Equipment and Building Systems," for more details.

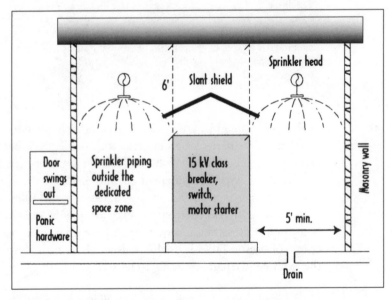

Figure 17.10 Section view of metal-enclosed medium-voltage equipment with no exposed live parts.

Basic Methods

Q24.   *Why is fault duty important and how is it determined?*

A.   While many nonelectrical professionals regard the issue of fault duty as a tool electrical professionals use to increase construction costs, fault duty is the single most important parameter the engineer must work with in order to make a building electrical system safe from violent electrical burndown. Section 110-9 says that equipment intended to interrupt current at fault levels must have an interrupting rating sufficient for the nominal circuit voltage and current that is available at the line terminals of the equipment. This says that the engineer must meet the **NEC** requirement by either designing a fully rated system or, as permitted by Underwriter's Laboratories, designing a series-rated system. Most electrical designers opt to design a fully rated system in order to avoid the complications inherent in coordinating fault duties in a mixed-vendor environment.

Fault duty is typically determined from information obtained from the serving utility. Quick and dirty estimates are possible for residential and small commercial services using a minimum amount of information about transformer impedance and full-load amperes and will likely yield a conservative estimate without the need for complex phasor arithmetic. When utility transformers begin to be as large as, say 500 kVA, an electrical professional should be retained to work through the details. A practical yet comprehensive treatment of this subject appears in Anthony, *Electrical Power System Protection and Coordination*, McGraw-Hill, 1995. An example appears at the end of Chap. 21.

Q25.   *What broad principles apply to the sizing of conductors?*

A.   There are four:

   ■   The overcurrent device must be sized for the sum of the noncontinuous load plus 125 percent of the continuous load. The overcurrent device may be rounded up to the next available standard size as long as the device does not exceed 800 A.

- The conductor must be sized so that the termination temperatures do not exceed 60°C for conductors smaller than 100 A or AWG No 1, and 75°C for conductors larger than 100 A or AWH No 1/0. This results in using the 60°C column of Table 310-16 for all conductor sizing of circuits 100 A and smaller, and the 75°C column for conductors larger than 100 A. Even if the higher-temperature insulation is used, *two* columns must be used to determine conductor sizes. (Many designers will specify 90°C conductors as a matter of preference, good practice, and convenience even though the specification exceeds **NEC** minimums.)

- Conductors and insulation must be specified on the basis of conditions of use. Considerations in determining the conditions of use should include whether the conductors are run parallel in the same raceway and whether the ambient temperature is high. High-temperature ambients are any temperatures that exceed 30°C or 86°F. Refer to the Solved Problem at the end of this chapter.

- The overcurrent device must protect the conductor under all conditions of use.

Q26.   *Are medium-voltage conductors selected the same way low-voltage conductors are selected?*

A.   No; medium-voltage conductors are handled differently because of the character of their construction and operation. Electrical contractors will often compare the conductor sizes on the contract documents with the uncorrected 75°C column in **NEC** Table 310-16 and conclude that the engineer has oversized the conductors to the detriment of the owner's budget. Table 310-16 applies only to conductors from 0 to 2000 V. Above this voltage, conductor ampacities must be selected by use of the IPCEA tables (an extract of which appears in Table 20.3) or derived from App. B of the **NEC**. A more complete discussion of this topic appears in Chap. 20.

Q27.   *How do we coordinate conductor and terminal temperatures?*

A.   The terminals of appliances and utilization equipment are based on the application of the 60° ampacity specified in Table 310-16 for circuits rated over 100 A or less and the 75°C ampacities for circuits rated over 100 A. If conductors of a higher temperature rating are used, the load applied to these conductors and equipment terminations is based on the 60 or 75°C ampacity from Table 310-16, whichever applies.

Q28.   *What broad principles apply to grounding of electronic systems?*

A.   A short introduction to the practical effect of **NEC** rules on the construction process is as follows.

■   When electronic systems are installed in a facility, the equipment installation technicians typically request an isolated ground not connected to any other ground in the building, often accompanied by the claim that their equipment will not work without the isolated ground. The conflict between the **NEC** requirement for connecting all grounding conductors to a single grounding electrode (Sec. 250-54) and the supplier's desire may be resolved by connecting the ground for the electronic system as close to the grounding electrode as possible. This meets the **NEC** requirement and gives the electronic systems their solid earth-ground connection. This makes minimum impedance between system ground and earth ground.

■   System technicians will request a voltage difference between neutral and ground *at the utilization equipment* of less than 1 V. The problem with electronic utilization equipment, however, is that it usually has switching power supplies, which generates significant harmonic currents that are impressed on the neutral of the system with a resultant increase in neutral current. The current through the constant impedance will cause a voltage drop and thus the difference between neutral and ground that the technicians see at their equipment terminals.

We take up the subject of grounding in more detail in Chap. 21. We place the question here to draw your attention to vastly rewritten Art. 640, which acknowledges formally what was once only used informally—"technical equipment ground." Sophisticated electronic utilization apparatus is becoming as prevalent in building systems as lights and motors and thus deserves special attention.

Q29. *What are the broad principles that are to be used in locating receptacles?*

A.   The rules for various kinds of facilities are concentrated in Chap. 2. The reason receptacles exist in the first place is that they are safer to use than extension cords. The 6-ft rule that applies in residential construction is not intended to make electrical contractors more money but rather to allow sufficient receptacles in a room so that the widespread use of extension cords will not cause a fire. Similar reasoning applies to receptacles built in the front and rear of residences. Holiday lighting and electrically driven garden tools would be more of a hazard without them.

Q30. *What broad principles apply to designing branch circuits for residences?*

A.   Add the ampere ratings of the noncontinuous load at 100 percent to the ampere ratings of continuous load at 125 percent. Fixed appliances, for example, are allowed to draw up to 50 percent of the rating of a branch circuit supplying two or more general-purpose outlets which serve lighting and receptacle loads.

The procedure for calculating the commercial receptacle load is determined by 180 VA per outlet. The first 10 kVA is to be calculated at 100 percent, if the receptacles are rated for noncontinuous duty. Continuous loads must be calculated at 180 VA and multiplied by 125 percent to determine the receptacle load.

The procedure for adding new circuits or extensions to existing electrical systems in dwelling units of 500 sq ft or less is determined by figuring the VA per square foot or the amperes per outlet method. Either method applies to a portion of the dwelling unit that has not been previously wired or an addition which exceeds 500 sq ft in area. Per Sec. 220-10, the OCPD does not have to be equal to the ampacity of the service conductors if two to six devices are located in a single enclosure or group of enclosures.

Q31. *How are branch circuits for appliances without motors applied?*

A.   Section 422-28e says when individual circuits supply only one nonmotor appliance rated 13.3 A or less, the overcurrent protective device is limited to 125 percent of the FLA rating of the appliance. If the overcurrent protective device trips, it is not to be increased up to 150 percent of the appliance rating. See **NEC** 430-28e.

For cases where 150 percent of the appliance rating can be utilized, the overcurrent protective device protecting only one nonmotor appliance rated at 13.3 A or more (13.3 × 125% = 16.7A) which is supplied by an individual circuit may be sized at 125 or 150 percent of the FLA of such an appliance.

Q32.   *What are the rules for designing branch circuits for receptacles in nondwelling units?*

A.   Section 220-13 says the procedure for calculating receptacle loads in industrial and commercial locations is computed at 100 percent for the first 10 kVA and all remaining VA at 50 percent. This reduction for all receptacles is on the basis that all receptacles are not used at the same time. The receptacle loads in hospitals, hotels, and warehouses may be lumped with the lighting load and the demand factors of Table 220-11 applied accordingly.

Solved Problem—Voltage Drop

Situation.

A 100-kW, 480/277-V load at 90 percent power factor is noncontinuous. The load is less than 50 percent ballast-type lighting. Length of feeder is 250 ft. Maximum allowable voltage drop is 2 percent.

Requirements.

Design circuit to comply with **NEC** requirements for voltage drop.

Solution:

100/0.832 = 134A.

  1. Determine amperes of 100-kW load.

    Line current = (100 kW × 1000)/($\sqrt{3}$ × 480 × 0.9) = 134 A

    Thus minimum ampacity is 134 A for a noncontinuous load.

  2. Minimum size for voltage drop:

    2 percent × 277 V = 5.54 V

    Line to neutral voltages are used in voltage drop calculations.

    Ampere feet = 134 × 250 = 33,500 = 33.5 × 1000

    Maximum voltage drop/1000 AF = 5.54/33.5 = 0.165

From a manufacturer's table (refer to Table 17.4) we look for a conductor that will yield a voltage drop of less than 0.165 per 1000 feet. Using the 90 percent power factor column, we can see that any conductor larger than AWG No. 1 (with a voltage drop of 0.162 per 1000 feet) will work.

Remarks.

(a) Sometimes you may need to raise conductor size to reduce voltage drop. When this happens, you will also need to raise the size of the overcurrent protective device. (b) Most wire and cable manufacturers supply tables like Table 17.4 to their customers, designers, and specifiers.

TABLE 17.4 MANUFACTURER'S TABLE OF COPPER RESISTIVITY

Size AWG or MCM	Magnetic Conduit or Armour			Nonmag. Conduit or Armour		
	80% P.F.	90% P.F.	100% P.F.	80% P.F.	90% P.F.	100% P.F.
14	2.540	2.790	3.067	2.535	2.780	3.060
12	1.570	1.749	1.917	1.565	1.749	1.923
10	.993	1.103	1.200	.987	1.103	1.201
8	.635	.699	.750	.629	.693	.751
6	.421	.462	.485	.461	.456	.485
4	.277	.300	.306	.271	.294	.306
2	.185	.196	.196	.179	.191	.191
1	.150	.162	.150	.150	.156	.150
1/0	.127	.133	.121	.121	.127	.121
2/0	.109	.110	.098	.098	.104	.092
3/0	.092	.092	.081	.081	.087	.075
4/0	.081	.075	.064	.069	.069	.057
250	.070	.070	.054	.064	.064	.051
300	.064	.064	.045	.056	.055	.042
350	.058	.055	.039	.051	.049	.036
400	.055	.051	.035	.047	.044	.032
500	.049	.045	.029	.042	.039	.026
600	.046	.041	.024	.038	.034	.022
700	.043	.038	.021	.036	.032	.019
750	.042	.037	.020	.034	.031	.017
1000	.038	.032	.016	.029	.025	.013

1. Values are per 1000 ampere-feet for three single conductors in conduit.
2. Values are based on three-phase, line-to-neutral voltages. For line-to-line voltage drops, multiply by a factor of 1.73. For single-phase circuits, multiply by a factor of 2.0.
Source: Courtesy of Canada Wire and Cable Limited.

Division 16100—Wiring Methods

Remarks. This CSI Division is one of the few divisions of the CSI Masterspec© format that has the same title as an **NEC** chapter. As Chap. 3 of the **NEC** it is, arguably, the heart of the **NEC**, showing in some of its language relics of wiring practice that was common over a hundred years ago. The reader may find interesting Appendix item A-29, which is an excerpt from the 1899 version of the National Electric Code. The focus of this chapter will be upon broad design and construction principles, rather than eccentric wiring conditions. Specific requirements for service, feeder, and branch circuit design for various facility types will appear in various questions in Part II.

*Q1.   What **NEC** requirements commonly appear in Division 16100?*

A.   The rules of Chap. 3 do not apply to the internal wiring of listed equipment; by implication, neither should it apply the internal wiring of unlisted equipment such as computers, audio systems, and appliances. Some variants of the CSI Masterspec© list the manner in which wires are to be labeled. Every organization may have its own system which helps distinguish 480-V, three-wire systems (using brown, yellow, orange) from say, 208-V, four-wire systems (using blue, black, red).

At the very least Sec. 310-11 of the **NEC** says that all conductors and cables must be marked to indicate the maximum rated voltage for which the conductor was listed, the manufacturer's name, trademark, or other distinctive marking by which the organization responsible for the product can be readily identified, the AWG size or circular mil area.

Q2.   Why are there so many cable and conductor types?

A.   The listing and marking of the various types of wires and cables are related to the general rules and give authorities enforcing the **NEC** the tools whereby they can judge whether the types of cable used substantially contribute to the hazards during fire conditions where fire fighters must of necessity be subjected to those products of combustion. Statistics show that most people die from smoke and products of combustion and not from the heat of fires. To the furthest extent possible, electrical materials which contribute to products of combustion should be held to a minimum.

A word to nonelectrical professionals on the distinction between a wire, a conductor, and a cable. The term *conductor* refers properly to the copper or aluminum wire that actually carries the electric current. An insulated conductor is one that is encased within electrical insulation material. The term *cable* refers to the complete wire assembly including the conductor, the insulation, and any shielding and/or outer protective covering where used. Cables can have just a single conductor, each separately insulated, but all enclosed in one overall jacketing.

Finally, when you see the word "ungrounded conductor" in the **NEC** in most cases it means the phase or "line" conductor, the conductor that is normally delivering energy to the load. The use of the word ungrounded conductor hews to the general principle of making the **NEC** as generally applicable as possible (to dc, 400 Hz, or unconventional transformer connections) but it comes at the cost of clarity to nonelectrical professionals.

Q3.   *What broad principles apply in the selection of electrical connections?*

A.   Refer to Art. 110. Because of different characteristics of copper and aluminum, devices such as pressure terminal or pressure splicing connectors and soldering lugs must be identified for the material of the conductor and must be properly installed and used. Conductors of dissimilar metals must not be intermixed in a terminal or splicing connector where physical contact occurs between dissimilar conductors (such as copper and aluminum, copper and copper-clad aluminum, or aluminum and copper-clad aluminum), unless the device is identified for the purpose and conditions of use. Materials such as solder, fluxes, inhibitors, and compounds, where employed, are suitable for the use and must be of a type that will not adversely affect the conductors, installation, or equipment.

One exception to this general rule appears in Sec. 346-3(a), which says that aluminum fittings and boxes may be used with steel rigid conduit, or steel fittings and enclosures with aluminum rigid conduit. Evidently, galvanic action between two dissimilar metals is not a problem.

Q4.   *Can conductors of the same circuit be split among different raceways?*

A.   Not unless the conductors are parallel conductors of the same phase (see Sec. 300-3). With few exceptions, all conductors of the same circuit and, where used, the grounded conductor and all equipment grounding conductors must be contained within the same raceway, cable tray, trench, cable, or cord. See Fig. 18.1. One common exception to the rule is for conductors in parallel. Some wire, such as THHN, is prohibited in its use outside of a raceway.

Q5.   *Can conductors of different systems be put in the same raceway?*

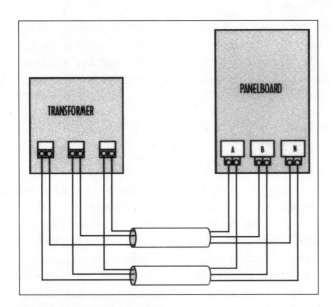

Figure 18.1 Parallel Conductor Wiring Method.

A.   Conductors of circuits rated 600 V or less are permitted to occupy the same equipment wiring enclosure, cable, or raceway. All conductors must have an insulation rating equal to at least the maximum circuit voltage applied to any conductor within the enclosure, cable, or raceway. Some types of grounding and bonding conductors can be run as single conductors on the exterior of a raceway. [see Sec. 300-5(i)]. It is the maximum circuit voltage in the raceway—not the maximum insulation voltage rating of the conductors in the raceway—that determines the minimum rating of conductors.

Q6.   *Can conductors of different voltages be put in the same raceway?*

A.   In general, no. Conductors of circuits rated over 600 V, nominal, cannot occupy the same equipment wiring enclosure, cable, or raceway with conductors of circuits rated 600 V, nominal, or less. All nonshielded conductors must have an insulation rating equal to at least the maximum circuit voltage applied to any conductor within the enclosures, cable, or raceway.

Q7.   *If a building has two different nominal voltages, what are the NEC requirements for identifying conductors for each system?*

A.   Requirements for identification of ungrounded conductors appear in Sec. 210-4(d). An example of this is where a 120/240-V 1-phase system and a 480/277-V, 3-phase 4-wire system. Section 210-4(d) does not, however, specify the colors to be used for phase (ungrounded) conductors but requires some means of identification. The means of identification must be permanently posted at each branch circuit panelboard.

Raceways and Boxes

Q8.   *How should conductors be grouped with metal enclosures or metal raceways?*

A.   See Sec. 300-20, where the **NEC** says that when conductors carrying alternating current are installed in metal enclosures or metal raceways, they must be so arranged as to avoid heating the surrounding metal by induction. To accomplish this, all phase conductors and, where used, the grounded conductor and all equipment grounding conductors must be grouped together. Where a single conductor carrying alternating current passes through metal with magnetic properties, the inductive effect must be minimized by (1) cutting slots in the metal between the individual holes through which the individual conductors pass or (2) passing all the conductors in the circuit through an insulating wall sufficiently large for all of the conductors in the circuit. The rule applies to medium-voltage circuits.

Q9.   *What kind of bending radius is required?*

A.   Section 300-34 says that a conductor cannot be bent to a radius less than 8 times the overall diameter for nonshielded conductors or 12 times the diameter for shielded or lead-covered conductors during or after installation. For multiconductor or multiplexed single-conductor cables having individually shielded conductors, the minimum bending radius is 12 times the diameter of the individually shielded conductors or 7 times the overall diameter, whichever is greater.

TABLE 18.1 MAXIMUM NUMBER OF CONDUCTORS AND FIXTURE WIRES IN ELECTRICAL METALLIC TUBING

Type letters	Conductor size AWG/kcmil	Trade sizes, in									
		1/2	3/4	1	1 1/4	1 1/2	2	2 1/2	3	3 1/2	4
THHN,	14	12	22	35	61	84	138	241	364	476	608
THWN,	12	9	16	26	45	61	101	176	266	347	443
THWN-2	10	5	10	16	28	38	63	111	167	219	279
	8	3	6	9	16	22	36	64	96	126	161
	6	2	4	7	12	16	26	46	69	91	116
	4	1	2	4	7	10	16	28	43	56	71
	3	1	1	3	6	8	13	24	36	47	60
	2	1	1	3	5	7	11	20	30	40	51
	1	1	1	1	4	5	8	15	22	29	37
	1/0	1	1	1	3	4	7	12	19	25	32
	2/0	0	1	1	2	3	6	10	16	20	26
	3/0	0	1	1	1	3	5	8	13	17	22
	4/0	0	1	1	1	2	4	7	11	14	18
	250	0	0	1	1	1	3	6	9	11	15
	300	0	0	1	1	1	3	5	7	10	13
	350	0	0	1	1	1	2	4	6	9	11
	400	0	0	0	1	1	1	4	6	8	10
	500	0	0	0	1	1	1	3	5	6	8
	600	0	0	0	1	1	1	2	4	5	7
	700	0	0	0	1	1	1	2	3	4	6
	750	0	0	0	0	1	1	1	3	4	5
	800	0	0	0	0	1	1	1	3	4	5
	900	0	0	0	0	1	1	1	3	3	4
	1000	0	0	0	0	1	1	1	2	3	4

From **NEC** Appendix, Table C1. Table 18.1 is one of the most handy tables in the **NEC**. It is most frequently used with Table 310-16.

Q10.   What is electrical nonmetallic tubing?

A.   See Sec. 331-1. Electrical nonmetallic tubing is a pliable corrugated raceway of circular cross section with integral or associated couplings, connectors, and fittings listed for the installation of electric conductors. It is composed of a material that is resistant to moisture and chemical atmospheres and is flame-retardant. A pliable raceway is a raceway that can be bend by hand with a reasonable force, but without other assistance. Electrical nonmetallic tubing is listed only when it is made of material that does not exceed the ignitability, flammability, smoke generation, and toxicity characteristics of rigid (nonplasticized) polyvinyl chloride.

Q11.   Where can electrical nonmetallic tubing be installed?

A.   It is permitted in any building not exceeding three floors above grade for exposed work, where not subject to physical damage, concealed within walls, floors, and ceilings. In any building exceeding three floors above grade, electrical nonmetallic tubing must be concealed within walls, floors, and ceilings where the walls, floors, and ceilings provide a thermal barrier of material that has at least a 15-min finish rating as identified in listings of fire-rated assemblies. The 15-min finish rated thermal barrier can be used for combustible or noncombustible walls, floors, and ceilings in locations subject to severe corrosive influences as covered in Sec. 300-6 and where subject to chemicals for which the materials are specifically approved.

Q12.   Where can intermediate metal conduit be installed?

A.   Section 345-3 says that use of intermediate metal conduit is permitted under all atmospheric conditions and occupancies. Where practicable, dissimilar metals in contact anywhere in the system must be avoided to eliminate the possibility of galvanic action. Intermediate metal conduit is permitted as an equipment grounding conductor. Intermediate metal conduit, elbows, couplings, and fittings are permitted to be installed in concrete, in direct contact with the earth, or in areas subject to severe corrosive influences where protected by corrosion protection and judged suitable for the condition. Intermediate metal conduit is permitted in or under cinder fill where subject to permanent moisture where protected on all sides by a layer of noncinder concrete not less than 2 in thick, where the conduit is not less than 18 in under the fill, or where protected by corrosion protection and judged suitable for the condition.

Q13.   *What is nonmetallic underground conduit with conductors?*

A.   Section 343-1 describes nonmetallic underground conduit with conductors as a factory assembly of conductors or cables inside a nonmetallic, smooth wall conduit with a circular cross section. The nonmetallic conduit is to be composed of a material that is resistant to moisture and corrosive agents. It must also be capable of being supplied on reels without damage or distortion and must be of sufficient strength to withstand abuse, such as impact or crushing, in handling and during installation without damage to conduit or conductors. The use of listed nonmetallic underground conduit with conductors and fittings is permitted for direct burial underground installation. For minimum cover requirements, see Tables 300-5 and 710-4(b) under rigid nonmetallic conduit.

Q14.   *What broad principles apply to the conduit bends?*

A.   Factory elbows may not be available to satisfy sidewall pressure limitations, so conduit sweeps produced in the field may be required. But the required bend radii may be prohibitively large. An alternate design should be carefully considered that contemplates the use of a concrete trench under switchgear, motor control centers, and other terminals, sufficiently wide from front to back to allow pulling the cable into the trench with sufficient cable to provide required makeup.

Per Sec. 370-71(a), minimum dimensions are set for high-voltage pull and junction boxes. Sharp bends in conduit simulate a coil. The induced voltage will decay but not before holding the voltage above earth potential.

Q15.   *What rules govern the selection of pull and junction boxes?*

A.   Section 370-71 says that pull and junction boxes must provide adequate space and dimensions for the installation of conductors. The length of the box cannot be less than 48 times the out-

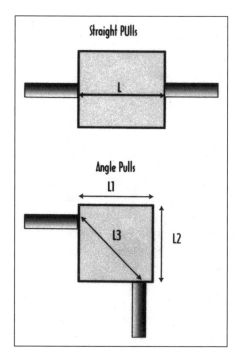

Figure 18.2 Straight Pulls and Angle Pulls.

side diameter, over sheath, of the largest shielded or lead-covered conductor or cable entering the box. The length cannot be less than 32 times the outside diameter of the largest nonshielded conductor or cable. For angle or U pulls the distance between each cable or conductor entry inside the box and the opposite wall of the box must not be less than 36 times the outside diameter, over sheath, of the largest cable or conductor. This distance must be increased for additional entries by the amount of the sum of the outside diameters, over sheath, of all other cables or conductor entries through the same wall of the box. Several exceptions apply. See Fig. 18.2.

Q16.   *What broad principles apply to the selection of manholes?*

A.   Judicious placement of manholes in a duct system is an important part of electrical design. Manholes are often included in a system that are not required and sometimes omitted where they are critically needed. The inclusion of a manhole between any two points adds substantially to the cost, complicates cable installation, and can produce cable damage. Whenever an intermediate manhole is included in a straight section of duct bank between two points, a design decision should be made about whether this intermediate manhole is to be a splice point or a point for feeding the cable in both directions. Slack should be provided in the event a splice should need repair. Some electric utilities install a complete loop of cable per long on one side of a splice. Splices should be staggered horizontally and be installed with adequate clearance vertically to facilities not only in the original installation but to provide enough cable for subsequent repair or replacement should this be necessary. More discussion of this subject appears in Chap. 20. Refer also to IPCEA Standard S-68-516.

Pulling cable through manholes should be avoided. It should be pulled at constant velocity (not to exceed 50 ft/min and not to be less than 15 ft/min). Lubricant should be applied liberally and continuously during the pull, and dynamometer readings should be taken periodically.

Q17.   *What broad principles apply to the specification of junction boxes?*

A.   Care must be given to the sizing and location of junction boxes. Junction boxes are not designed to pull through or pull around but are intended for junctions or splices. If an enclosure is to be used as a pull box, then careful attention should be given to its design based on whether the installation will require pulling to the box or feeding into the box. If a junction box is installed in a run and then cable is pulled through it, it is possible that the box might have been superfluous in the design.

Conductors and Cables

Q18.   *What broad principles apply to the selection of conductors?*

A. Conductor ampacity is usually the single most important parameter. The most commonly used table in the **NEC** is 310-16, which has been reprinted in Table 18.2. Insulated cables frequently fail because of the mechanical damage due to improper raceway and terminal design or to improper installation techniques. The trend in the industry is toward reduced insulation thickness, which makes it important to observe the established limitations on the physical treatment of these cables, sidewall pressure, jam ratio, minimum bending radii, pulling tension.

It is highly desirable to make these cables damage-proof, but that is beyond the state of the art, particularly when there are a number of other criteria these cables must satisfy: long service life under voltage stress and heat aging, flame retardance, thermal stability, dielectric strength. Cables installed in nuclear power plants are expected to have some measure of radiation resistance during loss of coolant accidents. In achieving these requirements some limits are imposed on the mechanical durability of the cable, and these limits must be recognized.

All ampacity table notes have been relocated for the 1999 revision (Table 18.2). General rules for medium voltage have been broken out and now appear in new Sec. 310-60.

Q19. *What factors determine allowable ampacity?*

A. There are many. The key is not to treat the wire and cable as a system in itself but as a component of the total electrical system. The terminations, equipment ratings, and ambient environment all affect the ampacity that can be assigned to the conductor.

The general rules for coordinating conductor termination ratings are as follows:

- Termination provisions for equipment rated 100 A or less, or equipment that is marked for No. 14 to No. 1 AWG conductors, are rated from use with conductors rated 60°C.

- Termination provisions for equipment rated 100 A or more, or equipment that is marked for No. 14 to No. 1 AWG conductors, are rated from use with conductors rated 75°C.

TABLE 18.2 Allowable Ampacities of Insulated Conductors Rated 0 through 2000 V, 60°C through 90°C (140°F through 194°F) Not More than Three Current-Carrying Conductors in Raceway, Cable, or Earth (Directly Buried), Based on Ambient Temperature of 30°C (86°F)

Size	Temperature rating of conductor (see Table 310-13)						Size
	60°C (140°F)	75°C (167°F)	90°C (194°F)	60°C (140°F)	75°C (167°F)	90°C (194°F)	
AWG or kcmil	Types TW, UF	Types FEPW, RH, RHW, THHW, THWN, THWN, XHHW, USE, ZW	Types TBS, SA, SIS, FEP, FEPB, MI, RHH, RHW-2, THHN, THHW, THW-2, THWN-2, USE-2, XHH, XHHW, XHHW-2, ZW-2	Types TW, UF	Types RH, RHW, THHW, THW, THWN, XHHW, USE	Types TBS, SA, SIS, THHN, THHW, THW-2, THWN-2, RHH, RHW-2, USE-2, XHH, XHHW, XHHW-2, ZW-2	AWG or kcmil
	Copper			Aluminum or copper-clad aluminum			
18	—	—	14	—	—	—	—
16	—	—	18	—	—	—	—
14*	20	20	25	—	—	—	—
12*	25	25	30	20	20	25	12*

10*	30	35	40	25	30	35	10*
8	40	50	55	30	40	45	8
6	55	65	75	40	50	60	6
4	70	85	95	55	65	75	4
3	85	100	110	65	75	85	3
2	95	115	130	75	90	100	2
1	110	130	150	85	100	115	1
1/0	125	150	170	100	120	135	1/0
2/0	145	175	195	115	135	150	2/0
3/0	165	200	225	130	155	175	3/0
4/0	195	230	260	150	180	205	4/0
250	215	255	290	170	205	230	250
300	240	285	320	190	230	255	300
350	260	310	350	210	250	280	350
400	280	335	380	225	270	305	400
500	320	380	430	260	310	350	500
600	355	420	475	285	340	385	600
700	385	460	520	310	375	420	700
750	400	475	535	320	385	435	750
800	410	490	555	330	395	450	800
900	435	520	585	355	425	480	900

TABLE 18.2 ALLOWABLE AMPACITIES OF INSULATED CONDUCTORS RATED 0 THROUGH 2000 V, 60°C THROUGH 90°C (140°F THROUGH 194°F) NOT MORE THAN THREE CURRENT-CARRYING CONDUCTORS IN RACEWAY, CABLE, OR EARTH (DIRECTLY BURIED), BASED ON AMBIENT TEMPERATURE OF 30°C (86°F) (*Continued*)

Size	Temperature Rating of Conductor (see Table 310-13)						Size
AWG or kcmil	60°C (140°F)	75°C (167°F)	90°C (194°F)	60°C (140°F)	75°C (167°F)	90°C (194°F)	AWG or kcmil
	Types TW, UF	Types FEPW, RH, RHW, THHW, THW, THWN, XHHW, USE, ZW	Types TBS, SA, SIS, FEP, FEPB, MI, RHH, RHW-2, THHN, THHW, THW-2, THWN-2, USE-2, XHH, XHHW, XHHW-2, ZW-2	Types TW, UF	Types RH, RHW, THHW, THW, THWN, XHHW, USE	Types TBS, SA, SIS, THHN, THHW, THW-2, THWN-2, RHH, RHW-2, USE-2, XHH, XHHW, XHHW-2, ZW-2	
	Copper			Aluminum or copper-clad aluminum			
1000	455	545	615	375	445	500	1000
1250	495	590	665	405	485	545	1250
1500	520	625	705	435	520	585	1500
1750	545	650	735	455	545	615	1750
2000	560	665	750	470	560	630	2000

Correction factors

Ambient temp., °C	For ambient temperatures other than 30°C (86°F), multiply the allowable ampacities shown above by the appropriate factor shown below						Ambient temp., °F
21–25	1.08	1.05	1.04	1.08	1.05	1.04	70–77
26–30	1.00	1.00	1.00	1.00	1.00	1.00	78–86
31–35	0.91	0.94	0.96	0.91	0.94	0.96	87–95
36–40	0.82	0.88	0.91	0.82	0.88	0.91	96–104
41–45	0.71	0.82	0.87	0.71	0.82	0.87	105–113
46–50	0.58	0.75	0.82	0.58	0.75	0.82	114–122
51–55	0.41	0.67	0.76	0.41	0.67	0.76	123–131
56–60	—	0.58	0.71	—	0.58	0.71	132–140
61–70	—	0.33	0.58	—	0.33	0.58	141–158
71–80	—	—	0.41	—	—	0.41	159–176

*See Sec. 240-3.

- Conductors with higher-temperature insulation may be terminated on lower-temperature rated terminations provided the ampacity of the conductor is based on the lower rating.

- The termination may be rated for a value higher than the value permitted in the general rules if the equipment is listed and marked for the higher-temperature rating.

See Solved Problem at the end of this chapter.

Q20. *Does the **NEC** require fireproofing?*

A. Not unless it is a requirement for listing, which, for medium-voltage cables, is generally the case. The AHJ will have to approve the field installation. For low-voltage or telecommunication wiring, Sec. 300-21 requires fire stopping.

Good practice for fireproofing should include any or all of the following:

- When the fireproofing around an existing cable has been modified due to new construction, the fireproofing must be replaced.

- All cables in electrical pullholes, vaults, and pullboxes should be fireproofed unless otherwise protected with metal or transition barriers.

- Fireproofing should be continuous, not interrupted at supports, and shall encircle three conductors when single-conductor cable is used, ending 1 in from the duct run entrance.

Q21. *Does exposure to sunlight pose hazards to electrical wiring?*

A. Yes. Requirements that appear in Sec. 310-8 recognize that the heat caused by sunlight can pose serious safety—as well as ampacity problems—for electrical systems. This requirement is consistent with the rules for thermal coordination of wires and terminations.

New to the **NEC** in the 1999 revision cycle is the requirement that "sunlight-resistant" conductors be specified where such conductors are so exposed. It is related to the generic rule that appears in Art. 100, which defines the basic types of locations (dry, wet, or damp) and requires that wiring be specified adequate to that purpose.

Solved Problem—Derating of Conductors

SITUATION.

An ambient environment of a panelboard in a mechanical room is 100°F—quite warm. Total load is 20 A.

REQUIREMENTS.

Determine the allowable ampacity and size OCPD for supply conductors.

SOLUTION.

From Table 310-16, select No. 10 AWG Cu=35 A.

Apply derating factors indicated in Table 310-16. Use the far right-hand column, which has temperature listed in Fahrenheit degrees.

$$35 \text{ A} \times 70\% = 24.5 \text{ A}$$

Apply derating factor again at 24.5 A $\times$ 82% = 20.09 A

Thus the allowable ampacity is 20.09. Select fuse or circuit breaker on this basis.

REMARKS.

This is very warm, but not uncommon ambient environment for many branch-circuit panelboards. The example illustrates a practical use of the conductor sizing derating factors.

Division 16200—Electrical Power

Remarks. Reliable electric power costs more than unreliable electric power. The same technological advances that reduced energy consumption have concurrently created a world much more dependent upon electric power. A "7 × 24 forever" operating requirement has become a standard and something of a buzzword in the business-critical enterprise industry where 100 percent power availability is necessary seven days a week, 24 hours a day. The literature on power reliability was quite thin until the early 1960s, when electrical engineers began to cast the problem in quantitative terms. Formal treatment of the reliability problem, its linkage to the cost of forced outages, and the proportional costs of power security switchgear now appears in the IEEE Gold

Book. Architects and engineers might find this reference helpful in designing on-site generation systems, the subject of this chapter. Electrical inspectors may find something in here that will give them insight into the degree to which an on-site generator may be more reliable than a utility service drop.

Q1.   *What NEC requirements commonly appear in Division 16200?*

A.   The generics appear in Art. 445; noteworthy modifications appear elsewhere. Transfer equipment appears in 230-83, emergency systems in 700, transfer equipment in 700-6, legally required systems in 701, optional standby systems in 702, interconnected electric power production sources in 705. If a fire pump is driven by a generator, refer to Art. 695, essential electrical systems for health care facilities in Art. 517 generally. In this chapter we lump together all equipment related to generators, including transfer equipment.

Q2.   *Are on-site generators any more reliable than a utility supply?*

A.   Refer to Tables 19.1 and 19.2. The data will show that utilities, particularly overhead utilities, are more prone to transients but obviously have no problem "starting." This data appears in an appendix of IEEE Standard 446, Recommended Practice for Emergency and Standby Power Systems for Industrial and Commercial Applications. The bottom line for architects and their associate engineers is that you need to consider the relative reliability of the local utility supply and the reliability of the best onsite generator the building owner can afford.

Q3.   *What is state of the art in distribution system reliability?*

A.   It appears that it is handled unevenly in the industry. California has some of the most aggressive reliability reporting requirements. It is a very sensitive issue because utility systems are, and should be, tied intimately into locality, its history of economic development, the weather patterns, geography, and the like.

The best data comes from the NERC as far as supply security is concerned. The data from distribution and delivery systems is much more difficult to follow. The California laws refer to SADI (System Average Interruption Duration Index) and other similar indices. None of this has any direct connection to the NEC except the extent to which the AHJ may use it to evaluate the independence and comparative reliability of an alternate source.

Q4.   *What is an emergency system?*

A.   An emergency system, as defined by the IEEE, is an independent reserve source of electric energy that, upon failure or outage of

TABLE 19.1 1980 GENERATOR SURVEY DATA

Equipment subclass	Average hours downtime per failure	Failure rate
Continuous service		
Steam turbine driven	32.7	0.16900 failure per unit year
Emergency and standby units		
Reciprocating engine driven	478.0	0.00536 failure per hour in use
Reciprocating engine driven	*	0.01350 failure per start attempt

*Small sample size; less than eight failures.
Appendix L of IEEE Std. 446 contains data from a recent survey of diesel and gas turbine generators, 600 to 1800 kW.

the normal source, automatically provides reliable electric power within a specified time to critical devices and equipment whose failure to operate satisfactorily would jeopardize the health and safety and personnel or result in damage for property.

Note the emphasis upon the work *automatic*.

By comparison, a standby power system is an independent reserve source of electric energy that, upon failure or outage of the normal source, provides electric power of acceptable quality so that the user's facilities may continue in a satisfactory operation.

A typical standby power arrangement might be a manually operated throw-over switch connected to a utility or to an on-site generator.

Q5.   *Is an on-site generator a service?*

A.   No. The definition in Art. 100 simply reads, "the conductors from the service point or other source of power to the service

TABLE 19.2 COMPARISON OF DIESEL AND GAS-TURBINE STARTING RELIABILITY STUDIES

Source	Number of units	Start attempts	Faniled starts	Starting reliability
Gas-turbine starting reliability studies				
ARINC Research Corporation	7	3.555	17	0.9952
Booz, Allen & Hamilton	34	12.316	80	0.9935
Kongsberg Dresser Power	38	17.749	141	0.9921
AT&T	28	13.644	106	0.9922
Diesel starting reliability studies				
ARINC Research Corporation	—	—	—	0.97
Electric Power Research Institute (EPRI)	155	22.320	83	0.9963
Consumers Power Company—Big Rock Point	2	669	12	0.9821
Northeast Utilities—Millstone	3	652	3	0.9954
Northeast Utilities—Connecticut Yankee	2	642	2	0.9969
Commonwealth Edison Company—Zion	4	1.693	30	0.9823
Consolidated Edison Company of New York, Inc.—Indian Point	6	424	4	0.9906
Institute of Nuclear Power Operations (INPO)	——Data not available——			0.9120
EPRI	——Data not available——			0.9829

disconnecting means." The reason for adding this phrase involves large industrial users that generate their own power on site in customer-owned generating plants. These operations are functioning as a private utility—as opposed to cooperatives and investor-owned utilities. This class of users typically do not have service points as they are defined in the Code in Art 100.

Q6.   *What is an optional standby power source?*

A.   Per **NEC** 702-2 second sources are installed purely for the owner's process protection. Consider double-ended substations that include provisions for throw-over tie breakers. Usually the motivation for installing these systems has to do with process continuity, not legal or life safety issues that qualify under 700 or 701. The alternate utility source becomes, technically, an alternate standby source. This could be disallowed and would effectively remove the provisions for optional standby sources to have separate services per Sec. 230-2.

Q7.   *What loads are permitted to be connected to the emergency generator?*

A.   Emergency generation systems should be designed primarily for providing standby electric service during outages for life and fire safety loads, including:

- Emergency white and exit lights
- Fire-alarm and fire-detection systems
- Fire sprinkler booster pumps
- Security systems and door locks
- Public address and telecommunication systems
- Closed circuit and master antenna television systems, elevators

Emergency white (EW) and exit lights are required at all egress points and changes in direction within corridors, as well as at main stairwell landings (nonintermediate), stairwell egress points, lobbies, and foyers. Exit light fixtures should have downlights, so they can be used in lieu of EW, and those exit lights delineating changes in egress direction require directional arrows.

Q8.   *On what basis should we consider other sources of emergency power?*

A.   Emergency lights should be supplied with 90-min, integral-standby-battery inverter-rectifier systems to sustain lighting for purposes of egress. Fluorescent lighting fixtures and exit lights equipped with integrally installed standby-power packages within their ballast channels are effective during transient loss of power. Exit lights also should use these emergency packages, which include floating battery sources. Used in conjunction with emergency generator sets, these allow facilities to ride through the 5- to 10-s delay for generator startup and in the event that generators fail to start altogether. See Fig. 19.1.

Q9.   *Can a normal source supply emergency loads?*

A.   Article 700 applies only to an emergency system when it is functioning in that mode. This means, for example, that a normal circuit to egress lighting need not comply with Art. 700 provided that the full electrical circuit pathway to that lighting is arranged so as to comply with Art. 700 during an outage. This might seem to support the view that since the tap ahead of the main is the normal source for a fire pump, it need not comply with Art 700. If the fire pump is legally mandated equipment for life-safety purposes (usually), and if the fire pump has no other supply (depends on facility design and local conditions), then Art. 700 will apply to the circuiting. The literal result will be to disallow the tap ahead of the main, because the rules in Chaps. 6 and 7 are of equal rank. In this case, the only way to comply with both chapters is to use a separate service or to use a standby generator that the system does not need in terms of its previously judged reliability. A more comprehensive treatment of this subject is given in *EC & M Magazine*, January 1997.

Q10.   *What are the broad principles that apply to high-rise building emergency power supplies?*

A.   In Sec. 700-9 the Code allows only listed electric circuit protective systems or feeder routed in spaces protected by sprinkler systems. The sprinkler provision now refers to a fully protected building instead of a fully protected space, a listed electric circuit protective system with minimum 1-h fire rating, a listed "thermal barrier system for electrical system components," a fire-rated assembly minimum rating of 1 h. This allows for multiple layers of sheet rock or other construction techniques that have been used for years to establish fire separations. Transfer equipment shall supply only emergency loads, per **NEC** 700-6.

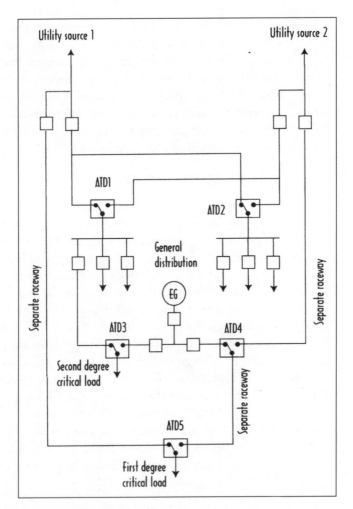

Figure 19.1 Two utility sources combined with an engine generator set to provide varying degrees of emergency power.

Packaged Engine Generators

Q11. *Are there any NEC rules governing the siting of an on-site generator?*

A. Some, one of the most important being the rule in Art. 700, Art. 695 and in the Life Safety Code that says you need to keep your

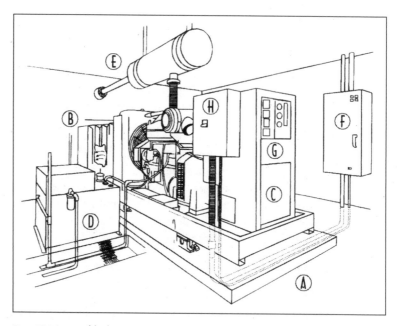

Figure 19.2 Isometric of diesel engine set room. (*Courtesy of EC&M Magazine.*)

Key to figure:

    A. Concrete foundation

    B. Radiator

    C. Cooling air inlet

    D. Fuel day tank and related supply and vent piping

    E. Muffler and exhaust pipe

    F. Transfer switch

    G. Control panel

    H. Circuit breaker

sources as independent as possible. Emergency feeders, distribution equipment, and generators must be in a 1-h fire-rated space or in a building fully protected by sprinklers. See Fig. 19.2.

Q12.   *What **NEC** rules govern the design of generator auxiliary equipment?*

A.   Section 700-12(b) says that if a generator requires a power-operated pump to move fuel from the main storage tank to the day tank for continued operation, the pump must be connected to the emergency power system. Thus, power-actuated dampers, ventilation, battery chargers and other auxiliary generator equipment may need to be connected to the emergency system.

It should be noted that the AHJ may judge that having the generator operate to failure is preferable to providing automatic means to shut it down which, in many cases, could present a greater hazard to personnel. An overload-sensing device would be permitted to be connected to an annunciator or an alarm instead of interrupting the generator and allow operating personnel to shut down load-side equipment in a safe and orderly fashion.

Q13.   *What type of generator should we specify?*

A.   The size and nature of the load, the space available, and the fuel type are among the principal considerations. The **NEC** does not

TABLE 19.3 COMPARISON OF GENERATOR-SET PRIME MOVERS

Considerations	Gas turbine	Reciprocating engine
Fuel	Diesel/natural gas	Diesel/natural gas
Starting time	60 s or longer	10 s
Ratings	250 kW and larger	15–2500 kW
Cooling	Air	Water
Cost	Higher than engine	Lower than turbine
Efficiency	Less efficient at full load than engine	More efficient at full load than turbine
Noise	Quieter than engine; high frequency	Noisier than turbine; low frequency
Transient-load response	Excellent	Good
Vibration	Low	High
Weight	Low	High

tell you which type, only how it must perform for a given facility class. See Table 19.3. Gas-turbine generators typically have more restrictive delays, coming on line in about 50 to 90 s. However, a new generation of turbine sets can assume loads within 8 to 10 s. They are also lighter and create less vibration than diesels. While suitable for rooftop mounting, gas-turbine units require substantially larger exhaust and muffler systems than diesels. Their exhaust stacks can be as large as 5 ft in diameter and must be up to +50 ft in length to permit adequate cooling of the exhaust-gas discharge, presenting architectural concerns. Turbine exhaust is also characterized by a piercing high-frequency whine, which cannot be completely muffled without sacrificing generator power and output.

Diesel-engine emergency generators require a 90-min fuel supply and a 72-h diesel tank supply in semicritical facilities such as nursing homes. In correctional and data-processing facilities, a 7-day fuel supply should be specified. Standby power should be fed to the load through a transfer switch.

Q14. *How do I select a backup generator?*

A. Sizing an engine-generator to carry its load involves many design parameters. If motors are involved, you must include them in the calculation to determine the generator set's starting kVA. Many generator manufacturers provide software to designers for making this determination based on the motor's code letter. A rule of thumb for estimating the inrush current when restarting the load under generator power: Use 2.3 times the normal load. You can reduce starting kVA by using soft starts, timed motor starts, and staggered load starts. A good practice is to divide the load into four equal parts. Then time the starts of each quarter about 20 s apart. Make sure the generator capacity allows you to start critical motors during emergency operations. If you have a UPS, you may still have an inrush. By design, the UPS will carry the load until your generator takes over.

When you cannot measure the starting kVA or dead-load pickup, then you must tabulate and calculate for each time of the expected load. Table 430-7(b) of the **NEC** lists locked-rotor kVA per horsepower for motors with specific code letters. For 24-h operation with varying loads, size generators to operate at no less than 50 percent of their rated load or as recommended by the manufacturer.

This prevents carbonization—sometimes called *wet-stacking*—of the engine. (Refer to *EC&M Magazine,* November 1997.)

Q15.   *How large should the fuel tank be?*

A.   Diesel-engine emergency generators require a 90-min fuel supply and a 72-h diesel tank supply in semicritical facilities such as nursing homes. In correctional and data-processing facilities, a 7-day fuel supply should be specified. Standby power should be fed to the load through a transfer switch. Refer to Table 19.4.

TABLE 19.4 FUEL-TYPE CONSIDERATIONS

Considerations	Natural gas	Diesel fuel
Engine generator size	Limited to 500–750 kW	2500 kW
Operations	Passive dependent on local utility	Active independent system
Source	Local gas utility	Fixed, requires monitoring and refueling because it is an independent source
Fuel-storage requirements	None	On site
Environmental issues	Limited	Fuel tank storage/site issues/disaster-recovery plan required
Control	Limited and dependent	Total control
Cost	Higher than diesel	Lower than natural gas
Emissions	Lower than diesel	Higher than natural gas
Transient-load response	Fair	Good

Q16. *Do generators require a disconnect?*

A. Yes. This is new to the 1999 **NEC**. For generators supplied by prime movers that cannot be readily shut down or that operate in parallel with another generator or other source voltage, a disconnect must be arranged between the load and "the generators and all protective devices and control apparatus," per 445-10.

Q17. *How do I connect a small gasoline generator for my house?*

A. The application would be considered an optional standby system unless an occupant's life depended upon reliable electric power. (This is a subject that can get complicated very quickly.)

For most single-phase residential applications, the safest method for connecting a generator is similar to Fig. 19.3. Here you see the basics of grounding and open transition switching. The safety of this installation should be confirmed by an electrician, the AHJ, the generator manufacturer, and the local utility. It is wise to have it "engineered" by a registered professional engineer to confirm its safety and economy.

Transfer Switches

Q18. *What are the broad principles that apply to the selection of transfer switchgear for on-site generators?*

A. Automatic transfer switches are used primarily for emergency and standby power-generation systems rated 600 V or less. These transfer switches do not normally incorporate overcurrent protection. They are available in ratings from 30 to 3000 A. An ATS is usually located in the main or secondary distribution bus that feeds the branch circuits. Because of its location in the system, the capabilities that must be designed into the transfer switch are unique as compared with the design requirements for other branch circuit and feeder elements. Special consideration should be given to the following characteristics of an ATS:

- Its ability to close against high inrush currents.

- Its ability to carry full-rated current continuously from normal and emergency sources.

- Its ability to withstand fault currents. In arrangements that provide protection against failure of the utility service, consideration should be given to

- An open circuit within the building area on the load side of the incoming service.

- Overload or fault condition.

- Electrical or mechanical failure of electric power distribution system within the building.

It is therefore desirable to locate transfer switches close to the load and to keep the operation of the transfer switches independent of overcurrent protection. It is frequently advantageous to apply multiple transfer switches of lower current rating located near the load rather than one large transfer switch at the point of incoming service. This adds cost and space but increases reliability.

Q19.   *Do I specify a three- or four-pole transfer switch?*

A.   This is important. Refer to Fig. 19.3. You can ground at either of two locations, at the service or at the generator. For the majority of generator applications: If you do it at the generator you will want a four-pole switch. If you do it at the service, you will want a three-pole switch. Once you've selected a four-pole switch, you must decide whether you want the neutral contacts to overlap in the transition. In most cases you do not. In hospital applications, you probably will. Switches may be specified with overlapping or broken transition of the neutral contacts.

Q20.   *Should I specify an electrically or mechanically held emergency transfer switch?*

A.   Mechanically held. Section 700-6(c) says that automatic transfer switches must transfer electrically but must mechanically latch into position. The rule applies to both switch positions. This protects against failure of the solenoid under load, which in turn would cause the transfer switch to drop away from the source prematurely.

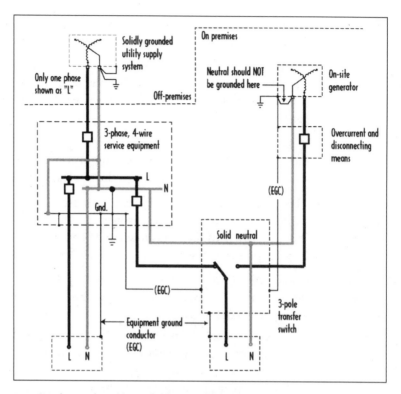

Figure 19.3 Phase-neutral-ground diagram of solidly interconnected neutral conductor grounded at service equipment and at the source of alternate power supply. Inset: pole transfer switch showing neutral contacts.

Q21.   *How should a transfer switch be programmed?*

A.   By a qualified person in complete cooperation with the local utility, with the Owner and the AHJ in the loop. It is risky to list a procedure because transfer procedures depend upon local requirements as well as **NEC** requirements.

The following procedure is something of an industry standard among automatic transfer switch manufacturers and may be used as a starting point for discussion among the various parties listed above. Assume that each of two engine sets (a lead and a non-lead) are paired with two automatic transfer switches (a primary and secondary).

1. Upon loss of utility power, an engine-start signal should be provided to the lead generator.

2. Begin timing sequence (say, 10–30 seconds) to allow the lead engine to reach an acceptable level of voltage and frequency.

3. If the lead generator is running properly, the primary ATS should transfer its assigned load to the lead engine source. All mission critical power is now supplied through the lead generator.

4. If the lead engine does not reach acceptable levels within the established time frame, or fails during operation, an engine-start signal should be provided to the nonlead engine.

5. This triggers the nonlead generator to start. Assuming the nonlead generator is running properly, a secondary ATS should transfer to the nonlead engine. The primary ATS will transfer power to the nonlead engine where it will remain.

6. Upon return of normal power, the primary ATS will transfer to the normal utility source. The engines should be shut down after all the cool down time delays expire.

The sequence can be modified to start both engines at the same time, but this practice is generally discouraged by engine-generator manufacturers because running a generator without load is detrimental to the engine-generator set in the long run. [Refer to *Pure Power Magazine* (supplement to *Consulting-Specifying Engineer*), Spring 1999.]

Q22.   *What other standards have bearing upon on-site generation?*

A.   Consider NFPA 54, the National Fuel Gas Code that sets minimum safety requirements for fuel gas piping systems, equipment that uses fuel gas, related accessories, and the safe venting of products of combustion. The key provisions of this code are as follows:

■ An emergency gas shutoff valve is required on the exterior of each building.

■ Before initial operation, all piping installation must be inspected and tested for pressure and leaks. The piping system must withstand the test pressure without showing any evidence of leaks or other defects.

- Operating instructions for gas-fueled equipment must be provided and left in a prominent place near the equipment for the consumer to use.

The National Fuel Gas Code is adopted by many states, and is used by most natural gas suppliers. When you select a natural gas generator, you are effectively selecting an on-site fuel source.

Solved Problem—Division 16200

SITUATION. A building with priority loads to be connected according to the following schedule:

$$
\begin{aligned}
100 \text{ hp} &= 124 \text{ A} \\
75 \text{ hp} &= 96 \text{ A} \\
60 \text{ hp} &= 77 \text{ A} \\
\text{Lighting} &= 62 \text{ A} \\
\text{Recepts} &= 25 \text{ A} \\
\text{Others} &= 122 \text{ A}
\end{aligned}
$$

REQUIREMENTS. Size the on-site generator kW.

SOLUTION.

1. Determine total running amperes=506 A (sum of full-load currents above).

2. Determine starting amperes by taking 100 percent of static loads, and 125 percent of motor starting loads.

Total starting current required:

$$
\begin{aligned}
124 \times 125\% &= 155 \text{ A} \\
96 \times 125\% &= 120 \text{ A}
\end{aligned}
$$

$$77 \times 125\% = 96 \text{ A}$$
$$62 \times 100\% = 62 \text{ A}$$
$$25 \times 100\% = 25 \text{ A}$$
$$122 \times 100\% = 122 \text{ A}$$
$$\text{Total} = 581 \text{ A}$$

3. Select running and starting kW by converting amperes to kilowatts.

$$506 \times 480 \times 1.732/1000 = 421 \text{ kW (running)}$$
$$581 \times 480 \times 1.732/1000 = 482 \text{ kW (starting)}$$

REMARKS.

The foregoing results may be converted to their horsepower equivalents. May want to figure in more conservative power factors. Motors may be started sequentially.

Division 16300—Transmission and Distribution

Remarks. One of the major changes in the NEC in the 1999 code cycle is the appearance of new code Art. 490, which pulls medium-voltage requirements from the old Art. 710. The workspace rules of Division 16000 apply here. The generic references to workspace, disconnects, overcurrent, and grounding still apply. Some of what is under the NEC umbrella falls under the NESC umbrella. The general requirements for installation are handled in 16000; we deal with medium-voltage circuit elements here.

Q1.   *What **NEC** requirements commonly appear in Division 16300?*

A.   Sometimes architect-engineers will label this section as high
voltage because it covers primary service, phasing, splices, and
terminations, and the construction installation sequence for
bringing new power on line. The specification for the unit substation is typically found in this division. If you are buying
above power at bulk primary rate (at a voltage above 1000 V
from a transformer larger than 1000 kVA, for example) then
the electric service construction requirements of Art. 230
appear here. Related sections in Division 3 are for concrete
work for housekeeping pad and concrete-encased duct.

Q2.   *What is the principal motivating factor in the selection of
medium voltage distribution?*

A.   The basic concept of load-center systems is to distribute power at
the highest economical voltage level to areas of concentrated electrical load centers where voltage is transformed to the lower utilization voltage and delivered to utilization equipment via relatively short secondary feeders. The size of secondary cables (often
500 kcmil and larger) is a design constraint that is familiar to all
electrical designers. The following is a comparison of cable costs
to distribute 1000 kVA at different voltage levels (assuming 440 V
as a reference voltage):

2400 V	25%
4160 V	15%
13,200 V	12.5%

Since distribution system planning has bearing on the sizing
and location of articulated unit substations within building a
few words about the planning process may be helpful for consultants. The planning process normally involves a partitioning
of the total distribution system planning problem into a set of
subproblems which can be handled by using available, usually
ad hoc, methods or techniques. The problem solution must
minimize the cost of substations, feeders and laterals to substations, and the cost of losses. In this process, however, we are
restricted by permissible voltage values, voltage dips, cable

ampacities, etc., as well as service continuity, reliability, and serviceability. In pursuing these objectives the planner ultimately has a significant influence on addition to and/or modifications of the primary distribution network, locations and sizes of substations, service areas of substations, location of breakers and switches, sizes of feeders and laterals, voltage levels and voltage drops in the system, the location of capacitors and voltage regulators, and the loading of transformers and feeders.

Q3.   *What are the broad principles that apply to medium-voltage switchgear?*

A.   You can find medium voltage requirements in parts of each of the following articles.

General Installation Requirements: 110-30

Services Over 600V: 230-200

Overcurrent Protection Over 600V: 240-100

Grounding Over 1 kV: 250-180

Outside Branch Circuits and Feeders: 225-50

Medium Voltage Cable: Article 326 and Section 310-60

Motors Over 600V: 430-121

New Article 490 pulls together all of the generic requirements. The equipment provisions of the former Article 710 have been relocated here.

Q4.   *What measures must be taken to protect building occupants from medium equipment?*

A.   Indoor electrical installations that are open to unqualified persons must be made with metal-enclosed equipment or must be enclosed in a vault or in an area to which access is controlled by a lock. Metal-enclosed switchgear, unit substations, transformers, pull boxes, connection boxes, and other similar associated equipment must be marked with appropriate caution signs. All doors should be labeled: DANGER—HIGH VOLTAGE—KEEP OUT.

Openings in ventilated dry-type transformers or similar openings in other equipment must be designed so that foreign objects inserted through these openings will be deflected from energized parts. In both cases, Section 230-203 requires signs with the words "Danger High Voltage Keep Out" must be posted in plain view where unauthorized persons might come in contact with energized parts.

Ventilating or similar openings in equipment must be so designed that foreign objects inserted through these openings will be deflected from energized parts. Where exposed to physical damage from vehicular traffic, suitable guards must be provided. Metal-enclosed equipment located outdoors and accessible to the general public must be designed so that exposed nuts or bolts cannot be readily removed, permitting access to live parts. Refer to Table 20.1. Where metal-enclosed equipment is accessible to the general public and the bottom of the enclosure is less than 8 ft above the floor or grade level, the enclosure door or hinged cover must be kept locked. Doors and covers of enclosures used solely as pull boxes, splice boxes, or junction boxes must be locked, bolted, or screwed on. Underground box covers that weigh over 100 lb will meet this requirement.

Q5.   *Under what conditions does the NEC require qualified maintenance personnel?*

A.   If secondary protection is not provided on a 2.5-MVA transformer on a medium-voltage system per Sec. 240-91 and Table 450-3(a).

Q6.   *Do I have much choice in the selection of a distribution system?*

A.   Typically not. In the vast majority of facilities, power is delivered to end users at utilization voltages such as 480, 240 or 208 V. That is because most residential, commercial, and even light industrial facilities can be supplied by a 100- to 1000-kVA transformer, pole-hung or pad mount. If you are working on behalf of an owner with a load in excess of 1000 kVA, then you may opt to purchase power at medium voltage. Even so, you may not know—or may not even need to know—all the details about how the local utility has wired the regional medium-voltage distribution system.

TABLE 20.1 MINIMUM CLEARANCE OF LIVE PARTS*

Nominal voltage rating, kV	Impulse withstand BIL, kV		Minimum clearance of live parts, in			
			Phase-to-phase		Phase-to-ground	
	Indoors	Outdoors	Indoors	Outdoors	Indoors	Outdoors
2.4–4.16	60	95	4.5	7	3.0	6
7.2	75	95	5.5	7	4.0	6
13.8	95	110	7.5	12	5.0	7
14.4	110	110	9.0	12	6.5	7
23	125	150	10.5	15	7.5	10
34.5	150	150	12.5	15	9.5	10
	200	200	18.0	18	13.0	13
46	—	200	—	18	—	13
	—	250	—	21	—	17
69	—	250	—	21	—	17
	—	350	—	31	—	25
115	—	550	—	53	—	42
138	—	550	—	53	—	42
	—	650	—	63	—	50
161	—	650	—	63	—	50
	—	750	—	72	—	58
230	—	750	—	72	—	58
	—	900	—	89	—	71
	—	1050	—	105	—	83

For SI units, 1 in = 25.4 mm.
*The values given are the minimum clearance for rigid parts and bare conductors under favorable service conditions. They shall be increased for conductor movement or under unfavorable service conditions or wherever space limitations permit. The selection of the associated impulse withstand voltage for a particular system voltage is determined by the characteristics of the surge protective equipment.

Figure 20.1 illustrates the classical medium-voltage distribution system types presented in order of increasing reliability and cost. Each arrangement is intended to supply power to a single facility. The salient points are as follows:

Simple radial. The entire load is supplied from a single source. Diversity among the loads makes it possible to minimize the installed transformer capacity.

Loop primary and primary selective radial. This system is most effective when two services are available from the utility. The electrical designer should work with the utility to determine the degree to which the incoming supply lines are "independent," i.e., whether they are themselves "looped" and can supply power from two geographically independent sources.

Loop selective and secondary selective. Normally operated as two electrically independent unit substations with the bus tie breakers open and with approximately half of the total load on each bus. This arrangement is commonly used to supply power to industrial and institutional facilities with large bulk distribution systems "nested" within a regional host utility. Transformer losses and initial costs are higher but so is the reliability.

Spot network. This regime offers maximum reliability to dense clusters of moderately-sized loads (100 to 1000 kW). It is used successfully, for example, in New York City. Aggressive analytic engineering of the network makes it possible for no single fault anywhere on the primary system to interrupt service to any load. Most faults will be cleared without interrupting service to any load.

The reader is referred to Cutler-Hammer-Westinghouse Consulting Guide for a comprehensive discussion of the various configurations shown here (Fig. 20.1).

Q7.   *What are the requirements for selecting and locating the service disconnect?*

A.   Part H of Sec. 230 contains the bulk of them. Each service disconnect must simultaneously disconnect all ungrounded service conductors that it controls and must have a fault-closing rating not less than the maximum short-circuit current available at its supply terminals. Where fused switches or separately mounted

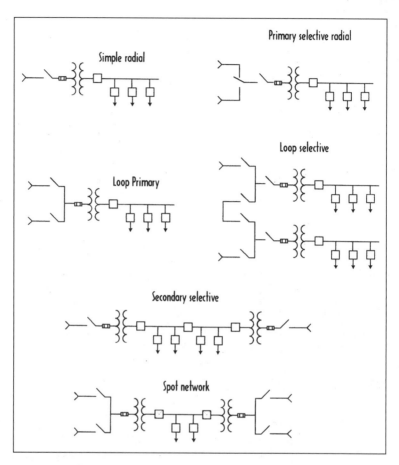

Figure 20.1 Types of distribution systems

fuses are installed, the fuse characteristics must be permitted to contribute to the fault-closing rating of the disconnecting means. Where the circuit breaker or alternative for it specified in Sec. 230-208 for service overcurrent devices meets the requirements specified in Sec. 230-205, they must constitute the service disconnecting means. Section 240-100 clarifies the rules for the application of overvoltage protective devices.

Q8.   *Does an interior substation supplied power from an underground service feeder need to have lighting arresters?*

A.   This is largely optional, though recommended [see Sec. 230-209, Surge Arresters (Lightning Arresters)]. Surge arresters installed in

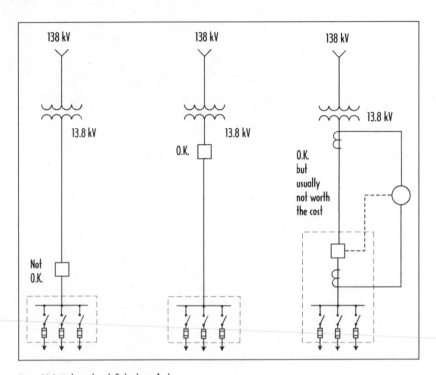

Figure 20.2 Medium-voltage bulk distribution feeders.

accordance with the requirements of Art. 280 are permitted on each ungrounded overhead service conductor. Where the voltage exceeds 15,000 V between conductors, they must enter either metal-enclosed switchgear or a transformer vault conforming to the requirements of Secs. 450-41 through 450-48.

Q9. *What are the rules for protection of medium-voltage service equipment?*

A. Article 490 says that pipes or ducts foreign to the electrical installation that require periodic maintenance or whose malfunction would endanger the operation of the electrical system cannot be located in the vicinity of the service equipment, metal-enclosed power switchgear, or industrial control assemblies. Protection must be provided where necessary to avoid damage from condensation leaks and breaks in such foreign systems. Piping and other facilities are not considered foreign if provided for fire protection of the electrical installation.

Q10. *What does the NEC have to say about medium-voltage cable?*

A. Article 326 contains a few very general, commonsense remarks which then refers to the tables of Art. 310. Those tables cover 0 to 2000 V. See Table 20.2 for a sample of an IPCEA cable ampacity table. All of the ampacity tables are based upon work by Neher-McGrath, researchers who did seminal work on the subject of cable ampacities.

Unit Substations

Q11. *What broad principles apply to selection of the incoming primary service switchgear?*

A. It is important to understand that a substation is not just a transformer. The need for a substation should be determined on the basis of the safety and sectionability that a substation adds to a *system*. See Fig. 20.3.

Section 240-100 requires the overcurrent device to be at the point of supply or otherwise located under engineering supervision. There has never been a medium-voltage overcurrent device location rule. The isolating switch for medium-voltage work can now be any device with visible break contacts, just not an air-break switch (Sec. 230-204). If service disconnect is oil switch or air, oil, vacuum, or SF6, then an isolating switch with visible contacts is required on the supply side of the disconnect. There are switches with visible break construction of other than air-break type that are acceptable per the NESC. A remote medium-voltage service disconnecting means at a multibuilding industrial installation under single management can be opened by a readily accessible remote-control device placed locally. See Fig. 20.2.

Q12. *What are the broad principles that apply to the transformers included as part of an articulated unit substation?*

A. Section 450-3 says that each transformer over 600 V, nominal, must have primary and secondary protective devices rated or set to open at no more than the values of transformer rated currents

TABLE 20.2 Medium-Voltage Cable Ampacity: Triplexed Concentric Stranded Rubber-Insulated Cable in Ducts

Copper conductor concentric strand
1 circuit 15 kV - 90°C conductor 20°C Ambient earth

Size	RHO-60				RHO-90				RHO-120				Delta TD for .0350 PF and RHO		
	30LF	50LF	75LF	100LF	30LF	50LF	75LF	100LF	30LF	50LF	75LF	100LF	60	90	120
2	180	176	169	162	178	173	164	155	177	170	160	150	0.76	0.84	0.91
1	206	201	193	185	204	197	187	176	202	194	182	170	0.80	0.89	0.96
1/0	235	230	221	211	233	225	214	201	231	222	208	194	0.84	0.93	1.01
2/0	269	262	251	239	267	257	243	228	264	252	236	219	0.88	0.98	1.07
3/0	308	300	287	273	305	294	278	260	302	289	269	250	0.92	1.03	1.12
4/0	353	343	328	311	349	336	317	295	346	329	307	283	0.97	1.09	1.19
250	388	377	360	341	384	369	347	323	380	361	336	310	1.01	1.13	1.24
350	470	456	434	410	465	445	418	387	460	436	403	371	1.09	1.24	1.36
500	572	553	525	494	566	540	504	465	559	527	486	444	1.19	1.36	1.49
750	707	681	643	603	698	663	616	565	689	647	592	538	1.32	1.51	1.67
1000	808	777	730	682	797	755	697	637	785	735	669	606	1.42	1.64	1.82

From the IPCEA Publication, Standard Cable Ampacities.

as noted in Table 450-3(a)(1). Electronically actuated fuses that may be set to open at a specific current must be set in accordance with settings for circuit breakers. Some relaxation of the rule is allowed where conditions of maintenance and supervision ensure that only qualified persons will monitor and service the transformer installation.

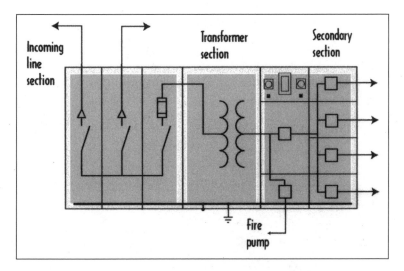

Figure 20.3 Basic components of a secondary unit substation.

Primary distribution, transformer section, and secondary breakers are sometimes called incoming section, and outgoing section. Load interrupter switches refer to ANSI C37 and C57 series. If the transformer is above 100 V, then it will be specified here along with related transformer controls, secondary breakers, and trip units. A new exception allows manufacturers to anticipate limited dual use of the grounding bus for both service and separately derived systems used in their equipment designs. The exceptions cover separately derived systems originating in listed equipment used for service purposes and allow for use of the service grounding electrode and internal grounding bus for both systems, assuming the sizes are appropriate [Sec. 250-30(a)].

Q13.   How much space can be saved by using a liquid-filled transformer?

A.   A little. Refer to Fig. 20.4, which illustrates the cost, footprint trade-off in a generic 1000-kVA substation using two common configurations. The temperature rise and basic impulse level classification of the transformer often determines the actual size of the metal cladding around switchgear. The higher the temperature rise (say, 150° rise) and the lower the BIL voltage, the smaller the cladding is likely to be. The size of the cladding around switchgear varies among manufacturers and depends upon the manner in which they modularize the components of their line of dry-type transformers.

This is not the only trade-off, however. Cheap transformers, though small and easy to fit into tight spaces, get too hot. All the money you want to save on the cost of the transformer you lose in building mechanical systems to keep the cheap, small transformer cool. On the other hand, expensive transformers stay cool, take up a lot of space, and practically last forever.

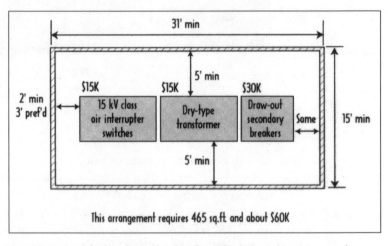

Figure 20.4 Comparison of medium-voltage switchgear layout for 1000-kVA, 480-V unit substation. Dry type transformer.

Figure 20.4 shows that cost and workspace square-footage requirements fall within a relatively narrow range. You want to have enough room for electrician safety, but not so much room that the users of the building will use the space for storage.

Q14.   *If the transformer is located indoors and doesn't meet the clearances called for in the listing of the liquid, what must be done?*

A.   Section 450-23(a)(2) says that a less flammable liquid insulated transformer specifically allows a violation of the liquid-listing conditions (often unachievable in retrofits) as a trade-off for installing automatic fire suppression and a liquid confinement area. Section 450-24 generally allows nonflammable fluid insulated transformers indoors with a liquid confinement area and vent.

Q15.   *If the transformer has a primary voltage of 40 kV, can the transformer be installed indoors?*

A.   Yes, but once you exceed 35 kV the transformer needs to go into a vault, regardless of the type of dielectric fluid.

Q16.   *With a 1-h fire rating, what are our options for siting a dry-type transformer?*

A.   Although transformers must generally be readily accessible, dry transformers are permitted in the open at greater heights. Dry transformers not over 50 kVA are generally permitted above suspended ceilings (not permanently closed in by structure) if the hollow space above the ceiling is fire-resistant, i.e., with at least a 1-hr fire rating. Note that any transformer ventilation requirements must still be met per Sec. 450-9. The area above the flush panel can be used for elements of the electrical installation, provided it is above the working clearance zone, per Sec. 384-4.

Q17.   *What broad principles apply to the protection of medium-voltage feeders?*

A.   Section 240-100 says that feeders must have a short-circuit protective device in each ungrounded conductor. The equipment used to protect feeder conductors must meet the requirements of Part K of Art. 250. The protective device(s) must be capable of detecting and interrupting all values of current that can occur at their location in excess of their trip setting or melting point. In no case can the fuse rating in continuous amperes exceed three times the ampacity of the conductor. Nor shall the long-time trip element setting of a breaker or the minimum trip setting of an electronically actuated fuse exceed six times the ampacity of the conductor. The operating time of the protective device, the available short-circuit current, and the conductor used will need to be coordinated to prevent damaging or dangerous temperatures in conductors or conductor insulation under short-circuit conditions. Some exceptions apply. Refer to Fig. 20.5. The left-hand side of this figure shows a cross section of a medium-voltage cable surrounded by a current transformer. Fault current is picked up by the current transformer and is used to close a contact on a breaker relay. The right-hand side of this figure shows the proper way to ground the shields of the cable so that this protection regime will operate properly.

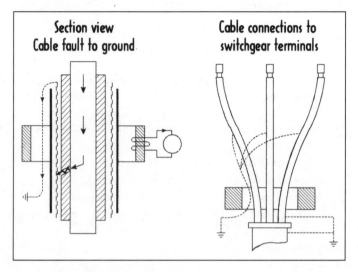

**Section view
Cable fault to ground**

**Cable connections to
switchgear terminals**

Figure 20.5 Medium-voltage cable ground fault protection.

Q18.   *What is "medium voltage," after all?*

A.   Nonelectrical professionals should be aware that the use of the terms low, medium, and high voltage will vary among electrical professionals depending on whether they follow the IEEE or the NEC definition of such voltage classes. The IEEE classifies any system above 115 kV as high voltage, and with medium voltage as ranging from 2400 V to 115 kV.

The use of "medium voltage" for the title of this CSI division is therefore not related to anything that appears in the NEC. As far as the NEC is concerned, anything over 600 V, nominal, is high voltage. The confusion is lamentable but typically does not change strategies for safe application of equipment operating at high potential.

Solved Problem—Division 16300

SITUATION.

A new unit secondary substation as shown in the Fig. 20.6. Of particular interest is the 480 motor load, which is approximately 25 percent of the substation load.

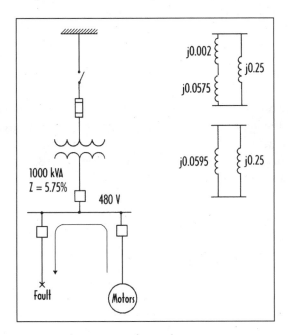

Figure 20.6 Fault current computation for a unit substation.

REQUIREMENT.

Select the interrupting capacity of the secondary feeder breakers.

SOLUTION.

Utility supply MVA = 500,000 kVA

Substation kVA = 1000 kVA

Source reactance on 1000-kVA base = 1000/500,000 = 0.002 pu

Combine positive sequence impedance networks. The total equivalent reactance is 0.0481 pu.

To convert this to equivalent amperes:

$$1/0.0481 \times (1000/1.732 \times 0.480)$$
$$= 24,969 \text{ A symmetrical fault current}$$

Select a breaker 30 to 42 kA.

Refer to Fig. 20.6.

REMARKS.

In many unit substations, where motors are a dominant component of the load, the feeder breakers may require AIC greater than the main breakers. To be conservative, many designers will simply size all breakers based on the maximum AIC. Practice in the field varies widely.

Division 16400—
Low-Voltage Distribution

Remarks. In a typical specification book this CSI division tends to be bulkier than all other divisions because it deals with the power back-bone within the building premises. It follows Division 16300 (Medium-Voltage Distribution) because it is reasonable to assume that the utility service is either already present or is built before the building premises wiring.

In this chapter we resume discussion of utility service with special focus on the low-voltage side of the transformer. We will dwell upon the notion of "service entrance" and its importance *as a legal artifice that determines where NESC requirements end and where **NEC** requirements begin. Some questions related to the subject of this chapter appear in Chapter 3, Site Requirements. We limit the discussion to single-occupancy facilities.*

Q1.   *What NEC requirements commonly appear in Division 16400?*

A.   Because building premise wiring is the central focus of the **NEC**, and because building premise wiring is largely low voltage, most of the information in Chaps. 2 and 3 of the **NEC** appears here. Because of the great quantity of design and construction issues that appear in this CSI division, we break the chapters down in the following subdivisions:

>   16400—Service, feeder, and branch circuit wiring
> 16450—Grounding
> 16460—Transformers
> 16470—Overcurrent protection
> 16480—Motors and motor control

Breaking up the subject matter in smaller, more manageable pieces will make easier the process of bringing forward the broad principles of **NEC** requirements for low-voltage distribution.

Q2.   *What are the broad contours of a building of low-voltage distribution system?*

A.   The **NEC** divides its own approach to the subject into three categories: services, feeders, and branch circuits.

Service entrance. This is the switchgear assembly for delivering energy from the serving utility to the premises being served. The conductors are terminated near their point of entrance into the building in the service equipment. The service equipment is the main control and means of cutoff for the supply for the building occupants and for the local fire department. Many requirements in the **NEC** are put there to make fire fighting easier. Article 230 contains the bulk of the **NEC** requirements.

Feeders. These are the conductors for delivering the energy from the service equipment to the panels which contain the final branch circuit fuse overcurrent devices (fuses or circuit breakers) which protect the end-user load equipment (lights or motors, etc.). Main feeders originate at the service equipment location. Subfeeders originate at bulk power panels (at the end of main feeders) and land on the panelboards which contain the overcurrent devices protecting end-user equipment. Subfeeders are necessary when a building is large and complex enough to have an additional bulk secondary distribution system. Otherwise most buildings require only main feeders. Article 215 contains the bulk of the **NEC** requirements.

Branch circuits. These are the conductors for delivering energy from the point of the final overcurrent device to the end-user utilization equipment. Article 210 contains the bulk of the **NEC** requirements.

When you are able to grasp this distinction, coupled with an understanding of the generic requirements of **NEC** Chaps. 1 through 4, and some general comfort with the basics of grounding, you will have mastered a large part of the National Electric Code.

Article 220 covers the rules for computing loads for each of the foregoing types of circuits. This section forms the basis for the types of load calculations that are necessary to pull a building permit. See Appendix Item A-3.

Service Entrance

Q3.   *What are the broad principles that apply to the design of service entrance?*

A.   Because the service entrance is the part of the building and involves equipment that is important to the utility, the choice of service equipment and service voltage should be a cooperative decision between the owner's engineer and the local electric utility. The operating voltage, ownership, maintenance, and burden of installation cost for this portion of the electric system will vary substantially from region to region.

An increasing number of utilities are offering medium-voltage services to only moderately sized facilities. Planning a medium-voltage distribution system within a building requires more time and effort but gives the design engineer a greater degree of flexibility in selecting equipment and designing electrical facilities within the building. The utility saves considerable capital investment in switching and transformation equipment which should be reflected in a lower electric rate.

When planning, it is important to route the incoming circuits to avoid clearance conflicts with existing or future underground or overhead structures. Poles located in areas subject to vehicular traffic may require curbs or barriers for protection. When open-wire circuits pass near buildings, adequate clearances must be provided. Some considerations for medium-voltage services are different from those for utilization voltage services, such as qualifications for operating and maintaining personnel and utility and code requirements.

In many states services are required by law to meet minimum construction standards. Underground and overhead services are normally built in accordance with local electric utility standards, which generally try to follow along the same lines as the **NEC**. If anything, local utility standards are likely to be more strict than **NEC** standards. The core group of standards is the NESC or state codes such as General Order 95 of California. The utility may assign responsibility for this construction to the commercial customer.

The physical arrangement of switchgear, primary switches (if any), transformers, and secondary distribution will vary considerably depending upon the type of distribution system at the utility interconnect and the type of facility. In some cases the utility will supply service from one or more transformer vaults located directly outside the building with bus stabs through the basement wall. This is the usual arrangement for buildings of moderate height in heavily loaded areas of many large cities. Transformer vaults are sometimes located within the building itself—in the basement and on the upper floors of tall buildings.

In all cases service entrance equipment rooms easily accessible to qualified persons should be dry and well lighted and should comply in all respects with the requirements of the electric utility and the AHJ. Plans should be made for possible future replacement of equipment. Provision for smoke exhaust should be considered in the event of an electrical fire.

In this chapter, however, we deal with low-voltage service entrances, those below 600 V. The bulk of the requirements appear in Art. 230. In Chap. 20, Division 16300, we discuss medium-voltage service entrances.

Q4.   How are the rules for medium-voltage service different from the rules for low-voltage service?

A.   Largely in the **NEC** generics of disconnect, access, and grounding. The generic rules for medium-voltage services appear in Part H of Art. 230. This section says that where oil switches or air, oil, vacuum, or sulfur hexaflouride circuit breakers constitute the service disconnecting means, an air-break isolating switch shall be installed on the supply side of the disconnecting means and all associated service equipment. Other features of medium-voltage service are as follows:

- Where fuses are of the type that can be operated as a disconnecting switch, a set of such fuses is permitted as the isolating switch.

- The isolating switch can only be accessible to qualified persons only

- Isolating switches must be provided with the means for readily connecting the load-side conductors to ground when disconnected from the source of supply. A means for grounding the load-side conductors is not required for any duplicate isolating switch installed and maintained by the electric supply company.

Q5.   Where is the service point for services where the utility owns the transformer?

A.   Many commercial and light industrial facilities with electric load on the order of 100 to 1000 kW operate under this kind of arrangement. The utility owns and maintains a transformer, on a pad or on a pole, which has a medium-voltage primary and a 480/277-V or 208/120-V secondary. Figure 21.1 or 21.2 indicates that the "service point" is defined to be in the underground pull box where a change in wiring method has taken place.

The service point is important because only those conductors on the load side of the service point come under the requirements of the **NEC**. Conductors on the line side of the service point come under the requirements of the NESC. The service point need not necessarily be in a below-grade junction or pull box. The service point may be an overhead secondary "makeup bus" where the customer's service entrance cable is landed. This makeup bus may be several feet in length, in which case the service "point" is not really a point at all.

When building a new service with a makeup bus it is worthwhile to come to an agreement with the utility about which point on the bus is the responsibility of the customer. It could be the termination point where the transformer cable and bus meet, where the makeup connects to the customer's secondary cable, or it could be the midpoint of the bus. The identification of service point will be critical in determining repair costs if service equipment is damaged.

Conductors on the load side of the service point are *service entrance* conductors and fall under **NEC** jurisdiction. All conductors upstream from the service entrance fall under NESC jurisdiction.

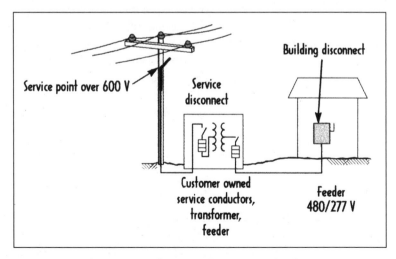

Figure 21.1 Service configuration. Service rated over 600 V supplying a customer-owned transformer.

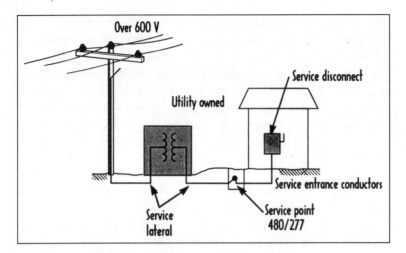

Figure 21.2 Service configuration. Service rated over 600 V supplying a utility-owned transformer.

Q6.   Where is the service point for a customer-owned transformer?

A.   Refer to Fig. 21.1, which shows a customer-owned and -maintained transformer with the service point up on the utility pole. All of the utility distribution equipment upstream from this point must meet NESC requirements. All of the customer-owned equipment downstream from this point must meet **NEC** requirements for medium-voltage equipment covered in Part H of Art. 230 and new Art. 490. The service disconnect is the fused transformer disconnect. The circuits originating from the secondary side of the transformer are considered to be feeders. The main service disconnecting means for the *building* is located as shown in the building. This is called the building disconnect; which is different from than the *service* disconnect.

Conductors on the load side of the transformer primary overcurrent protection are feeders.

Q7.   What are the protection requirements?

A.   Section 230-208 says that a short-circuit protective device shall be provided on the load side of, or as an integral part of, the service disconnect and shall protect all ungrounded conductors that it supplies. The protective device shall be capable of detecting and interrupting all values of current in excess of its trip setting or melting point that can occur at its location. A fuse rated in continuous amperes not to exceed three times the ampacity of the conductor, or a circuit breaker with a trip setting of not more than six times the ampacity of the conductors, shall be considered as providing the required short-circuit protection.

Overcurrent devices shall conform to the following:

(a) Equipment type. Equipment used to protect service-entrance conductors shall meet the requirements of Art. 710, Part C.

(b) Enclosed overcurrent devices. The restriction to 80 percent of the rating for an enclosed overcurrent device on continuous loads shall not apply to overcurrent devices installed in services operating at over 600 V.

Q8.   *Does the **NEC** govern where in the building the service equipment should be?*

A.   No, only where it cannot be (in a bathroom, for example). Some discussion of this issue appears in Chap. 17, where we covered workspace and access. One of the fundamentals that appear in Art. 230 is the requirement for concrete encasement of the service entrance conductors. When service entrance conductors have to pass through the building to the service equipment, a safety hazard exists because that part of the circuit in the building is not generally protected against short circuits, overloads, or arcing faults. However, when the distance is as short as possible, the hazard is minimized.

Longer circuits from the point of entrance through the building wall to the service equipment should be installed in a raceway encased in at least 2 in of concrete. Additional protection is provided by the use of metallic conduit suitably encased in concrete which is considered by the **NEC** as "outside the building." This protects the building by confining any fire or arcing (because of a short circuit) within the concrete envelope. The concrete-encased raceway may be installed along ceilings, under the basement floor, or on the roof.

Q9.   *Under what conditions does the **NEC** allow a building to have more than one service?*

A.   Section 230-2 offers a list of situations where it is allowed:

1.   For a fire pump

2.   For emergency, legally required standby, optional standby, or parallel power-production systems

3.   Multiple-occupancy buildings

4.   To meet capacity requirements

5.   For buildings of large area

6.   For services of different characteristics, such as for different voltages, frequencies, or phases, or for different uses, such as for different rate schedules.

Where a building or structure is supplied by more than one service, or any combination of branch circuits, feeders, and services, a permanent plaque or directory must be installed at each service disconnect location denoting all other services, feeders, and branch circuits supplying that building or structure and the area served by each .

Q10.   *How many services do we need?*

A.   This is a design question, not a question about what the **NEC** permits. The answer is: Design as few as possible. One service is preferred by electricians because it reduces the risk in not getting all the power cut off for maintenance and repairs. There are commonsense reasons why there must be more than one service, however. Among the obvious reasons why more than one service would be necessary, services are needed for systems of different characteristics, such as voltages, frequencies, phases, or different uses, such as different rate schedules. These may be planned for without the inspector's approval, and as long as they are built according to code, the design should pass inspection.

There are a few other reasons the **NEC** allows more than one service:

1.   The degree of reliability required for the installation with respect to the reliability of the power source. When supply security is important, multiple services or standby service, with load-transfer arrangements between the various parts of the building distribution system, may be indicated. In some cases economic considerations may indicate acceptance of reduced service availability and the interruption of nonessential loads during an emergency. If more than one service is required by a client, an additional charge may be assessed by the utility.

2.   The magnitude of the total load. Since the capacity of an individual service is limited by the utility to a maximum current value, additional services may be required to meet building demands.

3.   The availability of more than one system voltage from the utility. If more than one voltage is available, the utility may, for example, supply 208Y/120 for lighting and receptacles at one or more service entrance points, and 480Y/277 for power.

4.   The physical size of the building or the distances separating buildings compromising a single facility. Tall buildings, occupying a large ground area, and widely separated smaller buildings will often be supplied from multiple services.

5.   Local code requirements for firewalls.

Q11.   *How far away from the building must the service conductors be?*

A.   Section 230-9 says that service conductors installed as open conductors or multiconductor cable without an overall outer jacket must have a clearance of not less than 3 ft from windows that are designed to be opened, doors, porches, balconies, ladders, stairs, fire escapes, or similar locations. Conductors run above the top level of a window are permitted to be less than the 3 ft requirement above. Overhead service conductors cannot be installed beneath openings through which materials may be moved, such as openings in farm and commercial buildings, and cannot be installed where they will obstruct entrance to these building openings. Service raceways in concrete are considered outside a building. (Refer to Fig. 21.3.)

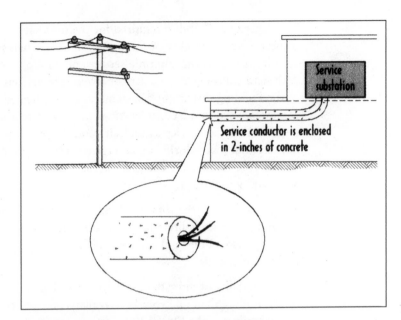

Figure 21.3 Encasement of service raceway will meet requirement of 230-70(a).

Q12. *Can you run service conductors through building A to get the building B?*

A.   No. Section 230-3 says that service conductors supplying a building or other structure may not pass through the interior of another building or other structure. You always want assurance that when you throw the switch to building A, it cuts out power only to building A. Likewise for building B. Any other regime compromises the safety of electricians and fire fighters. Now you may install service conductors along the *exterior* of one building to supply another building. Service entrance conductors installed *beneath* a building (encased in 2 in of concrete) or concealed in a raceway within a building (encased in 2 in of concrete) or installed in a transformer vault are considered outside the building.

In either case, each service, feeder, or branch circuit at the point where it originates shall be legibly marked to indicate its purpose unless located and arranged so the purpose is evident.

Q13. *If the second building is remote from a dwelling unit, what are the NEC disconnect requirements?*

A.   Article 225 contains the following requirements: A readily accessible disconnect is required at the remote building, located either outside or nearest the point of entrance inside Sec. 225-32. No more than six disconnects can be mounted in a single enclosure, or up to six separate enclosures (Sec. 230-71). The disconnects must be grouped, and each disconnect must be marked to indicate the load served.

Service-Entrance Conductors

Q14.   *What are the general rules for srvice entrance conductors?*

A.   Section 230-23 says that conductors must have sufficient ampacity to carry the current for the load as computed in accordance with Art. 220 and must have adequate mechanical strength. The conductors cannot be smaller than No. 8 copper or No. 6 aluminum or copper-clad aluminum. For installations to supply only limited loads of a single branch circuit, such as small polyphase power, controlled water heaters, and the like, they cannot be smaller than No. 12 hard-drawn copper or equivalent. The grounded conductor cannot be less than the minimum size as required by Secs. 230-42 and 230-79.

Q15.   *How do you size service entrance conductors for dwellings?*

A.   Service entrance conductors shall not be smaller than No. 6 unless in multiconductor cable. Multiconductor cable shall not be smaller than No. 8. Per 230-79, line conductors shall have an ampacity of not less than

1.   100 A, three-wire for a service to a one-family dwelling with six or more two-wire branch circuits.

2.   100 A, three-wire for a service to a one-family dwelling with an initial net computed load of 10 kVA or more.

3.   Grounded conductors. The grounded conductor shall not be less than the minimum size as required by Sec. 230-42.

Q16. *Give a summary of the service disconnect requirements, such as location, number, grouping, and access to occupants.*

A. Some of the **NEC** rules that apply to service disconnects are as follows:

■ A plaque is required at each service location to show the location of the other service.

■ A readily accessible disconnect is required to be located either outside or nearest the point of entrance inside.

■ There can be no more than six disconnects mounted in a single enclosure, or up to six separate enclosures.

■ The disconnects must be grouped [230-72(a)] and each disconnect must be marked to indicate the load served.

■ Fire pump, emergency, and standby power service must be grouped and located "sufficiently remote" from the normal service disconnecting means. In a multiple-occupancy building, each occupant must have access to his or her disconnecting equipment except where electrical maintenance is provided by continuous building management.

Q17. *What kind of equipment is allowed to be connected to the supply side of service disconnect?*

A. In general, nothing. There are a few exceptions, however. Section 230-82 allows for the following:

1. Cable limiters or other current-limiting devices.

2. Fuses and disconnecting means or circuit breakers suitable for use as service equipment, in meter pedestals, or otherwise provided and connected in series with the ungrounded service conductors and located away from the building supplied.

3. Meters nominally rated not in excess of 600 V, provided all metal housings and service enclosures are grounded in accordance with Art. 250.

4. Instrument transformers (current and voltage), high-impedance shunts, surge-protective devices identified for

use on the supply side of the service disconnect, load management devices, and surge arresters.

5. Taps used only to supply load management devices, circuits for emergency systems, standby power systems, fire pump equipment, and fire and sprinkler alarms if provided with service equipment and installed in accordance with requirements for service entrance conductors.

6. Solar photovoltaic systems or interconnected electric power production sources. See Art. 690 or 705 as applicable.

7. Where the service disconnecting means is power operable, the control circuit shall be permitted to be connected ahead of the service disconnecting means if suitable overcurrent protection and disconnecting means are provided.

8. Ground-fault protection systems where installed as part of listed equipment, if suitable overcurrent protection and disconnecting means are provided.

Service Equipment Overcurrent Protection

Q18. *What broad principles apply to overcurrent protection at the service entrance?*

A. Section 230-90 says each line must have overload protection; typically a fuse or a circuit breaker. This device must be an integral part of the service disconnecting means or shall be located immediately adjacent thereto. The fuse or circuit breaker must have a rating or setting not higher than the allowable ampacity of the conductor. Not more than six circuit breakers or six sets of fuses are permitted. The sum of the ratings of the circuit breakers or fuses are, however, permitted to exceed the ampacity of the service conductors.

Q19. *Will ground-fault protection on the service main cause nuisance tripping?*

A. It will if the overcurrent protection in the facility is not coordinated. Section 230-95 is one of the **NEC** requirements which enjoys a lively debate. It says that ground-fault protection of equipment must be provided for solidly grounded wye electrical services of more than 150 V to ground, but not exceeding 600 V phase-to-phase for each service disconnect rated 1000 A or more.

■ The rating of the service disconnect is considered to be the rating of the largest fuse that can be installed or the highest continuous current trip setting for which the actual over-current device installed in a circuit breaker is rated or can be adjusted.

■ The ground-fault protection provisions do not apply to a service disconnect for a continuous industrial process where a nonorderly shutdown will introduce additional or increased hazards.

■ The ground-fault protection provisions of this section shall not apply to fire pumps.

■ The ground-fault protection system must operate to cause the service disconnect to open all ungrounded conductors of the faulted circuit. The maximum setting of the ground-fault protection shall be 1200 A, and the maximum time delay shall be 1 s for ground-fault currents equal to or greater than 3000 A.

■ If a switch and fuse combination is used, the fuses employed must be capable of interrupting any current higher than the interrupting capacity of the switch during a time when the ground-fault protective system will not cause the switch to open.

Q20.   *Does the NEC require additional measures for maintenance of the ground-fault protection on the service main?*

A.   Yes. Because the main service equipment must be reliable, the NEC requires that the ground-fault protection system be performance tested when first installed on site. The test shall be conducted in accordance with instructions provided with the equipment. A written record of this test shall be made and be available to the authority having jurisdiction.

Q21. *What is so special about service entrance cable?*

A. Refer to Sec. 338-1. Service-entrance cable is a single conductor or multiconductor assembly provided with or without an overall covering, primarily used for services, and is of the following types:

- Type SE. Type SE, having a flame-retardant, moisture-resistant covering.

- Type USE. Type USE, identified for underground use, having a moisture-resistant covering but not required to have a flame-retardant covering. Cabled, single-conductor, type USE constructions recognized for underground use may have a bare copper conductor cabled with the assembly. Type USE single, parallel, or cabled conductor assemblies recognized for underground use may have a bare copper concentric conductor applied. These constructions do not require an outer overall covering.

- One uninsulated conductor. If type SE or USE cable consists of two or more conductors, one shall be permitted to be uninsulated.

Q22. *Is a grounding conductor required to be run to the second building?*

A. Article 225 does not contain any requirements for grounding, but Sec. 250-24 contains the requirements for remote buildings and structures. The basic rule is that a ground wire is not required if the grounded conductor (neutral) is used to ground the separate building disconnect. However, a ground wire is permitted to the separate building.

Q23. *What is new in the 1999 NEC on overcurrent protection?*

A. A service overcurrent protection device rating to be sized according to Note 3 of Table 310-16 for dwelling units. Per Sec. 230-90 service conductors shall have overload protection provided for each ungrounded service conductor. The general rule is that the overcurrent protection device setting must not be greater than the allowable ampacity of the conductor. A new Exception 5 was added to editorially correlate the rule that permits the overcurrent protection device to be sized in accordance with Note 3 of Table 310-16 for dwelling units.

Although busway is sometimes used for service entrance conductors, it is very difficult to provide protection for arcing faults. Neutral grounds normally employed at both the main switchboard and the service transformers make fault detection complicated.

Allied trade professionals should try to exceed **NEC** requirements by routing service cable systems to avoid high ambient temperature caused by steam lines, boiler rooms, UPS systems, etc. Also, avoid running raceways over the roof where conductors are subject to direct sunlight or reradiated or convective heat. Where cable is lead-sheathed, duct run through cinder beds should be avoided unless the duct system is encased in a sufficiently thick envelope of concrete to make it impervious to the acid condition that is prevalent in cinder beds. The lead sheath should also be jacketed to protect against corrosion. Precaution should be taken with polyethylene, cross-linked poly, and other organic jacketed cables to prevent chemical degradation of the jacketing when hydrocarbons may be present as in fueling areas, marshland, landfill areas, and similar locations. Cable systems should be protected from oils and chemicals that are used as preservatives in wood poles by suitable barriers on the riser pole, and enclosed in a raceway at a suitable distance from the pole.

Spare ducts should be considered to provide for the contingency in which a faulted cable becomes frozen in a duct and cannot be removed for replacement. This also simplifies installation of future cables that may be required for load growth. A duct system should not be laid in the same trench with gas or sewer service.

Miscellaneous

Q24.   *How close can wires be to windows?*

A.   Overhead conductor clearances from windows are now the same as service drop conductors. Section 225-19(d) relates to conductor clearances from windows and other areas; it was reworded to be similar to the requirements of service conductors; see Sec. 230-9. The rule requires that feeder and branch circuit conductors must be kept not less than 3 ft from windows that are designed to be opened, doors, porches, balconies, ladders, stairs, fire escapes, or similar locations. Service conductors installed as open conductors must have a clearance of not less than 3 ft from windows that are designed to be opened, doors, porches, "balconies, ladders, stairs," fire escapes, or similar locations. See Sec. 225-19(d) for similar requirements for branch circuits and feeders (Fig. 5).

Q25.   *How low can the service drop be?*

A.   First of all, use common sense. If a wire—any energized wire—looks too low to you, or if it looks as if someone could touch it without much of a stretch, then it is in violation of the **NEC**. Get the AHJ to throw the book at the installation.

That much said, the Sec. 230-24(b) of the **NEC** requires that service drop loop conductors maintain a minimum clearance of 10 ft. Service-drop conductors, where not in excess of 600 V nominal, must have a minimum clearance of 10 ft from final grade. The change to this section clarifies that the 10-ft vertical clearance above grade for service drop conductors also applies to the lowest point of the service drop loop conductors.

Q26.   *What are the requirements for the overhead conductor, such as drip loops, point of attachment, and clearances above roofs?*

A.   Article 225 contains the same requirements for outside overhead conductors as Art. 230 for overhead service conductors: The minimum clearance for the point of attachment is 10 ft but it may need to be higher to maintain the minimum clearances specified in Sec. 225-18. Overhead conductors must maintain clearances from ground and roofs according to Secs. 225-18 and 225-19.

Q27.   *What are the basics for installing service laterals underground?*

A.   Section 230-30 says that

- Service-lateral conductors shall withstand exposure to atmospheric and other conditions of use without detrimental leakage of current. Service-lateral conductors shall be insulated for the applied voltage.

- Service-lateral conductors shall have sufficient ampacity to carry the current for the load as computed in accordance with Art. 220 and shall have adequate mechanical strength.

■ Underground service-lateral conductors shall be protected against damage in accordance with Sec. 300-5. Service-lateral conductors entering a building shall be installed in accordance with Sec. 230-6 or protected by a raceway wiring method identified in Sec. 230-43.

Division 16450—Grounding

Remarks. In 1747, Sir William Watson of Great Britain astonished the scientific world by sending an electric current through a wire 2 miles long, using the earth as a conductor. Electrical people have been struggling with the meaning of this ever since. Even though its practical manifestation seems simple enough— the driving of rods in the ground, the bonding of metals to metal—it is the most subtle of all subjects in the electrical power industry.

We distinguish between the practice of grounding (which is mechanical bonding of conductors) and ground-fault protection. Ground-fault protection is a form of overcurrent protection which will be more fully discussed in Division 16470. This division deals principally with broad principles. We shall use the terms conductivity and resistance interchangeably. In most cases we are seeking high conductivity in the grounding regime, which is simply the reciprocal way of saying low ground resistance to the flow of fault current.

Q1. *What NEC requirements commonly appear in Division 16450?*

A. This is another **NEC** article with a title shared by a CSI division. The bulk of the requirements for grounding appear in Art. 250, which has undergone extensive reorganization in the 1999 **NEC,** so much so that the NFPA has put together Appendix E as a cross-reference guide. It is best just to read it from start to end as if it were a completely new document and sign up for a grounding training course. It is such a large and important body of information to understand and apply that we can only treat the main points here. Some related topics appear in Division 16600 (Lightning Protection) and Chap. 6 (Metals).

Grounding is no less an aspect of electrical systems than phase conductors. The core requirements are in Art. 250 and stated as exceptions elsewhere. It should be no surprise because the terminology for the grounded part of any circuit requires focus because slight differences in phraseology are big differences electrically. "Grounding *electrode* conductor" vs. "grounded *service* conductor" is a good example: two similar-sounding phrases and yet the distinction makes a big difference electrically. The sidebar, adapted from Art. 250 of the 1999 **NEC** itself, puts all the terminology side by side so that you can see that it will take time to become comfortable with the fine distinctions in grounding terminology.

GROUND:

A conducting connection, whether intentional or accidental, between an electrical circuit or equipment and the earth, or to some conducting body that serves in place of the earth. A "ground"—as often heard in casual conversation—as often heard in casual conversation—is actually a phenomenon, an occurrence, or a condition. The important word here is "intentional." Electrical systems are frequently intentionally grounded; electrical equipment is frequently accidentally grounded. The ground conductor is the actual physical copper wire that connects the earth to the equipment or system.

GROUNDED:

Connected to earth or to some conducting body that serves in place of the earth.

GROUNDED, EFFECTIVELY:

Intentionally connected to earth through a ground connection or connections of sufficiently low impedance and having sufficient current-carrying capacity to prevent the buildup of voltages that may result in undue hazards to connected equipment or to persons.

GROUNDED CONDUCTOR:

A system or circuit conductor that is intentionally grounded.

GROUNDING CONDUCTOR:

A conductor used to connect equipment or the grounded circuit of a wiring system to a grounding electrode or electrodes. The grounding electrode should be as near as possible to the grounding electrode conductor connection to the system.

GROUNDING CONDUCTOR, EQUIPMENT:

The conductor used to connect the non-current-carrying metal parts of equipment, raceways, and other enclosures to the system grounded conductor, the grounding electrode conductor, or both, at the service equipment or at the source of a separately derived system.

GROUNDING ELECTRODE CONDUCTOR:

The conductor used to connect the grounding electrode to the equipment grounding conductor, to the grounded conductor, or to both, of the circuit at the service equipment or at the source of a separately derived system.

GROUND-FAULT CIRCUIT-INTERRUPTER:

A device intended for the protection of personnel that functions to de-energize a circuit or portion thereof within an established period of time when a current to ground exceeds some predetermined value that is less than that required to operate the overcurrent protective device of the supply circuit.

GROUND-FAULT PROTECTION OF EQUIPMENT:

A system intended to provide protection of equipment from damaging line-to-ground fault currents by operating to cause a disconnecting means to open all ungrounded conductors of the faulted circuit. This protection is provided at current levels less than those required to protect conductors from damage through the operation of a supply circuit overcurrent device.

Q2.   *What are the broad principles of grounding that apply to the electrical systems in buildings?*

A.   The fundamental idea is to provide a low-impedance path for fault current to return to the earth before it can harm persons or do damage to electrical apparatus. It is a law of nature: electric current will instantly seek lowest potential. The effect is analogous to the behavior of water under the influence of the force of gravity (and just as immutable).

Per **NEC** 250-2(d), the earth must not be used as the only equipment grounding conductor of fault current path. Recommended *maximum* ground resistance for various facility types is as follows: Residential—25 ohms maximum. Light industrial—5 ohms. Heavy industrial or substation—1 ohm. When a significant lightning threat is expected, lightning arresters should be grounded with a resistance no more than 1 ohm.

In many cases, fortuitous grounds are established by underground water pipes, well casings, metallic building frames, and concrete piers. Experimental data indicate that metallic underground water systems, metallic underground sewers, or underground metallic gas-pipe systems have a ground resistance of less than 3 ohms. Wooden water pipes, plastic pipes, and nonconductive gaskets, however, provide circuit interruption. A great deal of Art. 250 of the **NEC** is devoted to wiring practice to ensure the continuity of a circuit that, under ideal operating conditions, should never carry any current. Most building systems in the United States are solidly grounded.

Q3.   *What is the most significant operating characteristic of the grounded system?*

A.   It provides for the highest level of fault current needed to drive overcurrent protective devices, and thus quick isolation of faulted circuits with a properly coordinated system. Grounded-wye systems, while having the disadvantage of a partial shutdown after the first ground fault, will limit system potential to ground. It provides the greatest maximum protection against system overvoltages due to lightning, switching surges, static, and contact with another high-voltage system. It limits the difference of electric potential between all uninsulated conducting objects in a local area. Refer to Fig. 21.4.

One of the key advantages of the grounded system as it is applied in building electrical systems is that it provides two voltage levels.

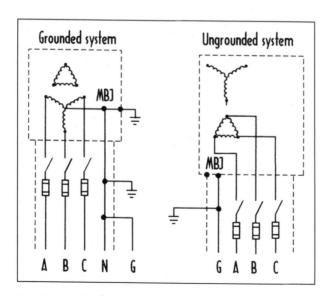

Figure 21.4 Comparison of grounded and ungrounded systems.

Single-phase loads such as lighting can be connected line-to-neutral. Three-phase loads such as motors can be connected line-to-line.

Q4. *What is the principal reason for operating a system ungrounded?*

A. You get an extra chance to find the fault. The first fault will not take down the circuit. If you have a ground annunciation circuit installed, it will give you a visual or audible alarm without opening the circuit breaker. This is sometimes important in process industries where the greater hazard may be in the sudden stop of a process. Thus, ungrounded systems are very common in industrial facilities.

A second fault on the same circuit, however, without having cleared the first, will open the circuit breaker. One key disadvantage of the ungrounded system is that high transient overvoltages will result when the first fault occurs. The second fault will typically open the circuit, but damage to cable insulation may be an accompanying result.

Q5. *How does water content affect the resistivity of the soil?*

A. Water actually improves ground current conductivity; electrolytic content can improve it as well. Very dry soils offer greater resistance to the flow of ground current, wet soils very little. Figure 21.5 shows soil resisitivities various parts of the United States. The map shows considerable variability. The practical effect of these differences for architects and their electrical engineers is that a "standard" ground mat design for a building in north Georgia may need to be larger than the ground mat for the same building in Marin County, California.

Q6. *What is the difference between grounding and a "ground"?*

A. Grounding is good. A "ground" is not. Ground*ing* refers to the electrode system, which may be the building steel or the water pipe, the metallic path that drains excess voltage into the earth, an *intentional* and planned-for conducting path. A "ground" is a somewhat more colloquial term for an accidental or unintentional and undesirable path that electricity will follow if the intended path is not restored. A ground may be a tree falling upon an overhead power line or the insulation wearing off the flexible cord to an appliance and making contact with the steel cladding, a washer

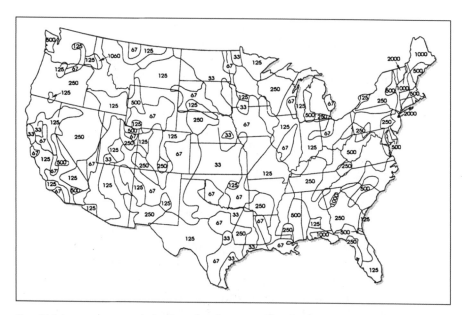

Figure 21.5 Average earth resistivity in the United States, showing how resistivity of the earth varies among regions.

or dryer, for example. Electrical people are always trying to get current to flow to ground before it flows through people.

Grounding and ground fault protection are not quite the same thing. We may speak of water pipes in connection with grounding, but when we say: fire pumps cannot have ground-fault protection OR ungrounded systems should have an audible or visual annunciation when there is a "ground," we are talking about different aspects of an electrical system that sound the same because the word "ground" appears in both.

Q7.   *What kind of circuit must be grounded?*

A.   The **NEC** makes a distinction between the kind of circuits that *must* be grounded and the kind of circuit that *may* be grounded. Per Sec. 250-21, they are (*a*) ac circuits of less than 50 V, (*b*) ac systems of 50 to 1000 V supplying premises wiring and premises wiring systems, (*c*) ac systems of 1 kV and over, and (*d*) separately derived systems. Sections 250-21 through 250-36 generally cover this topic.

Q8.   *How do grounding hazards develop?*

A. Shock hazard results when a human being is the intermediary between a region of high and low voltage. Thus, by connecting all conducting bodies together the potential equalizes and there is no potential difference. One of the principal ways of ensuring this is by making sure the impedance of the equipment grounding conductor is kept as low as possible. This will ensure proper operation of protective devices and minimize any hazardous differences of potential under fault conditions.

Where an ac system operating at less than 1000 V is grounded at any point, the grounded conductor cannot be smaller than the required grounding-electrode conductor specified in Table 21.1. For service entrance conductors larger than those listed in Table 21.1 the grounded conductor must be not less than 12.5 percent of total area of the largest set of ungrounded conductors. When parallel service raceways are involved, you apply the table based on the size of the ungrounded phase conductors in each of the parallel runs but never smaller than AWG 1/0, even when the neutral current in the grounded conductor is likely to be low.

Q9. *How is* **NEC** *Table 250-66 (Table 21.1) applied to select the size of the grounding electrode conductor?*

A. This is one of the most commonly used tables in the **NEC** and is given special focus in this book. It is keyed directly to the size of the service entrance conductors. Say you have two parallel 500-kcmil XHHW copper conductors for the phase service entrance conductors. Find that row in the far left column and read AWG 2/0 as the minimum size of the electrode conductor. Note that copper must match with copper. Refer to Table 21.2.

Q10. *How is* **NEC** *Table 250-122 applied to select the size of the bare conductor that must be run with the phase conductors in a new feeder circuit?*

A. This is another one of the most commonly used tables in the **NEC**. Say you have a trip unit rated 350 A and the branch feeder to the distribution panel is enclosed in a nonmetallic raceway. On the basis of the size of the rating of the upstream OCPD you would select an AWG No. 3 copper conductor for the equipment ground conductor. Refer to **NEC** Table 250-122.

TABLE 21.1 GROUNDING ELECTRODE CONDUCTOR FOR ALTERNATING-CURRENT SYSTEMS (NEC TABLE 250-66)

Size of largest service-entrance conductor or equivalent area for parallel conductors[1]		Size of grounding electrode conductor	
Copper	Aluminum or copper-clad aluminum	Copper	Aluminum or copper-clad aluminum[2]
2 or smaller	1/2 or smaller	8	6
1 or 1/0	2/0 or 3/0	6	4
2/0 or 3/0	4/0 or 250 kcmil	4	2
Over 3/0 through 350 kcmil	Over 250 kcmil through 500 kcmil	2	1/0
Over 350 kcmil through 600 kcmil	Over 500 kcmil through 900 kcmil	1/0	3/0
Over 600 kcmil through 1100 kcmil	Over 900 kcmil through 1750 kcmil	2/0	4/0
Over 1100 kcmil	Over 1750 kcmil	3/0	250 kcmil

[1]Where multiple sets of service-entrance conductors are used as permitted in Sec. 230-40. Exception 2, the equivalent size of the largest service-entrance conductors, shall be determined by the largest sum of the areas of the corresponding conductors of each set.

[2]Where there are no service-entrance conductors, the grounding electrode conductor size shall be determined by the equivalent size of the largest service-entrance conductor required for the load to be served.

[1]This table also applies to the derived conductors of separately derived ac systems.

[2]See installation restrictions in Sec. 250-64(a).

TABLE 21.2 MINIMUM SIZE EQUIPMENT GROUNDING CONDUCTORS FOR GROUNDING RACEWAY AND EQUIPMENT (NEC TABLE 250-122)

Rating or setting of automatic overcurrent device in circuit ahead of equipment, conduit, etc., not exceeding, amperes	Size (AWG or kcmil)	
	Copper	Aluminum or copper-clad aluminum*
15	14	12
20	12	10
30	10	8
40	10	8
60	10	8
100	8	6
200	6	4
300	4	2
400	3	1
500	2	1/0
600	1	2/0
800	1/0	3/0
1000	2/0	4/0
1200	3/0	250
1600	4/0	350
2000	250	400
2500	350	600
3000	400	600
4000	500	800
5000	700	1200
6000	800	1200

Where necessary to comply with Sec. 250-2(d), the equipment grounding conductor shall be sized larger than this table.
*See installation restrictions in Sec. 250-120.

The basic idea is to make sure that the conductor circular mil area is large enough to carry fault current to ground without being destroyed by it; thus the linkage with the rating of the overcurrent device. The conductor should be large enough to sustain the flow of fault current without causing excessive voltage drop. Excessive voltage drop may destroy the utilization apparatus itself during the fault. The **NEC** requires that special consideration be given to motor circuits, multiple circuits, and conductors in parallel. The rules for selecting the grounding conductor for supply circuits with parallel conductors with ground fault protection are more rigorous per Sec. 250-122(f)(2).

Q11.   *Where do you terminate an isolated ground?*

A.   The relevant section of the **NEC** is Sec. 250-146. An isolated ground may be terminated at any panel (within the same building between the protected circuit and the service), at the service, or derived service. You cannot go wrong by taking the isolated ground back to the original source, although you will likely need to increase the wire size depending on the length of the run. "Isolated" does not mean totally separate, only that the circuit ground is isolated from other circuit grounds and metal enclosures back to the source. You can never just drive a ground rod and call it an isolated ground.

Every code cycle, it seems, some new conducting body becomes a candidate for grounding. Table 21.2 tells you how large the conductor must be. For cables in parallel the EGC in each can be sized to the trip setting of the upstream GFPE if the GFPE is listed for the purpose. GFPE must not exceed the ampacity of a single ungrounded conductor.

Q12.   *When is a supplemental electrode required?*

A.   When the first made electrode runs above 25 ohms, Sec. 250-56 requires an additional electrode of any type specified in Sec. 250-118. Made electrodes do not actually supplement the water pipe. The usual resistances and dissipation capability of water pipe electrodes is so superior to that of made electrodes that a made electrode serves little initial purpose. Supplemental electrodes have one principal purpose: to provide an electrode for the system in the event that the metal water pipe is removed or made unavailable due to an insulating joint. (*EC&M*, June 1998, p. 54). Supplemental ground rods must meet the resistance requirements of made electrodes. Plate electrodes must be at least 30 in below grade per Sec. 250-52(d).

Q13. *What special considerations must be given to grounding service entrances when the conductors are in parallel?*

A. For service-entrance phase conductors larger than 1100 kcmil copper or 1750 kcmil aluminum, the grounded conductor cannot be smaller than 12.5 percent of the area of the largest service-entrance phase conductor. Where the service-entrance phase conductors are installed in parallel, six of the grounded service conductors are required to be based on the equivalent area of the ungrounded service conductors.

Figure 21.6 is an example of bonding electrodes together form the grounding electrode system. Water pipe electrodes for separately derived systems must follow the same distance-into-building constraints as similar electrodes for services.

Q14. *How are grounding electrodes formed by water pipes?*

A. Refer to Fig. 21.6. This is how the **NEC** allows it to be done. Even though the water pipe may be a suitable grounding electrode, you must provide at least one more grounding electrode and it must be bonded to the water pipe. The pipe must be in direct contact with the earth for more than 10 ft. The requirements are detailed in new Sec. 250-50.

For many years people have been concerned that grounding electrodes formed by water pipes will cause damage when electricity flows through them during faults. Effectively, an entire building raceway system "jumps" electrically for a short instant before the OCPD opens the circuit. A work group was formed by the International Association of Electrical Inspectors in 1944 to study the issue, and the outcome of the investigation did not result in prohibiting the use of water pipes as ground electrodes. The National Institute of Standards and Technology has since monitored the electrolysis of metal systems with a similar finding that water pipe grounding solves many more problems that it creates.

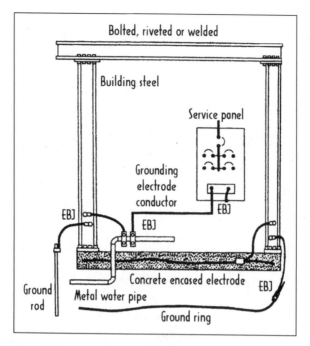

Figure 21.6 Water pipe grounding. (*Adapted from* **NEC** *Handbook.*)

Q15.   *Can underground metal gas piping be used as a grounding electrode?*

A.   Absolutely not [see Sec. 250-52(a)]. What may appear to be the use of the pipe as an electrode is actually the pipe itself being grounded in order to reduce the hazard of an electrical arc igniting natural gas. New Sec. 250-104(b) says that above-ground portions of gas piping systems *upstream* from the equipment shutoff valve must be electrically continuous and bonded to the grounding electrode system. Figure 21.7 shows the point where the gas pipe, typically inside your house, should be grounded. The requirement for bonding applies to all aboveground portions of gas piping whether it is located indoors or outdoors.

A more complete discussion of this subject appears in NFPA 54—the National Fuel Gas Code. You should consult the AHJ and/or the local natural gas company with further questions.

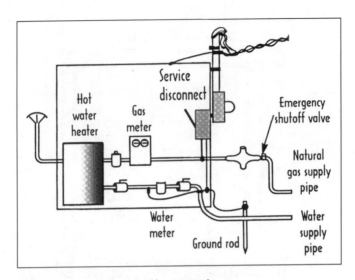

Figure 21.7 A gas pipe grounding. *(Adapted from 1999 NEC Changes.)*

The grounding conductor indicated in Fig. 21.7 will keep the gas pipe at zero potential. An authoritative source (*EC&M*'s Fred Hartwell) suggests sizing it according to new Table 250-66.

Q16.   *How does the grounding system for a typical building operate?*

A.   Most residential, commercial, and light industrial buildings are solidly grounded 480/277 or 208/120 V. This means that if the grounding system has been designed, built, and maintained properly, there will be virtually negligible resistance for fault current to flow from the point of fault to the driven rods located in the earth on the exterior of the building. In the solidly grounded system, the ground fault current flows through the equipment grounding conductors (usually the bare fourth wire pulled in a raceway) from a ground fault anywhere in the system to the bonding jumper between the equipment grounding conductors and the system ground-*ed* conductor. A normally operating utility+service equipment circuit is shown in the phase-neutral-ground schematic of Fig. 21.8. (The phase-neutral-ground diagram is a very common method of analyzing three-phase ac circuits because, by showing only one phase, all of the current flow fundamentals can be observed without cluttering the drawing. The magnitude of the other two phases is assumed to be equal to the first.)

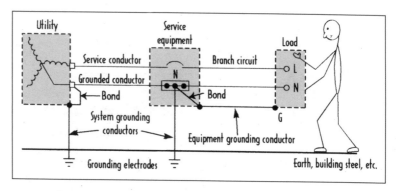

Figure 21.8 Phase-neutral-ground diagram of solid grounding regime operating normally.

The main bonding jumper and the grounding electrode conductor connections to the grounded system conductor have to be made at the same point. The objective is to keep (fault-related) grounded conductor current confined to electrical conductors.

Q17. *How is a fault cleared?*

A.   By getting enough of the fault current to flow through the OCPD (either fuse or circuit breaker). The faulted circuit in shown in Fig. 21.9. This condition may exist for example, when an internal power wire comes loose in an appliance or machine and strikes the metal housing. Note how the ground fault current return path is completed through two grounding conductors: through the bonding jumper in the service equipment and through the grounded service conductor to the supply transformer. The person touching the conductive enclosure in which the fault exists will be protected from shock injury if the equipment grounding conductors provide a shunt (parallel) path of sufficiently low impedance to limit the current through the person's body to a safe magnitude (much less than 5 mA, if possible). In solidly grounded systems, the ground fault current actuates the circuit protective devices to automatically deenergize a faulted circuit and thus remove the shock exposure.

Q18. *Will grounding the system at several places reduce electrical shock hazard?*

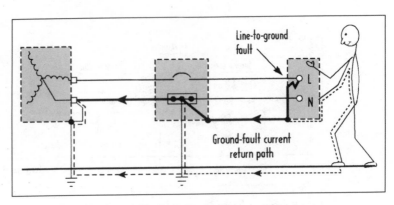

Figure 21.9 Phase-neutral-ground diagram of solid grounding regime with a line-to-ground fault.

A. No, additional grounding may bleed off the fault current that is necessary to drive overcurrent protective devices. If a grounded (neutral) circuit conductor is connected to the equipment grounding conductors at more than one point, or if it is grounded at more than one point, stray neutral current paths will be established. These stray neutral currents will be present in the system even during normal operation and are known to interfere with the proper operation of equipment, devices, or systems that are sensitive to electromagnetic interference such as telecommunication and computer systems.

Figure 21.10 illustrates the problem of stray neutral current due to multiple grounding of the grounded service conductor. (This figure is similar to Fig. 21.8. Stray neutral current may not be objectionable where there is a single overhead service drop connected to only one set of service entrance conductors.) This is a normally operating circuit except with another grounding conductor connected at the service equipment. Note the "loop" of neutral current flowing between the utility supply and service equipment. In situations where there is a single overhead service drop connected to only one set of service entrance conductors, the stray neutral current that is circulating may be tolerable, though it is not preferred.

Q19. *How does supplemental grounding work?*

A. Supplemental equipment bonding contributes to equalizing the potential between exposed non-current-carrying metal parts of

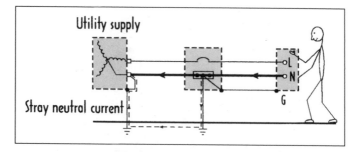

Figure 21.10 Phase-neutral-ground diagram of solid grounding regime showing stray neutral current due to multiple grounding of grounded service conductor.

the electric system and adjacent grounded building steel when ground faults occur. The inductive reactance of the ground fault circuit will normally prevent a significant amount of ground fault current from flowing through the supplemental bonding connections. The ground fault current path that minimizes the inductive reactance of the ground fault circuit is through the equipment grounding conductors that are required to run with or enclose the circuit conductors. Therefore, practically all of the ground fault current will flow through the equipment grounding conductors, and the ground fault current through the supplemental bonding connections will be no more than required to equalize the potential at the bonding locations.

Objectionable stray neutral currents are frequently caused by unintentional neutral-to-ground faults shown in Fig. 21.11. (This figure is similar to Fig. 21.9, but note the supplemental bonding. Objectionable stray neutral currents are frequently caused by unintentional *neutral-to-ground* faults, sometimes called *ground loops,* as shown here.)

Q20. *How is livestock affected by stray neutral currents?*

A.   The phenomenon is formally called "step potential." If a current-carrying conductor, even though nominally at ground potential, is connected to earth at more than one location, part of the load current will flow through the earth because it is then in parallel with the grounded conductor. Since there is impedance in both the conductor and the earth, a voltage drop will occur along

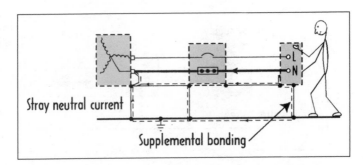

Figure 21.11 Phase-neutral-ground diagram showing stray neutral current due to unintentional grounding of a grounded circuit conductor.

both the earth and the conductor. Most of the voltage drop in the earth will occur in the vicinity of the point of connection to earth. Because of this nonlinear voltage drop to earth, most of the earth will be at a different potential than the grounded conductor owing to the load current flowing from this conductor to earth.

An equipment grounding conductor connected to the same electrode as the grounded load conductor will also have a potential difference from most of the earth owing to the potential drop caused by the load current. In most instances, the potential difference will be too low to present a shock hazard to persons or affect operation of conventional electrical load equipment. However, in many instances it has been of sufficient level to be detected by livestock, either by coming in contact with non-current-carrying enclosures to which an equipment grounding conductor is connected or where sufficient difference in potential exists between the earth contacts of different hoofs. Although potential levels may not be life-threatening to the livestock, it has been reported that as little as $1/2$ V of "step potential" can affect milk production.

Q21.   *What is the purpose of a ground mat?*

A.   In building systems the ground mat establishes the zero, or reference voltage level, and it provides the return-to-earth path for fault current. The electrodes that we see in this chapter are, in many cases, a mat—or a system of electrodes driven deep into the ground and connected with each other laterally. In utility systems, the ground mats are intended to protect linesmen and other switchyard personnel from dangerous step and touch potentials during normal operation, and to protect them under fault conditions.

A rigorous design of a ground mat is essentially an optimization problem involving some truly gnarly equations. The reader is referred to Meliopolous for insight into the basic ground electrode conductivity equations. There are several variables you must work with: soil resistivity; the radius, length, and material of the ground conductors; the depth at which the conductors are buried; their orientation; and the "shape" or area which is enclosed by the mat. To design a ground mat to measure a resistance below 5 to 25 ohms requires manipulation of very complicated equations using each of the foregoing input variables. The formal computation of ground resistance is virtually impossible to do in longhand in all but a few extremely rudimentary and impractical configurations. A commercial ground mat design software package is recommended.

You will find that the variable that will have the greatest effect upon lowering ground resistance will be the number of electrodes you use and the area which is enclosed by them. In many cases rigorous, formal computer-aided design is not necessary if enough is known about the performance of the grounding regimes in adjacent buildings. In other situations—especially where soil resistivity may not be known and where there is very little area to enclose, a rigorous design should be undertaken.

Q22. *Does the NEC require GFCI at construction sites?*

A. Yes. Since 1996 all branch circuits used by construction personnel must be GFCI protected except those used on small lighting generators and in industrial establishment (Sec. 305-6a for 1999 *NEC*). Construction site GFCI requirements have been extended to 30-A 125-V receptacles. The cord set allowance for portable GFCI protection now includes devices that do not include cords, such as self-contained GFCI units. These devices must be identified for portable use. The requirement for identification as suitable for portable use means that the device has what most people refer to as open neutral protection. GFCI breakers may be used in temporary panelboards supplying standard receptacles in portable boxes. Only receptacles used under temporary job conditions require GFCI protection. Other circuits of higher amperage, higher voltage, and three phases are not required by the Code to have GFCI protection. Thus a three-phase compressor or welder would be protected by standard overcurrent devices. Every multiwire branch circuit must have a disconnect means that simultaneously opens all ungrounded wires of the temporary circuit.

Q23.   *What is an assured equipment grounding conductor program?*

A.   Other receptacles not covered are permitted to have ground-fault circuit-interrupter protection for personnel, or a written procedure must be continuously enforced at the site by one or more designated persons to ensure that equipment grounding conductors for all cord sets, receptacles that are not a part of the permanent wiring of the building or structure, and equipment connected by cord and plug are installed and maintained in accordance with the applicable requirements of Arts. 210, 250, and 305.

The electrical contractor has the responsibility for the temporary wiring on any job site and is the one to develop, write, and supervise the assured equipment grounding program where that option is chosen as an alternative to use of GFCI protection. The approach mimics the OSHA approach to branch circuit safety on construction sites.

Q24.   *Can we substitute an assured equipment grounding conductor program in place of widespread GFCI?*

A.   You may if (*a*) it is acceptable to the AHJ, (*b*) you can stand the paperwork. The assured equipment grounding conductor program applies to all trades, applies to all receptacle outlets, and requires some person in charge of construction to make sure that all the cords on site are documented. It may be more practical. It is worth noting that OSHA began mandating GFCI on construction sites because 70 workers died on construction sites during the period from 1970 to 1975. More recently, the International Brotherhood of Electrical Workers data indicates that 106 workers were electrocuted on construction sites in 1992, for example. They died despite the use of GFCIs.

Q25. *Assuming the absence of lightning protection, should the ground ring for the transformer be tied to the building grounding system?*

A. Yes, this will prevent the flow of objectionable ground current over the equipment and service grounding conductor by keeping the two grounding systems at the same potential. This equipment and service conductor, which is run with the building entry conductors, also connects the transformer ground ring to the building grounding system. This type of current flow, which is due to the potential difference between the two grounding systems, often causes data corruption in digital end use devices.

Q26. *How do NEC grounding requirements relate to requirements for lightning protection?*

A. The basic rule, which now appears in Sec. 250-60, says that air terminal conductors for lightning protection systems are not substitutes for grounding wiring systems and equipment. (Air terminals are the "spikes" you will often see installed on the tops of buildings to "capture" lightning by providing a low resistance to ground).

One of the major changes to Art. 250 involves the removal of bonding and spacing requirements for lightning protection and relocating them in NFPA 780, Standard for the Installation of Lightning Protection Systems. The requirement that metal raceways, enclosures, frames, and other noncurrent-carrying metal parts of electrical equipment be kept at least 6 ft from lightning rod conductors is effectively stated through a Fine Print Note in Sec. 250-106.

Q27. *Do multiple-building facilities complicate grounding strategies?*

A. Yes, but the grounding problems should never prohibit architects and their associates from proposing, or adding to, existing multi-building facilities. There are grounding strategies that the NEC permits which can be solved by the electrical people and should not have a major impact on construction budgets, the concern for shared, hidden, or inadvertent metallic paths notwithstanding.

The relevant section of the code is Sec. 250-32, which covers the topic of two or more buildings supplied from a common service.

There are basically two choices for grounding at the second building which is supplied from a common ac grounded service. You may do one of two things:

1. Leave building 1 as a feeder and enter building 2 as a feeder. If you opt for this strategy the equipment grounding conductor and the grounded conductor remain isolated and the panelboard neutral block is not connected to the panelboard enclosure. See **NEC** Sec. 250-32(b)(1).

2. Leave building 1 as a feeder and enter building 2 as a service. If you opt for this strategy the conductors are treated like service conductors and the neutral block is bonded to the panelboard enclosure and the grounded conductor is used to ground any equipment or structural framework, etc.

In either case you must install grounding electrodes according to the rules of Part C of Art. 250. Considerable attention should dwell upon how many branch circuits are run from one building to another. One circuit is the least complicated arrangement. Considerable attention should dwell upon whether pipes, metal conduit, telephone, or CATV lines create inadvertent metallic paths between buildings.

A comprehensive treatment of this subject is beyond the scope of this book and the reader is referred to any one of the core references for a comprehensive discussion of the fine points of this subject.

Solved Problem—Ufer Ground

SITUATION.

A concrete pier is to be used for a building grounding system. Refer to the Ufer ground fashioned from the concrete and steel pier shown in

Fig. 4-6. Soil resistivity is 15,800 ohm-cm. The equation for determining this resistance is $(\rho/2\pi)[\ln(4L/a-1)]$, where L is the driven length of the rod, a is the diameter, and ρ is soil resistance (see Remarks). Based upon use of this equation, a 10-ft copper-clad steel rod is known to have 64.9 ohms resistance when its driven length is 8 ft.

REQUIREMENTS.

Determine the ground resistance of one or multiple concrete piers each formed of four rebars with spacer rings.

SOLUTION.

A reinforced concrete pier of four rebars will have approximately one-half the resistance of a simple driven rod of $5/8$ diameter of the same length. Use the earth resistance of the reference 10-ft rod and divide it by 2. Thus, the earth resistance of one reinforced concrete pier is 32.4 ohms.

In the case of multiple piers or footings divide the resistance of a single pier by half the number of outside piers. We have eight exterior piers. Thus the resistance is $32.4/(1/2)(8)=8.1$ ohms. This is significantly below the **NEC** minimum of 25 ohms. (Refer to Fig. 21.12.)

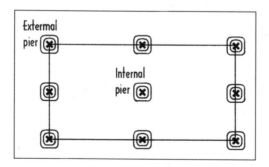

Figure 21.12 Multiple concrete pier arrangement.

REMARKS.

The broad principle illustrated here is how the actual foundation of a building itself may be used as a ground and how the area enclosed by the piers figures into the equation. Less well covered is the nitty-gritty arithmetic of determining the ground resistance of the driven rod. The equations, developed in 1936 by H.B. Wright (AIEE, vol. 55), involve a metric-English conversion in the use of ρ. Detailed calculations are beyond the scope of this book. The reader is referred to Meliopoulos for a very readable treatment of the theory of this subject.

Division 16460—Transformers

Remarks. The development in 1885 by William Stanley of a commercially practical transformer, coupled with Nicola Tesla's work, gave impetus to establishment of the ac systems as a standard for the electric power industry in North America. One hundred years later, we are still left with a passive and very efficient device, very much the same device. Manufacturers have since learned how to manipulate core size, core material, winding conductor size and material, insulation thermal, and dielectric types to affect efficiency and cost in a fashion that is as much art as science.

While transformers in the 100- to 1000-MVA range which are applied on the bulk transmission grid require a one-of-a-kind engineering approach and cost millions of dollars each, the transformers applied on the bulk distribution grid and in buildings are pretty much off-the-shelf items. Costs per kVA for a typical power transformer are in the $5 to $50 range. Transformers obviously allow safe and economical delivery of power, but a much less *appreciated fact is that transformers are a friend to protection engineering specialists because the impedance they add to any given circuit limits fault current and makes selectivity possible between overcurrent devices. The effect of this impedance is demonstrated in NEC Tables 450-3(a) and (b).*

Q1. *What NEC requirements commonly appear in this division?*

A. The bulk of them appear in Art. 450, Transformers and Transformer Vaults. There is some related material on this subject in other CSI Divisions—Division 3, where we covered the subject of concrete vaults; and Division 2, where we covered transformers from the standpoint of site layout. In Division 16100 we deal with the particular interior workspace issues that apply to transformers and other equipment.

In this chapter we focus on a few basic requirements that are electrical in nature. Article 450 of the **NEC** has been greatly improved in the 1999 revision cycle. The mother standard for all dry-type transformers is ANSI/IEEE C57 series of documents. Of particular interest in the discussion here will be the requirements that appear in C57.12.01-1979—Standard General Requirements for Dry-Type Distribution and Power Transformers.

Q2. *What broad principles apply to the design and installation of transformers?*

A. There several that are relevant to architects and space planners that appear in Art. 450:

■ You do not need to design a separate room for a transformer unless it is at least $112^1/2$ kVA. Section 450-21 covers dry-type transformers installed indoors. The break point is $112^1/2$ kVA. Below this kVA rating you need only provide at least a 12-in separation from combustible material or provide a fire-resistant heat-insulated barrier.

■ You do need to build a vault for transformers operating at 35 kV and above.

■ The degree of transformer overcurrent protection is directly linked to whether or not the installation is under supervision. This is a rather remarkable statement, one of the few requirements of the **NEC** where facility management and engineering overlap. The rule applies only to transformers operating over 600 V. The linkage is made in Tables 450-3(a) (Maximum Rating or Setting of Overcurrent Protection for Transformers over 600 V). The **NEC** defines a supervised location as one in which conditions of maintenance and supervision ensure that only qualified persons will monitor and service the transformer installation.

Q3. *How do cost factors figure into the selection of a transformer?*

A.   First cost relationships with respect to transformer insulation and coolant type—which usually affects construction and location selection. Exterior mineral oil, silicone-filled and high-molecular-weight hydrocarbon (HMWH) transformers are the least expensive, in order of increasing relative cost. Sealed interior dry-type transformers are the most expensive, typically by more than 200 percent the cost of a fluid-filled transformer. Ventilated interior dry-type transformers—the kind in most common use in low-voltage distribution systems in buildings—fall somewhere in between these extremes depending upon the temperature rise.

Q4. *Why are dry-type transformers preferred for building interior use?*

A.   Mineral insulating oil, because of its superior dielectric properties, has long been used as an insulating and cooling medium for transformers. However, oil-filled transformers create a severe fire and explosion hazard and as such have to be mounted either outdoors or in expensive fireproof and explosionproof vaults if installed indoors. To get around this requirement, askerel-filled transformers were developed some years ago. Synthetic askerel has the same excellent insulating qualities as mineral oil, but it will not burn or explode. The problem with this, however, is that askerel contains PCBs, which the U.S. government controls as a compound hazardous to the environment.

Q5. *What are the broad principles for locating transformers indoors?*

A.   Transformers of more than 112.5-kVA rating should be installed in a transformer room of noncombustible construction unless they have class B, class F, or class H insulation and are separated from combustible material by a distance of not less than 6 ft horizontally and 12 ft vertically, or by a noncombustible heat-insulating barrier. The barrier or distance separation is not necessary in sprinklered areas. Sealed tank gas-filled transformers are fire- and explosion-resistant and need no special safeguards.

In dust or corrosive atmospheres, less flammable dry transformers should be used in preference to the air-cooled dry type. Such atmospheres adversely affect windings and metal cores of air-cooled units, as well as the dielectric strength of insulating surfaces between bus bars and between exposed copper connections, causing insulation breakdown and arcing. Encapsulated transformers may offer an economic advantage if their application results in avoided costs associated with space preparation and/or a specialty HVAC apparatus.

When an air-cooled transformer is used in an area containing dusts or corrosive vapors, install the transformer or the unit substation of which it is part in a separate pressurized room and provide adequate, clean, filtered air for cooling.

Section 450-21 has application rules for general-purpose low-voltage dry-type transformers. It covers the largest class of transformers installed in buildings: the stepdown or stepup units that typically transform voltage from 480 to 208 or vice versa. The rules diverge when transformers exceed 112.5 kVA. The **NEC** says that a transformer up to 112.5 kVA must have a separation of at least 12 in from combustible material unless it is separated from the combustible material by a fire-resistant, heat-insulating barrier. A transformer larger than 112.5 kVA must be installed in a transformer room of fire-resistant construction unless it is separated from combustible material either by a fire-resistant, heat-insulating barrier or by free space not less than 6 ft horizontally and 12 ft vertically. So goes the broad sweep of it; exceptions and refinements abound. Underlying all of it is common sense about dispensing transformer heat. Per Sec. 450-9, ventilation should be adequate to ensure that fundamental operating characteristics are not lost.

Low-voltage transformers up to 50 kVA can be installed in hollow though inaccessible spaces as long as the ventilation and distance from combustible surface requirement is met. Spaces do not have fire ratings; only partitions have such ratings.

Q6.   *How must transformers be physically protected?*

A.   Section 450-8 says that transformers must have mechanical protection. Appropriate provisions must be made to minimize the possibility of damage to transformers from external causes where the transformers are exposed to physical damage. Dry-type transformers must be provided with a noncombustible moisture-resistant case or enclosure that will provide reasonable protection against the accidental insertion of foreign objects. Switches or other equipment operating at 600 V, nominal, or less, and serving only equipment within a transformer enclosure may be installed in the transformer enclosure if accessible to qualified persons only. All energized parts must be guarded in accordance with Secs. 110-26 and 110-34. Finally, the operating voltage of exposed live parts of transformer installations must be indicated by signs or visible markings on the equipment or structures.

Efficiency is the ratio of useful energy to energy consumed, expressed as a percentage. Therefore, one must know the transformer's losses in order to determine its efficiency. There are basically two types of losses. The core (iron losses) are constant (not affected by loading) whenever the transformer is energized. However, the winding (load) losses vary as the square of the change in loading. For this reason efficiency must be expressed at a given level of loading (usually full load). The winding losses are usually three to six times greater than the core losses and are primarily the result of the relative component of the impedance. The resistive component is relatively large in the smaller units. Therefore, efficiency is generally poorer in smaller transformers. Efficiency figures are on the order of 97.6 percent for 150° rise transformers and on the order of 98.9 percent for 80° rise transformers.

Thermal Considerations

Q7.   *What does temperature rise of the transformer have to do with its physical size?*

A.   An 80° transformer will be the most efficient and will stay "cool," but the design and the materials of the core construction require more physical space.

The temperature rise and basic impulse level classification of the transformer often determines the actual size of the metal cladding around switchgear. The higher the temperature rise (say 150° rise) and the lower the BIL voltage, the smaller the cladding is likely to be. The size of the cladding around switchgear varies among manufacturers and depends upon the manner in which they modularize the components of their line of dry-type transformers.

Q8.   *Does the NEC have anything to say about the temperature rating of the transformer?*

A.   No, but the temperature rating indirectly affects the space requirements for which the **NEC** has a great deal to say.

Bear in mind that there are two parameters in transformer construction which dominate the relationship between impedance, regulation and efficiency in power transformers: the insulating material itself, and the way the actual metal windings (copper or aluminum) are arranged (and ventilated) on the core. Current, heat, and resistance are interdependent effects.

The average rise of a winding is determined by the measurement of the resistance of the winding. The dry-type transformers were originally made for 80° average rise. Later the 115° system and the 150° systems came along. At present, unless otherwise specified, dry-type transformers are made as 150° rise units.

Dry-type transformer temperature rise classifications are 80, 115, and 150°C average temperature rise. Ambient conditions are 40°C maximum with 30°C average during any 24-h period.

The limiting temperature for 80° rise system is 150°.

The limiting temperature for 115° rise system is 185°.

The limiting temperature for 150° rise system is 220°.

To the extent that the **NEC** requires that the transformer circuit be designed adequate for the purpose, the increased heating effects of nonlinear load currents must be taken into consideration when determining the load on transformers.

Q9.   *How are the limiting temperatures of transformers determined?*

A.   The limiting temperatures for each system as determined as follows:

Hottest spot temperature rise maximum:

> 40° + average temperature rise of insulation class
> 130° (called hottest spot rise allowance)

Average hottest spot temperature rise:

> 30° + average temperature rise of insulation class

On the basis of the above:

Insulation rise class	Max. hot spot	Avg. hot spot
80	150	140
115	185	175
150	210	220

These figures are the crux of the life expectancy of a dry type transformer.

Q10. *How do I specify temperature rise?*

A.   A specifier can request that a transformer be built with 150° rise material but the temperature rise for rated kVA is limited to 80° or 115°. Such a transformer (1) will be able to carry more kVA load than its nameplate rating when operating at the limiting temperature of its insulation class. (2) Its relative life will be considerably longer when operated at rated kVA and under-rated ambient conditions, for example, if a 500-kVA dry-type transformer with 150° rise insulation operated under rated ambient temperature conditions and kVA load conditions has 100 percent life expectancy. A similar transformer with a specified 80° average rise will have a life expectancy for over 10,000 percent.

It should be plain that a transformer, properly applied, can last virtually forever. Improvements in materials associated with their thermal efficiency or overall circuit cost effectiveness tends to drive their replacement more than the end of their life. That much said, lower-temperature rise transformers will generally offer the longest life. Low-loss, high-efficiency, dry-type transformers can be specified with 115 or 80°C rise. A 115°C transformer has a life expectancy of about 10 times greater than that of a 150°C rise transformer. Dry-type transformers of 115° and 80°C rise also have the added advantage of having an emergency overload capability of approximately 15 and 30 percent, respectively. The reader is referred to the C57 series of ANSI standards for details.

Q11.   *What are the trade-offs with respect to temperature rise?*

A.   An 80° rise transformer wound with 150° rise insulation has lower I^2R losses than a transformer designed for 150° rise at the same kVA rating. It is due to the fact that in order to achieve the lower temperature rise the transformer manufacturer will use more conductors and make the ventilation ducts larger. On the other hand, since now the core length increases and the manufacturer does not want to increase the core material in order to keep the cost increase to within limits the core losses increase.

Ideally, as long as the load is *less than 40 percent* of rated kVA the 150° rise transformer will have fewer losses. *Greater than 40 percent* load the 80° rise transformer will have fewer losses. The actual value will depend upon the load cycle. An 80° rise transformer vs. 150° rise transformer may present advantage, especially in replacement situations, because the extra capability is made available at no increase in the secondary fault duty. If a 750-kVA unit were to be replaced with a 1000-kVA unit at 500 MVA available primary fault, the secondary fault will increase.

The advantage is apparent if the increase results in breaker replacement. On the other hand, in replacement situations or even when new equipment is specified, it may not be possible to utilize the extra capacity if the increased voltage regulation on the 80/150 rise rated transformer will not be adequate to the most remote load.

Remember that when a transformer set at 97.5 percent primary operates at 100 percent voltage its core losses go up. The same holds true for the magnetizing current. For the same rating the enclosure of 80° rise transformer may be larger than that of the 150° rise transformer of the same rating.

An 80° rise with 150° system is a good selection. Make sure that all the components of an 80° rise unit can carry the increased capacity while operating at 150°. Compare prices.

Do not forget to ventilate the room for the increased capability of the 80° rise transformer with 150° insulation system. The latest form of dry types encapsulated with resin transformers have limiting temperature rise of 185° because of the resins. If you intend to use an 80° transformer at its 150° rating, make sure the transformer is ordered with a hot spot temperature meter. If the transformer will be used with 120/208 or 277/480 distribution, then three such meters may be required.

Q12.   *What is askerel coolant?*

A.   Askerel (polychlorinated biphenyls). Askerel is the generic name for a class of liquid dielectric used in many transformers manufactured between 1929 and 1977. Askerel has excellent fire-resistant properties, high dielectric strength, and high heat capacity. For these reasons it was often used in transformers located in or near buildings. IEEE Standard 76-1974 provides guidance concerning the acceptance and maintenance of transformer Askerel. Askerel is typically composed of more than 50 percent PCBs. Under EPA regulations fluids containing less than 50 ppm PCB are classified as non-PCB. Fluids containing 50 to 500 ppm are classified as PCB contaminated. Any fluid containing more than 500 ppm is classified as a PCB fluid.

Since production of Askerel was discontinued in the United States, several manufacturers have developed replacement fluids with suitable electrical and flammability requirements. To date, none of the new fluids which have been developed match the performance of Askerel. Less flammable fluids used in the replacement of Askerel fall into one of the following classes of materials: polydimethyl-isiloxane (PDMS or silicone oils), polyalphaolafins (PAO), saturated paraffinic oils, and chlorinated benzenes.

Q13.   *What are the requirements for oil-insulated transformers installed indoors?*

A.   Oil-insulated transformers installed indoors must be installed in a vault constructed as specified in part C of Art. 450. Where the total capacity does not exceed 112½ kVA, the vault specified in part C of this article must be permitted to be constructed of reinforced concrete not less than 4 in thick. Where the nominal voltage does not exceed 600, a vault is not required if suitable arrangements are made to prevent a transformer oil fire from igniting other materials, and the total capacity in one location does not exceed 10 kVA in a section of the building classified as combustible, or 75 kVA where the surrounding structure is classified as fire-resistant construction.

Q14.   *What are the rules for siting oil-insulated transformers installed outdoors?*

A.   This answer should be considered in tandem with a similar question that appears in Division 2, Site Requirements. Section 450-27 says that combustible material, combustible buildings, and parts of buildings, fire escapes, and door and window openings must be safeguarded from fires originating in oil-insulated transformers installed on roofs, attached to, or adjacent to a building or combustible material. Space separations, fire-resistant barriers, automatic water spray systems, and enclosures that confine the oil of a ruptured transformer tank are recognized safeguards. One or more of these safeguards must be applied according to the degree of hazard involved in cases where the transformer installation presents a fire hazard. Oil enclosures must be permitted to consist of fire-resistant dikes, curbed areas or basins, or trenches filled with coarse, crushed stone. Oil enclosures must be provided with trapped drains where the exposure and the quantity of oil involved are such that removal of oil is important.

Q15.   *What is recommended practice for flammable fluid-filled transformers installed outdoors?*

A.   Transformer fluids may be used for outdoor installations. In locations where less flammable fluid filled transformers are installed outdoors in an isolated location and present no exposure hazard to important structures, no fire protection is necessary with the exception of the usual hydrant and hose protection.

■ Roof-mounted transformers should be installed within welded steel panels or curbed concrete mats having dimensions equal to the minimum curbed area recommended for indoor installations. The sides of the pan or curbs should be of sufficient height to contain the entire fluid content of the transformer. Whenever practical, the pan or curbed area should be drained to a safe location acceptable to the authority having jurisdiction. Space of multiple roof-mounted transformers should be in accordance with recommendations for indoor transformers.

■ Transformers which expose important buildings or other structures should be provided with the curbing and crushed stone or drained basin normally recommended for outdoor oil-filled transformers. However, means should be provided to prevent direct flame impingement on the wall. This can be accomplished by allowing a minimum clearance of 5 ft between the inside edge of the containment area and the building wall.

■ Install outdoor transformers on a concrete foundation or pad for stability. A basin is needed to contain fluid. The basin should be formed around the transformer by installing a minimum 6-in-high noncombustible curb. This basis should be filled with coarse crushed stone about 1.25 in in diameter to a minimum depth of 6 in. The volume of the basin should be of sufficient capacity to contain the volume of stone plus the oil contents of the transformer plus the discharge from any water spray and/or hose streams for the expected duration of the fire. The stone fill should be loosened and turned as necessary to prevent the clogging of drainage by dirt, dust, and silt. This may be needed as often as monthly where the surrounding area is dusty, such as near foundries and cement plants and after snowstorms, which may leave the basin filled with ice. An acceptable alternative to providing a large-capacity basin would be to drain the minimum 6 in deep basin through a trap to a safe location, acceptable to the authority having jurisdiction, such as a holding tank. Provisions should be made to prevent the drain openings from freezing. Where drainage is such that burning oil will be quickly removed from the basin, the crushed stone may be eliminated.

■ Water-spray systems, where provided, should be designed to provide a discharge of 0.25 gal/min per sq ft over the entire tank in accordance with Factory Mutual Data Sheet 4-1N.

The waterspray should be designed for a minimum duration of 60 min. When planning new installations, consider the location of the nearest water-spray supply or provide an on-premises water supply.

■ Cables, isolated bus duct, or cable trays penetrating an exposed wall should be sealed with a fire barrier or stop.

■ Ventilation louvers should be relocated to an unexposed area.

■ Overhanging combustible eaves should be removed or protected by fireproofing or an automatic sprinkler or water-spray system. Sprinkler or water-spray systems for protection of exposed walls and roof should be piped separately from transformer water-spray systems.

■ Where transformers expose more than one building they may be placed in a fire-resistive vault and protected as shown in FM Table IV as an alternative to the exposure protection.

Q16.   *How shall we design the cooling system for the transformer?*

A.   **NEC** requirements are stated in Sec. 450-9. Refer to Chapter 16 for in-depth treatment (450-9, Ventilation). The ventilation must be adequate to dispose of the transformer full-load losses without creating temperature rise that is in excess of the transformer rating.

Transformer Overcurrent Protection

Q17.   *What are the broad principles of transformer overcurrent protection?*

A.   The rules are stated in Sec. 450-3(a) for medium-voltage transformers, and Sec. 450-3(b) for low-voltage transformers. It is important to understand that, while these rules apply only to the transformer, they need to be applied in conjunction with the overcurrent protection rules that appear in Art. 240.

These are described in Fig. 21.14. If installing fuses, the overcurrent protection device for the primary side of a transformer rated over 600 V must be rated not greater than 250 percent of the rated primary current of the transformer with individual overcurrent

protection on the primary. In a transformer in a nonsupervised location with a secondary voltage of 600 V or less, the OCPD and conductors on the secondary side must be sized at 125 percent of the
FLA rating with primary and secondary protection. In a transformer
of 600 V or less, nominal, where the rated secondary current of a
transformer is between 2 and 9 A, an overcurrent protection device
rated or set at no more than 167 percent of secondary current may
be used. In a transformer in a supervised location with a secondary
voltage rated 600 V or less, the OCPD on the secondary side must
be sized at not more than 250 percent of the FLA rating.

Q18.   *Does the overcurrent protection specified in Sec. 450-3 provide
protection for the primary and secondary conductors?*

A.   Not entirely. Where polyphase transformers are involved, primary
and secondary conductors will usually not be properly protected.
The rules of Sec. 450-3 are for transformer protection alone. The
primary overcurrent device provides short-circuit protection for
the primary conductors and a degree of overload protection for the
transformer. Secondary overcurrent devices prevent the transformer
and secondary conductors from being overloaded. The transformer
is considered the point of supply, and the conductors it supplies are
required to be protected in accordance with their ampacity.

Miscellaneous

Q19.   *How do I read transformer nameplate data?*

A.   Symbols used in transformer nameplates are the dash(-), the
slant(/), and the cross(x). The dash(-) is used to separate voltages
of different windings, the slant(/) is used to separate voltages of
the same winding, and the cross(x) is used to indicate a series-
multiple connection. To designate a winding with a midtap
which will provide half the full winding kilovoltampere rating at
half the full-winding voltage, the full-winding voltage is written
first, followed by a slant, and then the midtap voltage. For
example, 240/120 is used for a three-wire connection to designate a 120-V midtap voltage with a 240-V full winding voltage.
A winding which is appropriate for series, multiple, and three-

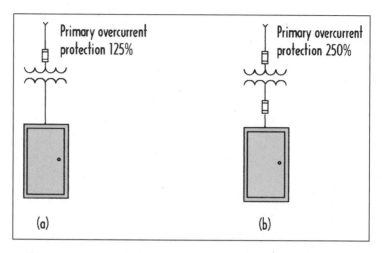

Figure 21.13 Rules for low-voltage transformer overcurrent protection. (a) Per Sec. 430-3, transformers are required to be protected by either a fuse or a circuit breaker on the primary side to protect the transformer primary. Notice the absence of a secondary overcurrent device. (b) Note the presence of the fuse on the secondary. With such fusing the primary overcurrent protection may be sized up to 250 percent of the rated current of the transformer primary.

wire connections will have the designation of multiple voltage rating followed by a slant and the series voltage rating; e.g., the notation 120/240 means that the winding is appropriate for 120-V multiple connection, for 240-V connection, and for 240/120 three-wire connection. When two voltages are separated by a cross(x), a winding is indicated which is appropriate for both multiple and series connection but not for three-wire connection. The notation 120x240 is used to differentiate a winding that can be used for 120-V multiple connection and for 240-V series connection but not for a three-wire connection.

According to NEMA and ANSI standards, the higher-voltage winding is identified by HV or H, and the lower-voltage winding is identified by LV or X. Transformers with more than two windings have the windings identified as H, X, Y, Z, or order of decreasing voltage. The terminal H1 is located on the right-hand side when facing the high-voltage side of the transformer. On single-phase transformers the leads are numbered so that when H1 is connected to X1, the voltage between the highest numbered H lead and the highest numbered X lead is less than the voltage of the high-voltage winding.

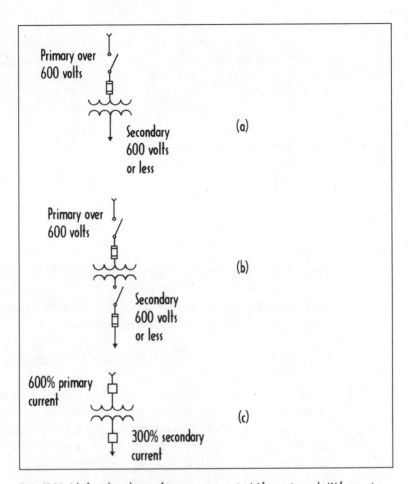

Figure 21.14 Rules for medium-voltage transformer overcurrent protection: (*a*) fuse on primary only; (*b*) fuses on primary and secondary; (*c*) breakers on primary and secondary.

On three-phase transformers the terminal H is on the right-hand side when facing the high-voltage winding, with the H2 and H3 terminals in numerical sequence from right to left. The terminal X1 is on the left-hand side when facing the low-voltage winding, with the X2 and X3 terminals in numerical sequence from left to right.

Q20.   *What are the special requirements of double-ended substations?*

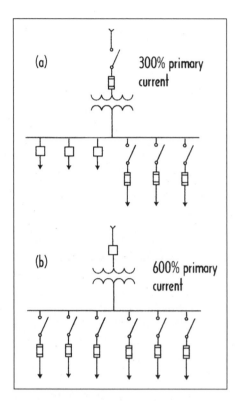

Figure 21.15 Rules for transformer overcurrent protection. (a) Breakers and fuses mixed; (b) all fuses.

A.   The Code cites only a few; the most notable is the requirement in Sec. 450-6 on the subject of secondary ties. A secondary tie is a circuit operating at 600 V, nominal, or less, between phases that connects two power sources or power supply points, such as the secondaries of two transformers. The **NEC** basics are as follows:

■   The tie may be one or more conductors per phase and must have overcurrent protection at each end as required in Art. 240.

■   Where all loads are connected at the transformer supply points at each end of the tie and overcurrent protection is not provided in accordance with Art. 240, the rated ampacity of the tie cannot be less than 67 percent of the rated secondary current of the largest transformer connected to the secondary tie system. This rule is particularly relevant in the sizing of breakers of double-ended substations.

■   Where load is connected to the tie at any point between transformer supply points, overcurrent protection is not provided in accordance with Art. 240, the rated ampacity of the tie cannot be less than 100 percent of the rated secondary current of the largest transformer connected to the secondary tie system.

Q21.   *What is a K-factor transformer and why should we consider it in our specifications?*

A.   K has been accepted by the transformer industry and recognized and listed by UL as the harmonic coefficient. Harmonic currents are integral multiples of the 60-Hz fundamental frequency and, when they accumulate in the neutral conducting path, can cause overheating of some elements of an electrical supply circuit, among other problems. K-factor transformers will mitigate this effect. In a proposed revision of C57-110-1986, IEEE Recommended Practice for Establishing Transformer Non-sinusoidal Load Currents, from a guide to a standard, this committee has decided to change the letter from K to F, since F is used in a converter standard in Europe.

Q22.   *What can we do to reduce the noise?*

A.   The noise comes from magnetostrictive deformation of the core, aerodynamic noise produced by cooling fans, and mechanical and flow noise from the oil-circulating pumps. The radiated core noise, consisting of a 120-Hz tone, is the most difficult to reduce and is also the noise that generates the most complaints from residents living near the transformer. Improved core-construction techniques and lower-loss cores both tend to reduce transformer core noise. If further reduction in core noise is needed, it can only be achieved by increasing the cross-sectional area of the core to reduce the flux density. This design will increase the construction cost and decreases core losses. However, a point of diminishing returns is reached at which the cost of increasing core size outweighs the savings in reduced losses.

Q23.   *How are transformer phases identified?*

A.   The American standard for designating terminals H1 and X1 on Y-D transformers requires that the positive-sequence voltage drop from H1 to neutral lead and the positive-sequence voltage drop from X1 to neutral by 30°, regardless of whether the Y or D winding is on the high-tension side. Similarly, the voltage from H2 to neutral leads the voltage from X2 to neutral by 30°, and the voltage from H3 to neutral leads the voltage from X3 to neutral by 30°. Electrical people should be aware, however, that this convention may not be followed by the local utility.

It is absolutely critical to make sure that whenever manual, automatic, or other emergency power supply circuits are built, that an investigation into phase rotation is undertaken to confirm that phases match up.

Solved Problem—Division 16460

SITUATION.

A 40-hp, 460-V fire pump requires a transformer. The transformer has 4160 V on the primary and 480 V on the secondary. Auxiliary equipment is 30 A total.

REQUIREMENTS.

What is the maximum OCPD permitted on the primary side of the fire pump?

SOLUTION.

Transformers for fire pumps may have OCPD on their primary side, but it must be capable of carrying the locked-rotor current of the fire pump plus accessory loads indefinitely.

1. Find LRA of the fire pump. 40 hp requires 312 A.

2. Calculate the primary to secondary ratio. 4160/480 = 8.67.

3. The sum of LRA and auxiliary equipment = 342 A.

4. Reflected ratio = 342/8.67 = 39.5.

5. OCPD may be 40 A on the 4160 side.

REMARKS.

A variant of this method is to simply oversize the fuse so that the transformer will run to destruction. This method will allow the transformer to survive when locked-rotor amperes are lower than the level needed to damage the transformer thermally.

Division 16470—Overcurrent Protection and Switchboards

Remarks. Every profession has its jargon. Much like the term "ungrounded conductor," the term "overcurrent device" is one of those terms that is so general that it serves the purpose of keeping people out of the electrical trades. Simply put, an overcurrent device is usually either a circuit breaker or fuse, a 50 percent savings in words. A switchboard, panelboard, power or lighting panel are names for what is essentially a rack of overcurrent devices. We treat the salient points together in this chapter.

Q1.   *What NEC requirements commonly appear in Division 16470?*

A.   The generic rules for overcurrent protection appear in Art. 240 with modifying conditions scattered throughout the **NEC**. Article 384 contains the requirements for switchboards and panelboards. Related passages appear in 110-26 (installation); in 230-G (when switchboards are used as service equipment), and in 517-14 (as applied in health care facilities). Mother standards for this class of equipment are NEMA PB1 panelboards, ANSI/UL 67 panelboards, and a whole family of ANSI C37 standards for fuses, circuit breakers, and related switchgear.

16470: Comparison of fuses versus circuit breakers

ADVANTAGES

1. Low initial cost

2. Simple, no parts to maintain

3. High current-interrupting capabilities

4. Provide current limitation, thus materially reducing or eliminating the possibility of conductors or equipment being damaged by mechanical or thermal stresses under fault conditions.

5. Can provide good selective coordination on electrical faults, thus eliminating unnecessary shutdowns.

6. Inherently fail-safe. If they fail, they open the circuit and make it safe.

DISADVANTAGES

1. Must be replaced after each operation; longer downtime

2. Can cause single phasing, which may damage motors without single-phasing protection.

3. Not adjustable; time-current characteristics are fixed

4. Affected by ambient temperature, which may cause nuisance opening of the protected circuits.

5. Can be replaced by another fuse with incorrect ratings or characteristics.

6. Requires the stocking of replacement fuses for each type used.

7. Cannot by themselves be used as an isolating switch for the circuit; they must be mounted in a combination with a switch.

Q2. *What are the broad principles of overcurrent protection?*

A. As applied in the low-voltage feeder and branch-circuit conductors must be protected by overcurrent-protective devices connected at the point the conductors receive their supply. We make a distinction between circuits that operate at 800 A or above. The next higher standard overcurrent protection device is permitted for conductors not part of a multioutlet branch circuit supplying receptacles for cord- and plug-connected portable loads and where the ampacity of the conductors does not correspond with the standard ampere rating of a fuse or a circuit breaker without overload trip adjustments above its rating. The basic rule for typical residential and commercial branch circuits is illustrated in Fig. 21.16.

Q3. *What are the basics for branch-circuit overcurrent protection?*

A. They are stated in Sec. 210-24 and summarized in Table 21.4. One of the practical applications of this table will result in a panelboard schedule (Fig. 21.17). In many jurisdictions, the electrical contractor is required to submit a schedule like this in order to obtain a building permit.

Q4. *Are there any benefits to using circuit breakers instead of fuses?*

A. There aren't any safety advantages to having circuit breakers in your home instead of fuses. The main difference is that circuit breakers can be reset while fuses operate only once and then must be replaced. If your breakers and fuses trip repeatedly, call an electrician because you may have a problem with your electrical system. Refer to the sidebar comparing the differences.

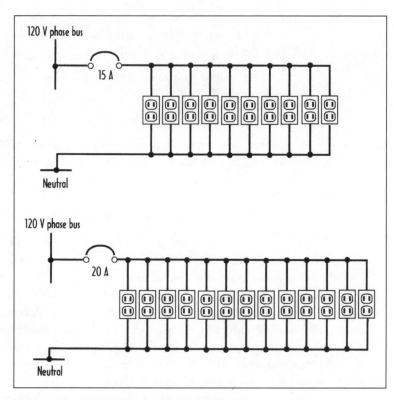

Figure 21.16 Basic branch circuits that meet NEC 180 VA requirement.

Q5.   *Why is ground-fault protection not required on circuits rated less than 1000 A?*

A.   Several studies were undertaken in the late 1960s and early 1970s which proved that 480-V circuits could sustain (by resonance) · very low level fault currents without tripping overcurrent devices. The **NEC** adopted the results of this research into its requirements for ground-fault protection and made it mandatory for circuits with line-to-neutral voltages greater than 150 V. Typically, the magnitude of the available fault current, even a current-limited fault, is high enough that conventional protection will take it out for devices rated below 1000 A. When you get into the higher ratings, a low-level arcing ground fault is in the overcurrent part of the device trip curve, not the instantaneous part.

Q6.   *Can we tap a feeder circuit with an overcurrent device?*

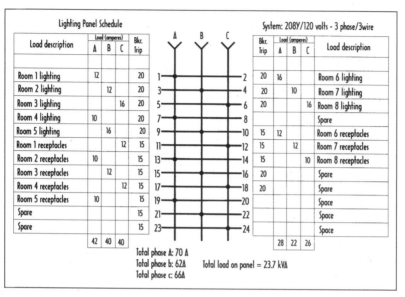

Figure 21.17 Example panelboard schedule.

TABLE 21.4 SUMMARY OF BRANCH-CIRCUIT REQUIREMENTS (**NEC** TABLE 210-24)

Circuit rating	15 A	20 A	30 A	40 A	50 A
Conductors (min size):					
Circuit wires*	14	12	10	8	6
Taps	14	14	14	12	12
Fixture wires and cords— see Sec. 240-4					
Overcurrent protection	**15 A**	**20 A**	**30 A**	**40 A**	**50 A**
Outlet devices:					
Lampholders permitted	Any type	Any type	Heavy duty	Heavy duty	Heavy duty
Receptacle rating†	15 max. A	15 or 20 A	30 A	40 or 50 A	50 A
Maximum load	**15 A**	**20 A**	**30 A**	**40 A**	**50 A**
Permissible load	See Section 210-23(a)	See Section 210-23(a)	See Section 210-23(b)	See Section 210-23(c)	See Section 210-23(c)

*These gauges are for copper conductors.
†For receptacle rating of cord-connected electric-discharge lighting fixtures, see Sec. 410-30(c).

A.   Yes, but only if the distance between the tap and the load will be no greater than 10 ft. Section 240-21 says that conductors are permitted to be tapped, without overcurrent protection at the tap, to a feeder or transformer secondary where all the following conditions are met:

1.   The length of the tap conductors does not exceed 10 ft.

2.   The ampacity of the tap conductors is not less than the combined computed loads on the circuits supplied by the tap conductors, and not less than the rating of the device supplied by the tap conductors, or not less than the rating of the overcurrent-protective device at the termination of the tap conductors.

3.   The tap conductors do not extend beyond the switchboard, panelboard, disconnecting means, or control devices they supply.

4.   The tap conductors are enclosed in a raceway, which must extend from the tap to the enclosure of an enclosed switchboard, panelboard, or control device, or to the back of an open switchboard.

5.   The rating of the overcurrent device on the line side of the tap conductors cannot exceed 10 times the tap conductor's ampacity (Fig. 21.18).

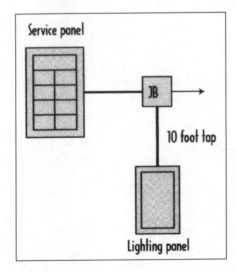

Figure 21.18 10-ft tap rule.

Q7. *How far can the outside tap conductors enter the building?*

A. The same distance that the service conductors are permitted to enter the building; see Sec. 230-70(a). Generally, this is established by local ordinance or by the inclination of the inspector. Many areas restrict the service conductor to a maximum of 5 ft indoors.

Q8. *Can a tap be made beyond 10 ft?*

A. Yes, but only under certain conditions: Conductors are permitted to be tapped, without overcurrent protection at the tap, to a feeder where the length of the tap conductors does not exceed 25 ft. The ampacity of the tap conductors is not less than one-third of the rating of the overcurrent device protecting the feeder conductors. The tap conductors terminate in a single circuit breaker or a single set of fuses that will limit the load to the ampacity of the tap conductors. This device is permitted to supply any number of additional overcurrent devices on its load side. The tap conductors are suitably protected from physical damage or are enclosed in a raceway. Refer to Figs. 21.18 and 21.19.

Q9. *Can feeder taps supplying a transformer be handled differently?*

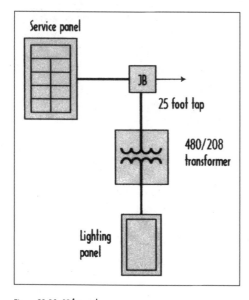

Figure 21.19 25-ft tap rule.

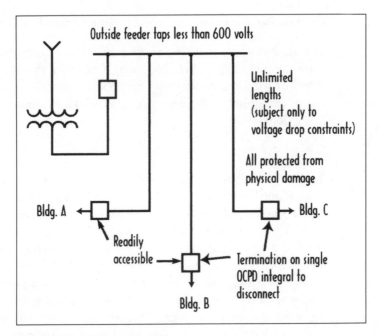

Outside feeder taps less than 600 volts

Unlimited
lengths
(subject only to
voltage drop constraints)

All protected from
physical damage

Bldg. A

Readily
accessible

Termination on single
OCPD integral to
disconnect

Bldg. B

Bldg. C

Figure 21.20 Outside feeder overcurrent protection.

A.   Yes. Conductors supplying a transformer are permitted to be
tapped, without overcurrent protection at the tap, from a feeder
where the conductors supplying the primary of a transformer have
an ampacity at least one-third of the rating of the overcurrent
device protecting the feeder conductors. The conductors supplied
by the secondary of the transformer must have an ampacity that,
when multiplied by the ratio of the secondary-to-primary voltage,
is at least one-third of the rating of the overcurrent device protecting the feeder conductors. The total length of one primary plus
one secondary conductor, excluding any portion of the primary
conductor that is protected at its ampacity, is not over 25 ft. The
primary and secondary conductors are suitably protected from
physical damage. The secondary conductors terminate in a single
circuit breaker or set of fuses that will limit the load current to
not more than the conductor ampacity that is permitted by Sec.
310-15. Some exceptions exist for this rule as applied to high-bay
manufacturing facilities. (Refer to Fig. 21.20.)

Q10.   *What are some of the possible dangers associated with replacing
current-limiting fuses with non-current-limiting fuses?*

A.   Current-limiting fuses can limit the available fault let-through current. Care must be taken when replacing current-limiting fuses; be sure to replace the fuse with the proper rating. Using the wrong interrupting rating fuse could literally blow the equipment off the wall. Current-limiting fuse enclosures have a special feature that keeps non-current-limiting fuses from being installed.

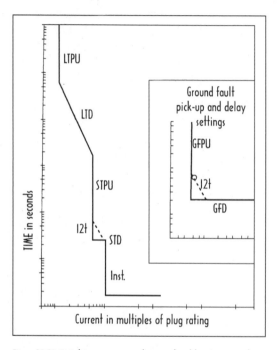

Figure 21.21 Typical time-current curve adjustments for solid-state trip unit with adjustable phase current settings.

Q11.   What is an overcurrent coordination study?

A.   Overcurrent coordination is defined as properly localizing a fault condition to restrict outages to the equipment affected, accomplished by choice of selective fault-protective devices. Section 240-12 says that where an orderly shutdown is required to minimize the hazard(s) to personnel and equipment, a system of coordination based on coordinated short-circuit protection and overload indication is based on monitoring systems or devices. The monitoring system may cause the condition to go to alarm, allowing corrective action or an orderly shutdown, thereby mini-mizing personnel hazard and equipment damage.

It should be plain from the foregoing discussion that overcurrent coordination is an important engineering problem to solve. An example of the finished product is shown in Fig. 21.21. Properly coordinated overcurrent devices is the signature of a well-engineered electrical power system. It is also one of the most difficult and time-consuming problems in all of electrical power engineering. Architects, mechanical engineers, and their subcontract consultants should allow sufficient time in the design budget for qualified engineers to do this work in professional fashion.

Switch and Panelboards

Q12.   *What is the difference between a panelboard and a switchboard?*

A.   A panelboard is a group of panel units designed for assembly into a single panel, including buses, automatic overcurrent devices, and/or switches, for the control of light, heat, and power circuits. The assembly is suitable for mounting in a cabinet, which can then be placed in or against a wall and which is accessible only from the front. A panelboard is a convenient method of groups together with those overcurrent devices that are fed from a common source and that provide circuit protection to a number of branch circuits.

The distinction between panelboard and switchboard is generally a matter of size, with panelboards being smaller and more commonly applied. Switchboards are larger than panelboard and tend to be free-standing but of a construction class below secondary distribution switchgear that can handle draw-out steel frame circuit breakers. The **NEC** further defines a lighting and appliance branch-circuit panelboard as one having more than 10 percent of its overcurrent devices rated 30 A or less *for which neutral connections are provided.*

Most panels installed for lighting and receptacle branch circuit fall within this definition and are therefore subject to the Art. 384 requirements. Section 384-15 restricts any one lighting panel to a maximum of 42 single-pole overcurrent devices or their equivalent.

Q13.   *What is a power panelboard?*

A.   New to the 1999 **NEC,** Art. 384 now includes a formal definition of power panelboard and some revised requirements. A power panelboard is one having 10 percent or fewer of its overcurrent devices protecting lighting and appliance branch circuits. A further change defines them as branch circuits that (1) use neutral connections and (2) have overcurrent protection rated 30 A or less. The power panelboard is exactly complimentary to the revised lighting and appliance branch circuit panelboard definition, which is one with more than 10 percent of its overcurrent devices protecting lighting and appliance branch circuits. The result is to exactly split the panelboard universe between power panelboards and lighting and appliance branch-circuit panelboards, depending on how they are used in the field [*EC&M,* August 97, p. 39].

Q14.   *What is the difference between a switch and panelboard?*

A.   Section 384-14 now classifies panelboards as either one of the following: Lighting and Appliance Branch Circuit Panelboard or Power Panelboard. Such a panel has over 10 percent of its overcurrent protection devices rated at 30 A or less with neutral connections.

Q15.   *What broad principles apply to panelboards?*

A.   Section 384-13 says that all panelboards must have a rating not less than the minimum feeder capacity required for the load computed in accordance with Art. 220. Panelboards must be durably marked by the manufacturer with the voltage and the current rating and the number of phases for which they are designed and with the manufacturer's name or trademark in such a manner as to be visible after installation, without disturbing the interior parts or wiring. All panelboard circuits and circuit modifications must be legibly identified as to purpose or use on a circuit directory located on the face or inside of the panel doors.

Q16.   *Where should panelboards be located?*

A.   They are typically installed flush-mounted into a wall in finished areas, or they can be surface-mounted on a wall in unfinished areas. The placement of panelboards should obviously be convenient and safe for the building occupants. Some designers prefer to locate them in small electrical closets at the same location on each floor. The panels can then be surface-mounted on the walls of the closet and the feeders supplying the panels can be run vertically up through the closet space. In industrial facilities is not as easy to find a suitable wall in an open manufacturing area. Some designers will either specify a free-standing switchboard or specify column-width panels that are manufactured to fit between the flanges of the structural steel columns.

Q17.   *How can a panelboard be used as service equipment?*

A.   With some care. Each switchboard, or panelboard, if used as service equipment, must be provided with a main bonding jumper sized in accordance with Sec. 250-79(d) or the equivalent placed within the panelboard or one of the sections of the switchboard for connecting the grounded service conductor on its supply side to the switchboard or panelboard frame. All sections of a switchboard must be bonded together using an equipment grounding conductor sized in accordance with Table 250-95. Terminals in switchboards and panelboards must be so located that it will not be necessary to reach across or beyond an ungrounded line bus in order to make connections.

Q18.   *What are the broad principles that apply to the installation of switch and panelboards?*

A.   Some of the rules are covered in the workspace discussion of Chap. 17. Section 384-4 says that in indoor installations space dedicated to the electrical installation must include a space equal to the width and depth of the equipment and extending from the floor to a height of 25 ft or to the structural ceiling, whichever is lower. No piping, ducts, or equipment foreign to the electrical installation must be located in this zone. A dropped, suspended, or similar ceiling that does not add strength to the building structure is not a structural ceiling. Sprinkler protection is permitted for the dedicated space where the piping complies with this section. The working clearance space must include the zone described in Sec. 110-26. No architectural appurtenances or other equipment must be located in this zone. Outdoor electrical equipment must be installed in suitable enclosures and must be protected from accidental contact by unauthorized personnel or by vehicular traffic or by accidental spillage or leakage from piping systems. Some exceptions apply.

Q19.   *How many circuit breakers can be installed on a panelboard?*

A.   By using listed equipment, you would not normally have to deal with this. The manufacturers already comply with NEMA rules. But, for the sake of understanding the NEMA rule, not more than 42 overcurrent devices (other than those provided for in the mains) of a lighting and appliance branch-circuit panelboard must be installed in any one cabinet or cutout box. A lighting and appliance branch-circuit panelboard must be provided with physical means to prevent the installation of more overcurrent devices than that number for which the panelboard was designed, rated, and approved. A two-pole circuit breaker must be considered two overcurrent devices; a three-pole circuit breaker is considered three overcurrent devices.

Q20.   *Does the panelboard itself require protection?*

A.   Yes. Section 384-16 says that each lighting and appliance branch-circuit panelboard must be individually protected on the supply side by not more than two main circuit breakers or two sets of fuses having a combined rating not greater than that of the panelboard. Panelboards equipped with snap switches rated at 30 A or less must have overcurrent protection not in excess of 200 A. The total load on any overcurrent device located in a panelboard must not exceed 80 percent of its rating where, in normal operation, the load will continue for 3 h or more. Where a panelboard is supplied through a transformer, the overcurrent protection must be located on the secondary side of the transformer.

Solved Problem—Division 16470

SITUATION.

The circuit is shown in Fig. 21.19. The nearest upstream circuit breaker is rated 600 A.

REQUIREMENTS.

We want to make a tap. The 25-ft tap rule allows smaller tap conductors to be made to larger-sized conductors and extended to a distance of over 10 to 25 ft provided the current-carrying capacity of the tap is at least 1.3 times the feeder overcurrent device.

SOLUTION

1.   Calculate minimum size tap 600/3=200 A.

2.   Select conductors from 310-16; 200 A requires AWG No. 3/0.

3.   Select circuit breaker. From 240-3, AWG No. 3/0 requires 200 A.

REMARKS.

This is a very general example. When sizing raceway and junction box you must size on the basis of the larger conductor. The same applies with the EGC.

Division 16480—Motor Controllers

Remarks. In a review of dozens of electrical specifications from consulting firms all over the United States electrical designers are leaving the selection of motor-circuit elements to the electrical contractor and to the mechanical trades. In the building industry it is quite common for bidding documents to be prepared with electrical designers having only have preliminary motor information. Thermal overloads for all motor starters may well be selected after final installed horsepower of a motor is determined. Contractors are sometimes expected to submit to the engineer a schedule of motor full-load amperes and thermal overload used, at the completion of the project. Architects who have subcontract relationships with electrical engineers should be attentive to this process.

Article 430 is one of the most frequently applied parts of the NEC and contains a great many useful motor supply circuit design tables.

*For reader convenience, we have reproduced a few—but not all—of them, courtesy of the NFPA. Some code answers that appear here will require reference to the original 1999 National Electric Code **NEC** text.*

Q1.   *What NEC requirements commonly appear in Division 16480?*

A.   The NEC governs just about everything having to do with
      motors, motor circuits, and motor control centers in Chap. 4.
      All work performed under this section is subject to all the
      requirements contained in CSI divisions 16000 and 16100 with
      which NEC Chaps. 2 and 3 are closely related.

      We will treat the subject of motor circuit design differently in this
      division than in the treatment we gave motors in Division 15. In
      Division 15, HVAC&R equipment is utilization apparatus. Here
      we look more closely at the details with which an electrical
      designer is typically concerned: the power backbone of the build-
      ing system as a working whole. Since there is so much overlap
      between Division 16480 and Division 15, we have made the arbi-
      trary choice to put the ASD material in Division 15 to help leaven
      the content of Division 16480. Adjustable-speed drives will be
      handled as utilization apparatus specific to mechanical trades.

Q2.   *Does the NEC assert requirements for the selection of motors?*

A.   Not explicitly; however, under the general statement that motors
      must be appropriate to the purpose, it effectively does. As indi-
      cated in Division 15000, motors are already selected by the man-
      ufacturers of the utilization apparatus: elevators, HVAC, and the
      like. Common practice is to select a motor so that the load is
      75 to 95 percent of its rated full load. This assures high efficiency.
      Given a unit of utilization apparatus, the contractor is expected
      to make the connection. It is the field connection where the NEC
      applies. Using listed equipment assumes that the equipment is
      suitable for the purpose. Motor starters in package control panels
      will be furnished by the mechanical trades.

Q3.   *Is there a horsepower assignment to distinguish a large motor*
       from a small motor?

A.   For motors greater than 100 hp, a general use or isolating switch
      may have to be used. When such a switch is used, it must be
      clearly marked *do not operate under load.* Another exemption
      from the horsepower rating requirement is for stationary motors
      rated at 2 hp or less and 300 V or less. The disconnecting means
      can then be a general use switch having an ampere rating not less
      than 200 percent of the full-load current of the motor.

TABLE 21.5 VOLTAGE RATINGS OF STANDARD MOTORS.

Nominal system voltage	Nameplate voltage
Single-phase motors	
120	115
240	230
Three-phase motors	
208	200
240	230
480	460
600	575
2,400	2,300
4,160	4,000
4,800	4,600
6,900	6,600
13,800	13,200

Q4.   *What are the fundamental considerations in motor design?*

A.   *Motor characteristics:* These include type, speed, voltage, horse-power rating, service factor, power factor rating, type of motor enclosure, lubrication arrangement of winding and their temperature limits, thermal capabilities of rotor and stator during starting, running, and stall conditions. Refer to Table 21.5. *Starting conditions:* Full voltage or reduced voltage, voltage drop and degree of inrush during starting, repetitive starts, frequency, and total number of starts, and others. *Ambient conditions:* Temperature maxima and minima, elevation, adjacent heat sources, ventilation arrangement, exposure to water, chemicals, exposure to rodents, and various weather and flood conditions. *Driven equipment:* Characteristics will influence chances of locked rotor, failure to reach normal speed, excessive heating during acceleration, overloading, stalling.

Q5.   *What is the difference between wound-rotor, squirrel-cage motors?*

A.   The squirrel-cage motor is the simplest in construction. There are no commutator bars, slip rings, brushes, and the like. This simple construction results in a much lower cost motor with fewer maintenance problems. The squirrel-cage motor is by far the most common electric motor in use today. When this motor first appeared at the end of the nineteenth century, it was very common to people to have squirrels for pets, thus the resemblance to the rotating cages used to provide the squirrels with exercise.

The wound-rotor induction motor does not have its rotor windings shorted out internally in the rotor like the squirrel-cage motor. The electrical connections to the rotor windings are instead brought out via slip rings and brushes to terminals on the frame of the motor. Thus the rotor circuit is accessible, and external resistance can be added to each phase as desired in order to change the starting torque and starting current or the full-load running speed of the motor.

Although this is not a practical circuit, Fig. 21.22 shows motors applied in listed HVAC utilization apparatus as a feeder, as a branch circuit, and in a motor control center. They each are comprised of these elements in a variety of configurations.

Q6.   *What are the basics for the environmental and mechanical protection of motors?*

A.   Section 430-132 says that exposed live parts of motors and controllers operating at 50 V or more between terminals must be guarded against accidental contact by enclosure or by location by either installing the motor in a room or enclosure that is accessible only to qualified persons, by installation on a suitable balcony, gallery, or platform, so elevated and arranged as to exclude unqualified persons, by elevation 8 ft or more above the floor. Exceptions exist for stationary motors having commutators. You should review the relevant OSHA regulations.

Q7.   *Does the NEC have rules for the location of motors?*

A.   See Sec. 430-14, Location of Motors. (*a*) Ventilation and maintenance. Motors must be located so that adequate ventilation is provided and so that maintenance, such as lubrication of bearings and replacing of brushes, can be readily accomplished.

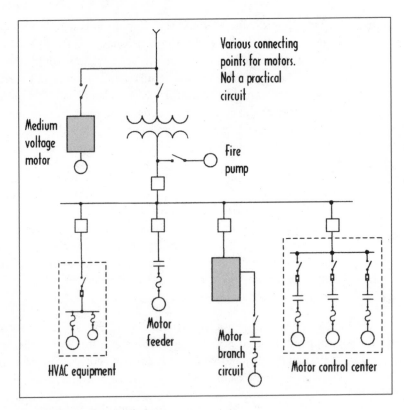

Figure 21.22 Variety of motor circuit configurations..

(*b*) Open motors: Open motors having commutators or collector rings must be located or protected so that sparks cannot reach adjacent combustible material, but this must not prohibit the installation of these motors on wooden floors or supports.

Q8.   *What are the basic characteristics of motors applied in buildings?*

A.   Motors are typically rated in horsepower output, while most other electrical equipment is rated in watts or voltamperes (VA) input. This typically requires a horsepower to watt conversion. Circuits supplying motors are sized according to the input to the motor (input equals the output plus losses of the motor). The losses are not the type of information found on the nameplate of the motor. It is important to understand that circuits that supply motors not rated in horsepower must still be sized according to the input of the motor, rated in amperes. Sizing circuits based solely upon kilowatt output will result in seriously undersized conductors and the improper application of overcurrent devices.

Q9. *What happens to motors that operate a voltages below nameplate rating?*

A. Motor voltages below nameplate rating result in reduced starting torque and increased full-load temperature rise. Motor voltages above nameplate rating result in increased starting torque, increased starting current, and decreased power factor. The increased starting torque will increase the accelerating forces on couplings and driven equipment. Increased starting current causes greater voltage drop in the supply circuit and increases the voltage dip on lamps and other equipment. In general, voltages slightly above nameplate rating have less detrimental effect on motor performance than voltages slightly below nameplate rating.

Q10. *Can I swap a standard motor for a premium efficiency motor?*

A. Sure, but make sure that if it is overall energy efficiency you want to increase, do not overlook the losses in the mechanical system itself. When you change motors, you should not assume that it will be as simple as a horsepower to horsepower replacement. If the torque characteristics of the replacement motor will drive the load, the controller and the disconnect may not be able to handle the new motor. See Part J, Art. 430 for sizing disconnects. You may find your energy savings eaten up with the cost of the next larger size disconnect.

Q11. *How should we apply premium-efficiency motors?*

A. Energy efficiency does not fall within the primary focus of the National Electric Code. Thus, requirements that pertain to premium efficiency motors are those that have bearing upon fire safety.

The general requirement, which appears in Sec. 430-109, is that a motor disconnect must have a horsepower rating not less than the horsepower rating of the motor. A more rigorous exception requires that the motor disconnect for design E motor (rated more than 2 hp, but not over 100 hp) must be either design E rated or the disconnect must have a horsepower rating of not less than 1.4 times the rating of motor. The disconnect for design E motors rated over 100 hp must be either design E rated or the disconnect must have a horsepower rating of not less than 1.3 times the rating of the motor.

Q12.   *How do efficiencies compare between standard and premium-efficiency motors?*

A.   The classical induction motor is design B, which is used in a wide range of utilization apparatus for simple general-purpose across-the-line starting duty. It has a relatively high starting torque for accelerating high inertia loads and can handle short-duration overloads to 200 percent of full load before reaching breakdown torque. NEMA standard MG-1 requires that if manufacturers want to label these motors energy-efficient, then nominal efficiency ratings need to range from 91.7 to 93.0 percent. By contrast, the design E version of the same motor must have a nominal efficiency of 95.4 percent and a minimum efficiency of 94.5 percent.

Properly evaluating a motor for a particular job requires that you have an understanding of the relationship between its speed and torque capabilities and current requirements. These characteristics can be controlled by changing the variables of steel lamination, slot shape, air gap, ventilation, turns in the winding, and manufacturing techniques to configure motors for specific applications such as for elevators, fans, compressors, and the like.

The most commonly used motor is the design B motor, a general-purpose motor with its torquing peak at about 80 percent of synchronous speed. Many manufacturers have already produced design B motors that exceed the NEMA values.

Disconnects, Starters, and Controllers

Q13.   *What is a motor starter?*

A.   Because a motor has more complicated dynamics than, say, a light bulb, it requires a special kind of "switch" in order to manage the current and voltage changes when it is operating normally (and abnormally). These switches range from very simple to very complex and are either called starters, controllers, or contactors. A list of the most common types of starters and contactors appears in Table 21.6.

TABLE 21.6 MAGNETIC STARTER RATINGS—600 V MAXIMUM—60, 50, 25 CYCLES

Continuous ampere rating			Maximum horsepower rating		Maximum horsepower rating for plugging service*	
NEMA size	General purpose	Volts	Single-phase	Three-phase	Single-phase	Three-phase
		115	1/3	—		
		200	—	1 1/2		
00	9	230	1	1 1/2		
		460	—	2		
		575	—	2		
		115	1	—	1/2	—
		200	—	3	—	1 1/2
0	18	230	2	3	1	1 1/2
		460	—	5	—	2
		575	—	5	—	2
		115	2	—	1	—
		200	—	7 1/2	—	3
1	27	230	3	7 1/2	2	3
		460	—	10	—	5
		575	—	10	—	5
		115	3	—	2	—
		200	—	10	—	7 1/2
2	45	230	7 1/2	15	5	10
		460	—	25	—	15
		575	—	25	—	15
		—		—		—
		200		25		15
3	90	230		30		20
		460		50		30
		575		50		30
		—		—		—
		200		40		25
4	135	230		50		30
		460		100		60

	575	100	60

TABLE 21.6 MAGNETIC STARTER RATINGS—600 V MAXIMUM—60, 50, 25 CYCLES (*Continued*)

Continuous ampere rating			Maximum horsepower rating		Maximum horsepower rating for plugging service*	
NEMA size	General purpose	Volts	Single-phase	Three-phase	Single-phase	Three-phase
		—		—		—
		200		75		60
5	270	230		100		75
		460		200		150
		575		200		150
		—		—		—
		200		150		125
6	540	230		200		150
		460		400		300
		575		400		300
		—		—		
		200		—		
7	810	230		300		
		460		600		
		575		600		
		—		—		
		200		—		
8	1215	230		450		
		460		900		
		575		900		
		—		—		
		200		—		
9	2250	230		800		
		460		1600		
		575		1600		

*An example is plug-stop or jogging (inching duty) which requires continuous operation with more than five openings per minute.

The most common controller is the across-the-line magnetic type starter with a control voltage of 120 VAC. An overload relay is placed on the load side of the contactor that has overload protection in each phase. Some overload relays have separate indicating contacts to which a light or other alarm can be wired to indicate tripping. Overload relays are trip-free; most of them should be reset manually, but some are available for automatic reset. Automatic reset is not acceptable on machines where automatic restart of the motor could be hazardous. The starters and contactors shown here are standard to NEMA. There are also three classes of overload relays. NEMA rated or NEMA type controllers are designed to provide a high level of performance over a wide set of application conditions. They are the form most commonly found in general use. Another form, IEC controllers, are rated based upon their performance in the laboratory. Typically, they are considerably smaller and provide a lower level of performance than similarly rated NEMA-type devices.

New to the 1999 **NEC**, a nonfused motor circuit switch is now permitted to have an ampere rating less than 115 percent of the load, provided its horsepower rating is at least equal to the motor rating or to the equivalent horsepower rating of a combination load.

Q14. *What is the difference between a motor disconnect and a motor controller?*

A.   A controller will start and stop a motor. Each controller must be capable of starting and stopping the motor it controls and must be capable of interrupting the locked-rotor current of the motor. A motor disconnect is put in the circuit to open the motor during an emergency. It is a safety feature; the controller is an operational feature associated with normal operation. The classic motor control circuit is shown in Fig. 21.23.

New to the 1999 **NEC**, each controller must have a horsepower rating not less than what would be required for the highest motor winding.

Q15. *What do I need to know to select a motor disconnect?*

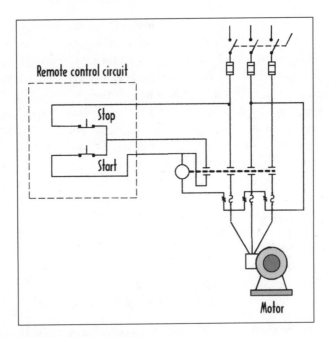

Figure 21.23 Classical motor control circuit.

A.   The locked-rotor amperes. If a motor is mechanically stalled, you
      want the conductors to withstand the heat until the stall is relieved.
      When a motor is stalled, it becomes, essentially, a transformer. (You
      may also think of a motor as a rotating transformer.) **NEC** Table
      430-7(b) gives you multipliers to use to determine this. For many
      years you could depend upon locked-rotor currents being on the
      order of 6 times the full-load amperes of the motor, but since motor
      designers have become so clever about getting more efficiency to
      the motor, motors have a wider variety of dynamic characteristics.

Q16.   *Where should the disconnect be located?*

A.   Section 430-102 says that a disconnecting means must be located
      in sight from the controller location and must disconnect the
      controller. When it is not possible to do this, the **NEC** offers
      some relief in Sec. 430-102(b). When a motor disconnect is not
      within sight of the motor, the disconnect for the controller must
      be capable of being individually locked in the open position. The
      new word "individually" was added to clarify that a separate
      locking means is intended and that a circuit breaker or switch

within a locked panelboard or locked equipment room does not satisfy the "individual" locking requirement.

Q17.   *How do I size the disconnect?*

A.   Section 430-110 says that the disconnecting means for motor circuits rated 600 V, nominal, or less, must have an ampere rating of at least 115 percent of the full-load current rating of the motor. Where two or more motors are used together or where one or more motors are used in combination with other loads, such as resistance heaters, and where the combined load may be simultaneous on a single disconnecting means, the ampere and horsepower ratings of the combined load must be determined from the sum of all currents, including resistance loads, at the full-load condition and also at the locked-rotor condition. The combined full-load current and the combined locked-rotor current so obtained is considered as a single motor for the purpose of this requirement.

New to the 1999 **NEC**, the listed manual controller additionally marked "suitable as motor disconnect" is permitted as a disconnecting means *if* located on the load side of the final short-circuit and ground-fault protection, or *if* for a stationary motor rated 2 hp or less, and 300 V or less, regardless of its location in the circuit.

Q18.   *Can a switch or circuit breaker be used as both controller and disconnecting means?*

A.   Yes. Section 430-111 says that a switch or circuit breaker complying with Sec. 430-83 is permitted to serve as both controller and disconnecting means if it opens all ungrounded conductors to the motor, if it is protected by an overcurrent device (which is permitted to be the branch-circuit fuses) that opens all ungrounded conductors to the switch or circuit breaker, and if it is either an air break, an oil switch, or an inverse time circuit breaker. Some exceptions apply.

New to the 1999 **NEC** is a group of coordinated controllers driving several parts of a single machine; the common disconnecting means no longer needs to be adjacent to the controllers and it and the controller need only be in sight from the machine. This could result in controller's being out of sight of its disconnect.

Q19. *What is the maximum number of motors that can be served by one controller?*

A. The general rule is one controller for one motor. What would a rule be without an exception, however. Section 430-87 gives the number of motors served by each controller. Each motor is provided with an individual controller. An exception exists for motors rated 600 V or less. A single controller rated at not less than the sum of the horsepower ratings of all of the motors of the group is permitted to serve the group of motors under any one of the following conditions:

a. Where a number of motors drive several parts of a single machine or piece of apparatus, such as metal and woodworking machines, cranes, hoists, and similar apparatus.

b. Where a group of motors is under the protection of one overcurrent device as permitted in Sec. 430-53(a).

c. Where a group of motors is located in a single room within sight from the controller location.

Motor Supply Circuit Design

Q20. *What information should appear on the nameplate?*

A. NEMA Standard MG 1, Sec. 10.38 requires the following: manufacturer, type, frame, horsepower, time rating, ambient temperature, rpm, frequency, phases, rated load amps, voltage, efficiency, either the locked-rotor amperes or the code letter, service factor, and insulation class. Motor-operated appliances marked with both horsepower and full-load current are permitted to have circuit characteristics based on the full-load current. Refer to Tables 21.5 and 21.8.

Q21. *How do I determine equivalent amperes for a motor?*

A. Convert horsepower to kilowatts, then kilowatts to amperes. Get the nameplate horsepower information and the operating voltage. This formula applies to motors of all horsepowers. For example,

TABLE 21.7 MAXIMUM RATING OR SETTING OF MOTOR BRANCH-CIRCUIT SHORT-CIRCUIT AND GROUND-FAULT PROTECTIVE DEVICES (NEC TABLE 430-152)

Type of motor	Percentage of full-load current			
	Nontime delay fuse*	Dual-element (time-delay) fuse*	Instantaneous trip breaker	Inverse time breaker†
Single-phase motors	300	175	800	250
AC polyphase motors other than wound-rotor				
Squirrel cage:				
Other than design E	300	175	800	250
Design E	300	175	1100	250
Synchronous‡	300	175	800	250
Wound rotor	150	150	800	150
Direct current (constant voltage)	150	150	250	150

Note: For certain exceptions to the values specified, see Secs. 430-52 through 430-54.

*The values in the nontime delay fuse column apply to time-delay class CC fuses.

†The values given in the last column also cover the ratings of nonadjustable inverse time types of circuit breakers that may be modified as in Sec. 430-52.

‡Synchronous motors of the low-torque, low-speed type (usually 450 rpm or lower), such as are used to drive reciprocating compressors, pumps, etc., that start unloaded, do not require a fuse rating or circuit-breaker setting in excess of 200 percent of full-load current.

a 30-hp, 208-V motor would draw amperes thus: 30 hp × 746 W/hp = 25,380 W. Then 25,380/(1.732 × 208) = 70.2 A. Many manufacturers have cardboard "motor calculators" that designers can use.

New to the 1999 **NEC** is a type of motor control apparatus, a "listed protected combination controller" that can now be used in lieu of other protective devices specified in Table 21.7. The apparatus bundles short-circuit and ground-fault protective functions in a single self-contained unit. The allowable settings are the same as for an instantaneous trip circuit breaker.

TABLE 21.8 FULL-LOAD CURRENT THREE-PHASE ALTERNATING-CURRENT MOTORS [*,†]

The following values of full-load currents are typical for motors running at speeds usual for belted motors and motors with normal torque characteristics. Motors built for low speeds (1200 rpm or less) or high torques may require more running current, and multispeed motors will have full-load current varying with speed. In these cases, the nameplate current rating shall be used. The voltages listed are rated motor voltages. The currents listed shall be permitted for system voltage ranges of 110 to 120, 220 to 240, 440 to 480, and 550 to 600 V.

Horsepower	Induction type squirrel-cage and wound-rotor, A						
	115 V	200 V	208 V	230 V	460 V	575 V	2300 V
1/2	4.4	2.5	2.4	2.2	1.1	0.9	—
3/4	6.4	3.7	3.5	3.2	1.6	1.3	—
1	8.4	4.8	4.6	4.2	2.1	1.7	—
1 1/2	12.0	6.9	6.6	6.0	3.0	2.4	—
2	13.6	7.8	7.5	6.8	3.4	2.7	—
3	—	11.0	10.6	9.6	4.8	3.9	—
5	—	17.5	16.7	15.2	7.6	6.1	—
7 1/2	—	25.3	24.2	22	11	9	—
10	—	32.2	30.8	28	14	11	—
15	—	48.3	46.2	42	21	17	—
20	—	62.1	59.4	54	27	22	—
25	—	78.2	74.8	68	34	27	—
30	—	92	88	80	40	32	—
40	—	120	114	104	52	41	—
50	—	150	143	130	65	52	—
60	—	177	169	154	77	62	16
75	—	221	211	192	96	77	20
100	—	285	273	248	124	99	26
125	—	359	343	312	156	125	31
150	—	414	396	360	180	144	37
200	—	552	528	480	240	192	49

TABLE 21.8 FULL-LOAD CURRENT THREE-PHASE ALTERNATING-CURRENT MOTORS* (*Continued*)

| Horsepower | Induction type squirrel-cage and wound-rotor, A | | | | | | |
	115 V	200 V	208 V	230 V	460 V	575 V	2300 V
250	—	—	—	—	302	242	60
300	—	—	—	—	361	289	72
350	—	—	—	—	414	336	83
400	—	—	—	—	477	382	95
450	—	—	—	—	515	412	103
500	—	—	—	—	590	472	118

*For 90 and 80 percent power factor, the figures shall be multiplied by 1.1 and 1.25, respectively.
†This table is an adaptation of **NEC** Table 430-150.

Q22.   *What is the difference between design letters and code letters?*

A.   As explained in Sec. 430-7(b), *design letters* reflect characteristics inherent in motor design (such as locked-rotor current, slip at rated load, and locked-rotor breakdown torque). *Code letters* are marked on motor nameplates to show motor input with locked rotor as shown in **NEC** Table 430-7(b).

Q23.   *How do I size the conductors to a single motor?*

A.   Much depends upon the type of wire that is suitable to the environment. Once you know the insulation type and the ambient temperature, you may size wire. It is fair to say that the **NEC** rule most widely applied by mechanical engineers appears in Sec. 430-22: "Branch-circuit conductors supplying a single motor shall have an ampacity not less than 125 percent of the motor full-load current rating." By sizing motor circuit conductors based on a minimum of 125 percent of the full-load current of the motor, you coordinate the wire size and the running overload protection, which allows the running overload protection to protect both the motor and the branch-circuit conductors from damage due to overloads. The ampere values are based on the horsepower rating and nominal voltage listed on the motor nameplate.

Q24. *What is the service factor all about?*

A. Service factor is the permissible overload a motor can handle within a defined temperature range without overheating to the point of damaging insulation. NEMA assigns the standard specification at 1.15 (for open motors), with 1.00 and 1.25. NEMA requires open-type designs A, B, and C motors to have service factors of 1.15; the design E motor may have a service factor of 1.0.

Q25. *Where do I apply service factor information?*

A. The higher the service factor, the more overload the motor can handle. When voltage and frequency are maintained at nameplate rated values, the motor may be overloaded up to the horsepower calculated by multiplying the rated horsepower by the service factor shown on the nameplate.

A continuous-duty motor with a marked service factor of not less than 1.15 or with a marked temperature rise of not over 40°C can carry a 25 percent overload for an extended period without damage to the motor. Other such types of motors are those with a service factor of less than 1.15 or those with a marked temperature rise of greater than 40°C that are incapable of withstanding a prolonged overload, where the motor overload protection device opens that circuit if the motor continues to draw 115 percent of its rated full-load current.

Where a motor is selected for duty-cycle service (intermittent, short-time, periodic, or varying), it can be assumed that the motor will not operate continuously, owing to the nature of the apparatus or machinery it drives. Therefore, prolonged overloads are not likely to occur unless mechanical failure in the driven apparatus stalls the motor; in this case, however, the branch-circuit protection device would open the circuit. The omission of overload protection devices for such motors is based on the type of duty and not on the time rating of the motor.

Q26. *How do I size the fuse or circuit breaker supplying power to a motor?*

A. The branch-circuit short-circuit and ground-fault protection device is sized to the percentages listed in Table 430-152. The revised exception to Sec. 430-52(c)(1) permits "the next size up" protection device, when the values derived from Table 21.7 do not correspond with the standard rating or setting of overcurrent protection devices as listed in Sec. 240-6(a). In the 1993 **NEC** the requirement was that we rounded down.

Q27.   *How will duty-cycle factors reduce the cost of the power-supply circuit?*

A.   When sizing the classification of service for a motor, the conductors are sized from the percentage nameplate current rating in amperes listed in **NEC** Table 430-22(b). Conductors are sized for supplying individual motors used for short-time, intermittent, periodic, or varying duty. (See Art. 100 for definitions of different duties.) Motors that operate on short-time duty cycles, with cooling periods in between, may have their conductors calculated at a percent less than 125 percent. This permissive rule is due to the conductors having a chance to cool during periods of driving the load.

Q28.   *Are the feeders to ac adjustable-voltage motors sized differently?*

A.   Yes. For motors used in alternating-current, adjustable-voltage, variable-torque drive systems, the ampacity of conductors, or ampere ratings of switches, branch-circuit short-circuit and ground-fault protection, etc., must be based on the maximum operating current marked on the motor or control nameplate, or both. If the maximum operating current does not appear on the nameplate, the ampacity determination must be based on 150 percent of the values given in **NEC** Table 430-150 (a portion of which is reproduced here as Table 21.8). The relevant section of the code for adjustable voltage drives is Sec. 430-6(c).

Q29.   *Is there a difference in sizing motor-circuit overcurrent elements?*

A.   Yes. You must use the **NEC** table to size the cable and disconnect fuses, and/or circuit breakers. The motor overload, however, must be taken from the motor nameplate itself, a big difference. The various tables list motor voltages at 460, but the footnote explains that these currents are permitted for voltage ranges on the order of 440 to 480. In some cases the nameplate current rating of each motor will be less than that in the table.

TABLE 21.9 TYPICAL MANUFACTURER MOTOR CONTROL CENTER SIZING WORKSHEET [*][†]

| NEMA Size | Maximum Horsepower | | | | | Switch Size | Unit Size | |
	208 V	240 V	380 V	480 V	600 V		Inches	× Space
Full Voltage Non-reversing—Fusible								
							12	2×
1	7.5	7.5	10	10	10	30	18	3×
							12	2×
2	10	15	25	25	25	60	18	3×
3	25	30	50	50	50	100	24	4×
4	40	50	75	100	100	200	36	6×
5	75	100	150	200	200	400	60	10×
							66	11×
6	150	200	300	400	400	600	72	12×
Full Voltage Reversing—Fusible								
1	7.5	7.5	10	10	10	30	24	4×
2	10	15	25	25	25	60	24	4×
3	25	30	50	50	50	100	30	5×
4	40	50	75	100	100	200	54	9×
5	75	100	150	200	200	400	72	12×
6	150	200	300	400	400	600	72	12×

[*]Adapted from Westinghouse-Cutler-Hammer *Consulting Guide.*
[†]This table can be used with the solved problem on p. 387.

The higher rating will allow the feeder to accommodate future changes. All motor feeder elements should be slightly oversized beyond the **NEC** amount. Remember, the **NEC** figure is a minimum standard. A slightly larger feeder design will reduce voltage drop, but you must be careful not to exceed the conduit size restrictions.

Q30.   *How are motor overloads specified?*

A.   In Sec. 430-22, the procedure for sizing the overload protection for motors rated more than 1 hp is found by multiplying the nameplate FLA of the motor's nameplate by 115 or 125 percent. The percentage selected is based upon service factor or temperature rise of the motor. A thermal protector built into a motor may be set with ratings greater than permitted for other forms of overload protection, which is designed to protect motor windings. The procedure for sizing overload protection for motors rated more than 1 hp is found by multiplying the nameplate FLA of the motor nameplate by 130 or 140 percent. The percentage selected is based upon service factor or temperature rise of the motor. If an overload relay fails to allow a motor to start and run due to inrush current, the size of the relay may be increased to the percentages listed in **NEC** 430-34.

New to the 1999 **NEC**, motors that have an internal thermal protector will restart the motor automatically after tripping. It must not be used if an automatic restart can result in injury.

Q31.   *Can we connect a motor to a mixed-use, general-purpose branch circuit?*

A.   Yes. Section 430-42 says that you may but that you must be careful with the overload protection. Overload protection for motors used on general-purpose branch circuits must be provided as specified in (*a*), (*b*), (*c*), or (*d*) below.

(*a*) *Not over 1 hp:* One or more motors without individual overload protection may be connected to a general-purpose branch circuit only where the installation complies with the limiting conditions. These conditions appear in Secs. 430-32(b) and (c) and Secs. 430-53(a)(1) and (a)(2).

(*b*) *Over 1 hp:* Motors of larger ratings than specified in Sec. 430-53(a) may be connected to general-purpose branch circuits only where each motor is protected by overload protection selected to protect the motor as specified in Sec. 430-32.

(*c*) *Cord- and plug-connected:* Where a motor is connected to a branch circuit by means of an attachment plug and receptacle and individual overload protection is omitted as provided in (*a*) above, the rating of the attachment plug and receptacle cannot exceed 15 A at 125 V or 15A at 125 or 250 V.

(*d*) *Time delay:* The branch-circuit short-circuit and ground-fault protective device protecting a circuit to which a motor or motor-operated appliance is connected must have sufficient time delay to permit the motor to start and accelerate its load.

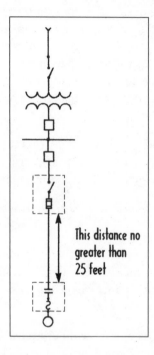

This distance no greater than 25 feet

Figure 21.24
Feeder tap rule for motors.

Protective devices (branch circuit, short circuit, and ground fault) for a branch circuit are located not more than 25 ft from the point where the conductors are tapped to the main feeder.

Q32. *Can I make things simpler by tapping the feeder?*

A. Yes, sometimes, but the rule appears in Sec. 430-28. The so-called tap rules are somewhat complicated and beyond the scope of this book. However, a brief familiarity with the rule is possible. In Sec. 240-21 the basic rule appears: All conductors must be protected at the point where they receive their supply. Tap conductors are permitted to be up to 25 ft long if they (*a*) have an ampacity equal to that of the feeders or (*b*) have an ampacity at least one-third that of the feeders and are protected from physical damage. (Refer to Fig. 21.24.)

Small Motors

Q33.   *Are the rules for motors less than 1 hp different?*

A.   Yes. These are commonly called *fractional horsepower motors*. In Sec. 430-22, three-phase motors rated 1 hp or less that are not permanently installed and not automatically started, such as bench grinders, drill presses, and portable electric tools, are not required to have overload protection and may be protected by the branch-circuit short-circuit fuse or circuit breaker. This type of equipment is usually attended by the operator, who can immediately shut off power to the motor should it overheat.

Q34.   *Do small motors require controllers?*

A.   No, as long as the motor is less than 1 hp. The **NEC** requirements rules for motors of this type appear in Sec. 430-32, which discriminates between automatically and nonautomatically started motors.

  ■   For a stationary motor rated at $^1/_8$ hp or less that is normally left running and is so constructed that it cannot be damaged by overload or failure to start, such as clock motors and the like, the branch-circuit protective device is permitted to serve as the controller.

  ■   For a portable motor rated at $^1/_3$ hp or less, the controller is permitted to be an attachment plug and receptacle.

Q35.   *How do I design the circuit for a group of motors?*

A.   The basic rule is to take the largest branch-circuit device and add the full-load currents of the remaining motors. When instantaneous trip circuit breakers are involved, you need to substitute the full-load current multiplied by the factor in Table 21.7 for the same type of overcurrent device to be used to protect the feeder.

Multiple motors can be connected in a single circuit provided the requirements of Sec. 430-52 have been met for each motor. The broad principles are as follows:

- Confirm inverse time circuit breaker or fuse specification and whether you round up or down on standard sizes of either. (Motor design letter does not figure into the specification of time delay fuses and conductor size.)

- The largest ampere current that a motor feeder will ever be required to carry occurs when the largest motor is started and all the other motors supplied by the same feeder are running and delivering their full-rated power. The general rule appears in Sec. 430-24. In addition, Sec. 430-63 clarifies that if the motor feeder conductors are sized larger than the minimum required, then the size of the overcurrent device for the feeder may be based upon the size of the feeder conductors. Apply the generic feeder rules of Arts. 210 and 220.

- Where the conductors are feeders, the highest rating or setting of the feeder short-circuit and ground-fault protection devices for the minimum-size feeder conductor permitted by this section is specified in Sec. 430-62.

Q36.   *How do you specify motor feeders to groups of motors?*

A.   Feeder circuit conductors supplying power to two or more motors that are utilized to serve duty-cycle loads must have the largest motor selected based upon their conditions of use. The conductors for feeder circuits supplying power to multimotor and combination equipment loads are required to be sized using the nameplate FLA is listed or applying the requirements of Secs. 430-25 and 430-24. In some motor installations there may be a special situation in which a number of motors are connected to a feeder circuit . Because of their operation, certain motors do not operate together and the feeder circuit conductors may be sized on the group that has the greater current rating.

New to the 1999 **NEC**, instead of summing the horsepower ratings of the motors into the group to determine the rating of the combined controller, the equivalent horsepower method is to be used instead.

Large Motors

Q37. *How do the rules for large motors differ from those for smaller motors?*

A.   Not by much; though architects and mechanical engineers should be mindful of the space requirements for both the motor and the motor controller—especially if the apparatus will operate at medium voltage. Frequently, advantages in energy savings are reduced by the need to prepare interior space for the equipment, especially in renovation projects.

This much said, Sec. 430-121, Part K, recognizes the additional hazard due to the use of higher voltages. It adds to or amends the other provisions of Art. 430. Other requirements for circuits and equipment operating at over 600 V, nominal, are in Art. 490.

What is different is that overload protection and fault-current protection may be provided by the same device. This recognizes that manufacturers can produce motor protection with microprocessor technology. Large motors have a higher relative transient inrush than smaller machines even though they meet NEMA maximum steady-state requirements. (Refer to Fig. 21.25.)

Q38. *Where do we locate the disconnecting means for a medium-voltage motor starter?*

A.   Section 430-102 states an exception to the general rules that a disconnecting means must be located in sight from the controller location and must disconnect the controller. One of the exceptions states that for motor circuits over 600 V, nominal, a controller disconnecting means capable of being locked in the open position is permitted to be out of sight of the controller, provided the controller is marked with a warning label giving the location of the disconnecting means.

Q39. *What are the most significant changes for motor circuit design and construction in the 1999 code cycle?*

A.   There are very few changes that would be of interest to trades other than the electrical trades. A few changes that would be of interest to electrical professionals are as follows:

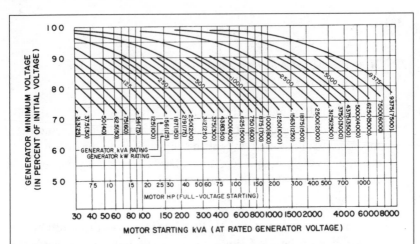

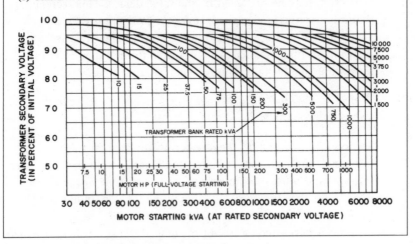

Figure 21.25 Motor starting kVA. (From IEEE Std 241; used with permission.)

- Motor-operated appliances marked with both horsepower and full-load current are permitted to have circuit characteristics based on the full-load current. This is something of a "truth in labeling" requirement, intended to protect the public from advertising hyperbole in the marketing of small, typically residential appliances.

- Minimum linear dimensions of motor terminal housing revert to 1993 sizes, but the 1996 volumes remain. Manufacturers will take care of this.

- Maximum 250-V receptacle rating for cord- and plug-connected motor loads without overload protection are raised to 15 A from 10 A.

- Motors that have an internal thermal protector will restart the motor automatically after tripping. It must not be used if an automatic restart can result in injury.

- Design B energy-efficient motors can now have their ground-fault protection and short-circuit protection set to the same allowances as design E motors.

- In addition to sizing the wires to the highest current winding, now each controller must have a horsepower rating not less than what would be required for the highest motor winding.

- Instead of summing the horsepower ratings of the motors into the group to determine the rating of the combined controller, the equivalent-horsepower method is to be used instead.

- Standard bus arrangements now apply only to the horizontal common bus, not to terminal placements or to internal interconnecting conductors.

- For a group of coordinated controllers driving several parts of a single machine, the common disconnecting means no longer needs to be adjacent to the controllers and it and the controller need only be in sight from the machine. This could result in having the controller out of sight of its disconnect.

- A listed self-protected combination controller now qualifies as a motor disconnecting means.

- The listed manual controller additionally marked "suitable as motor disconnect" is permitted as a disconnecting means *if* located on the load side of the final short-circuit and ground-fault protection, or *if* for a stationary motor rated 2 hp or less, and 300 V or less, regardless of its location in the circuit.

■ A nonfused motor-circuit switch may now have an ampere rating less than 115 percent of the load, provided its horsepower rating is at least equal to the motor rating or to the equivalent horsepower rating of a combination load.

This is only a partial list. The reader is referred to the core references on 1999 **NEC** changes.

Motor Control Centers

Q40.   What are the broad principles that apply to motor control centers?

A.   A motor control center is an assembly of one or more enclosed sections having a common power bus and principally containing motor control units. They contain all necessary buses, incoming and outgoing lines, and safety features for convenience and safety. There are very few specialty rules for their design and installation, only the **NEC** generics regarding workspace, grounding, and overcurrent protection. The safe manufacture and field installation of motor control centers depend heavily upon conformance to NEMA and UL standards.

The mother standard for motor control centers is NEMA ICS-2, Industrial Control Devices, Controllers, and Assemblies. The reader is referred to the solved problem at the end of the chapter for a general introduction to the procedure for designing a motor control center.

Q41.   What particulars should show up in the MCC specification?

A.   Information about shipping splits, equipment weights, mounting and alignment, and special installation requirements. Mechanical bracing should be sufficient to withstand fault currents on the order of 42 to 100 kA. A ground bus must be provided for the full length of the control center assembly. All terminal connections should be accessible from the front of the structure.

The enclosure should conform to ANSI and NEMA ICS 6, standards for motor control centers.

Motor starter installation should conform to NEMA PB 1 standards for motor control centers.

Install overload relays in motor starters to match installed motor characteristics.

Q42.   *Can a motor control center be used as service equipment?*

A.   Yes. Section 430-95 says that where used as service equipment, each motor control center must be provided with a single main disconnecting means to disconnect all ungrounded service conductors. A second service disconnect is permitted to supply additional equipment. Where a grounded conductor is provided, the motor control center must be provided with a main bonding jumper, sized in accordance with Art. 250, within one of the sections for connecting the grounded conductor, on its supply side, to the motor control center equipment ground bus.

Q43.   *What broad NEC principles apply to the motor control center unit compartments?*

A.   Electrical designers and engineers should always show on the blueprints both the schematic and the pictorial (elevation view) of the motor control center. The schematic should show starter conductor and conduit size, and overcurrent device information. This helps secure good bids from equipment suppliers. The specifications should include a few, if not all, of the following typical requirements: MCCs should consist of a fusible visible blade, switch-type combination starter (wherever disconnects are required) and spaces to receive future units. Capacitors should not be provided for motors controlled by variable-speed frequency drive units. Wherever possible, control circuits must be wired so that motors automatically restart after a power disruption. Provide time delays to allow large fans and pumps to coast down before restarting after a momentary power disruption and to prevent all of the large motors from restarting simultaneously. Structures must contain a horizontal wireway at the top, isolated from the horizontal bus and readily accessible. Each structure must contain an isolated vertical wireway with cable supports, accessible through hinged doors and a horizontal wireway at the bottom.

Starter and feeder tap unit doors must be interlocked mechanically with the unit disconnect device to prevent unintentional opening of the door while energized and unintentional application of power while the door is open. An interlock between the unit disconnect device and the structure will prevent removal or reinsertion of a unit when the disconnect is in the *on* position. Means must be provided for releasing the interlock for intentional access and/or application of power. Padlocking arrangements must permit locking the disconnect device *off* with at least three padlocks with the door closed or open. Means must be provided to padlock the unit in a partially withdrawn position (test) with the stabs free of the vertical bus.

Overload relays must be reset from the outside of the enclosure by means of an insulated button. Motor control centers may require control power transformers for the combination starter units. These control circuits are 120 to 480 V and the transformers are 50 to 100 VA.

An external operator handle must be supplied for each switch or breaker. This mechanism must be engaged with the switch or breaker at all times regardless of unit door position to prevent false-circuit indication. The operator handle must have a conventional up-down motion and must be designed so that the position will indicate the unit is *off*. For added safety, it must be possible to lock this handle in the *off* position with up to three padlocks. A defeater mechanism must be provided for the purpose of defeating this interlock by a deliberate act of an electrician should one desire to observe the operation of the operator handle assembly or the unit components.

Q44.   *What are the rules for installing capacitors in motor control centers?*

A.   Capacitors are usually incorporated into motor control centers. Typically they are applied to all constant-speed motors 10 hp and above. The horsepower limit is sometimes determined by a company technical standard. Capacitors for power-factor correction must be provided and installed by the electrical contractor for all individual motors rated 10 hp and larger. The kVAR rating is frequently left to the capacitor manufacturer. A match is made with a motor of a given horsepower. A correction of power factor to 86 percent minimum when the motor is loaded to 75 percent of nameplate rating. Capacitors for package equipment where motor starters are provided by mechanical trades as factors installed will be provided and installed by the mechanical trades.

Solved Problem—Chapter 21 (Division 16480)

SITUATION.

A group of 480-V motors for HVAC and elevators in a generalized building requires a motor control center. A motor list, produced by the mechanical designer, is given to the electrical designer for the purposed of determining how much space the architect should block out for the motor control center.

MOTOR LIST

4 motors of 7.5 hp each [full voltage, nonreversing (FVNR)]

5 motors of 10 hp each (FVNR)

2 motors of 20 hp each (FVNR)

2 motors of 30 hp each (FVNR)

2 motors of 40 hp each [full voltage, reversing (FVR)]

1 motor of 100 hp [reduced voltage, nonreversing (RVNR)]

REQUIREMENT.

Estimate the size of the motor control center.

Solution

Step 1. *Determine the ampacity of the bus.* Use the table below.

Motor HP	Full-load motor amperes			
100	124	125 percent of largest motor	1.25 × 124	155
7.5	11		4 × 11	44
10	14		5 × 14	70
20	27		2 × 27	54
30	40		2 × 40	80
40	52		2 × 52	104
		2 × 30 kW heater	2 × 38	76
			Total	583 A

Therefore, the motor control center bus should be sized at 600 A, minimum.

Step 2. *Consult with manufacturer regarding space requirements.* For typical manufacturer information given in Table 21.9. The first column in "unit size" is the depth of the motor control center. Note that the 100-hp motor determines maximum depth at 54 in. The depth of other starters may be as shallow as 12 in. The NEMA starter size and the number of spaces required for each starter is tabulated below.

We determine the space factors for the 30-kW heating load thus:

$$I = [30 \times 1000]/[\sqrt{3} \times 480] = 38 \text{ A required for each heater.}$$

A 60-A disconnect switch is adequate for each heater. Allow 1 space factor for each 60-A disconnect and 1 space factor for the main incoming feeder. Thus the total number of space factors is 46+2+1=49.

Step 3. *Determine the number of space factors per vertical section.* This information is also available from manufacturers' catalogs. A typical value for the number of space factors for a standard height motor con-

Table 21.10 MCC Example

Motor Starters			Number of Space Factors		
Motor HP	**Starter Type**	**NEMA Size**	**Each Unit**	**Quantity**	**Space Total**
7.5	FVNR	1	2×	4	8
10	FVNR	1	2×	5	10
20	FVNR	2	2×	2	4
30	FVNR	3	4×	2	8
40	FVR	3	5×	2	10
100	FVNR	4	6×	1	6
				Starter *Total*	46

trol center is 10 to 12. Use 12. Thus, we may estimate the minimum number of vertical sections to be 49/12 = 4.1 vertical sections. We round up to 5 sections in order to allow some room for spare spaces.

Figure 21.26 is a "roughed-out" layout of the motor control center sufficient for schematic design purposes. All of the NEC generics regarding working space about electrical apparatus apply.

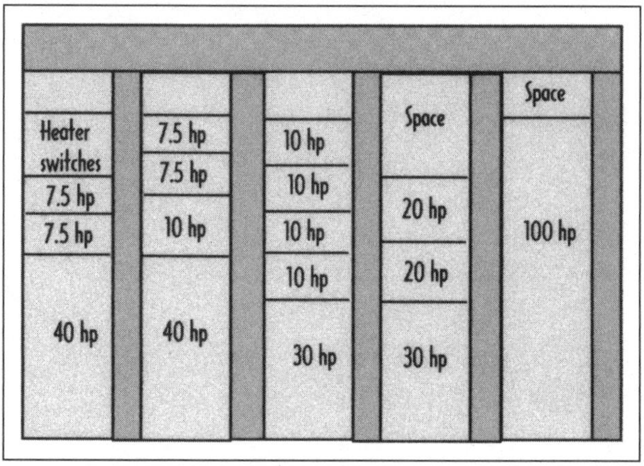

Figure 21.26 Motor control center layout.

REMARKS.

This example is intended to show the basic steps. If you need to play a game of inches—often the case in renovation projects—manufacturer application engineers are a handy resource for determining dimensions and cost trade-offs.

Division 16500—Lighting

Remarks. After rotating machinery, lighting is the most common application of electrical energy. For many electrical engineers, the design of lighting systems provides the only outlet for their creative energies, for merging their flair for art and their skill in engineering. It is a tenuous privilege, however, since every-one responds to light in an individual way.

Q1.   *What NEC requirements commonly appear in Division 16500?*

A.   Construction requirements appear in Art. 410, which covers installation, location, grounding, support, and wiring of lighting fixtures and associated auxiliary equipment. The design and engineering of branch-circuit requirements appear in Art. 220, and these rules overlap with the rules for receptacle design. A lighting receptacle is a switched receptacle intended to be used by a lamp, typically by a flexible cord. The rules governing lighting panels appear in Art. 384. All the general rules for selecting wiring appear in Chap. 3, and the rules for grounding appear in Art. 250.

It is worth noting that the international term for a lighting fixture is "luminaire" and is defined as a complete lighting unit consisting of a lamp or lamps together with the parts designed to distribute the light, to position and protect the lamps, and to connect the lamps to the power supply. Some of the basic nomenclature for the lighting industry is listed in the sidebar.

Nomenclature for lighting

The measure of how well electricity is converted into light is called *luminous efficacy* and is expressed in lumens per watt. If the energy in a light source could be converted without loss into yellow-green light, the efficacy of the source would be 684 lumens per watt. Practical commercial lighting sources fall well short of this.

Quantity	Quantity is a measure of	Symbol	Unit SI	English	Definition of unit
Luminous intensity (candlepower)	Ability of source to produce light in a given direction	*I*	Candela (cd)		Approximately equal to the luminous intensity produced by a standard candle

Quantity	Quantity is a measure of	Symbol	Unit SI	Unit English	Definition of unit
Luminous flux	Total amount of light	φ	Lumen (lm)		Luminous flux emitted in a solid angle of 1 steradian by a 1 candela uniform point source
Illuminance (illumination)	Amount of light received on a unit area of surface (density)	E	Lux (1x)	Footcandle (fc)	One lumen equally distributed over one unit area of surface
Luminous exitance	Density of light reflected or transmitted from a surface	M	lm/m²	lm/ft²ª	A surface reflecting or emitting 1 lumen per unit of area
Luminance (brightness)	Intensity of light per unit of area reflected or transmitted from a surface	L	cd/m²	cd/in²	A surface reflecting or emitting light at the rate of 1 candela per unit of projected area

1 meter (m) = 3.28 ft; 1 cd/m² = 3.14 lm/m²
1 m² = (3.28)² = 10.76 ft²; 1 cd/in² = 452 lm/ft²
1 fc = 10.76 1x
ªFormerly "footlambert," which is no longer a preferred term.

*Q2.   What are the broad **NEC** principles that apply to lighting systems?*

A.   As with all **NEC** requirements, the underlying concern is for fire safety. Fire is a risk if excessive heat is produced by improper installation or use of lighting fixtures. Incandescent fixtures require a higher wattage to obtain the same amount of light as electric discharge fixtures; thus the heat produced by incandescent lamps is usually greater than for electric discharge fixtures. Recessed lighting fixtures, if not installed properly, can create a fire hazard. Thus, the **NEC** requires recessed incandescent fixtures to be thermally protected. Electric discharge fixtures usually have a ballast which is a source of heat in addition to the lamp. Ballasts for fluorescent fixtures are required by Sec. 410-73?? to be thermally protected unless they are remote HID ballasts and they are not recessed.

Mother standards for lighting systems are UL Standards 1570, 1571, and 1572. Article 220-3, which, among other subjects, deals with branch circuits that typically supply power to lighting fixtures, was overhauled in order to correlate it to changes made to Art. 210. A typical lighting branch circuit is shown in Fig. 22.1, and Table 22.1 through Table 22.3 are included here (pp. 396‑398) for handy design reference. We will cover only the **NEC** generics for lighting and follow up with a solved problem at the end of this chapter.

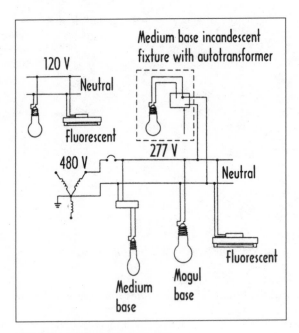

Figure 22.1 Typical lighting branch circuit construction.

Q3.   *Does the* **NEC** *require lighting fixtures to be installed for final inspection?*

A.   No. Where the Code requires a lighting outlet to be installed, this means that an outlet intended for the connection of a lighting fixture is required, not that a lighting fixture is required! Lighting outlets that do not have a fixture canopy, lampholder, or similar device installed must be provided with a nonmetallic or metal cover (Secs. 370-25 and 410-12). Where metal covers or plates are used, they must comply with the grounding requirements of Sec. 250-110.

Q4.   *What are the general rules for wiring lighting circuits?*

A.   Branch circuits for lighting are restricted to a maximum of 20 A unless supplying heavy-duty lampholders. Branch circuits for the lighting in dwellings and guest rooms are restricted to a maximum of 120 V. Branch circuits for medium-base, screw-shell incandescent lampholders are restricted to a maximum of 120 V. Branch circuits for electric discharge lamps may be 277 V nominal. Designers should try to design a switching regime to encourage the occupants to save energy.

The reader should be aware that the Illuminating Engineering Society of North America's energy management committee, along with the American Society of Heating, Refrigerating and Air-Conditioning Engineer's 90.1 Lighting Subcommittee, have been involved in achieving consensus in the ASHRAE 90.1-89R standards for lighting systems. The federal EPACT legislation requires states to match the provisions of the 90.1-89R standard for their building codes in the future. This is another standard whose provisions for additional metering may have bearing on electrical design.

Q5.   *What are the rules for lighting work space?*

A.   Illumination of working spaces around service equipment, switchboards, panelboards, and motor control centers installed indoors and operating at 600 V or less is required by Sec. 110-26(d). Where electrical equipment mentioned in Sec. 110-34(a) operates at voltages above 600, illumination must be provided for the working spaces. Lighting outlets and fixtures must be located and arranged so that persons changing lamps or making repairs will not be endangered by live parts of other equipment.

Q6.   *How are lighting circuits supplied power?*

A.   From a lighting panel. A lighting panel is a panel that has more than 10 percent of its overcurrent devices rated at 30 A or less for which a neutral connection is provided. A lighting panel is restricted to a maximum of 42 single-pole overcurrent devices. It

TABLE 22.1 PRESCRIPTIVE UNIT LIGHTING POWER ALLOWANCE (ULPA) (W/FT²)—GROSS LIGHTED AREA OF TOTAL BUILDING

Building type or space activity	0– 2000 ft²	2001– 10,000 ft²	10,001– 25,000 ft²	25,001– 50,000 ft²	50,001– 250,000 ft²	>250,000 ft²
Food service						
Fast food/cafeteria	1.50	1.38	1.34	1.32	1.31	1.30
Leisure dining/bar	2.20	1.91	1.71	1.56	1.46	1.40
Offices	1.90	1.81	1.72	1.65	1.57	1.50
Retail*	3.30	3.08	2.83	2.50	2.28	2.10
Mall concourse	1.60	1.58	1.52	1.46	1.43	1.40
Multiple-store service						
Service establishment	2.70	2.37	2.08	1.92	1.80	1.70
Garages	0.30	0.28	0.24	0.22	0.21	0.20
Schools						
Preschool/elementary	1.80	1.80	1.72	1.65	1.57	1.50
Jr. high/high school	1.90	1.90	1.88	1.83	1.76	1.70
Technical/vocational	2.40	2.33	2.17	2.01	1.84	1.70
Warehouse/storage	0.80	0.66	0.56	0.48	0.43	0.40

*Includes general, merchandising, and display lighting.
This prescriptive table is intended primarily for core-and-shell (i.e., speculative) buildings or for use during the preliminary design phase (i.e., when the space uses are less than 80 percent defined). The values in this table are not intended to represent the needs of all buildings within the types listed.
From ASHRAE/IES 90.1-1989, Energy Efficient Design of New Buildings Except New Low Rise Residential Buildings. Used with the permission of ASHRAE.

must be individually protected on its supply side by its own overcurrent device. Lighting loads must be reasonably balanced between the phases.

Q7.   *Where do I put switched lighting outlets?*

A.   In every habitable room. Section 210-70 contains requirements for the installation of lighting outlets. In dwelling units, part (a) requires wall switch-controlled lighting outlets in every habitable room: bathrooms, hallways, stairways, attached garages, and detached garages with electric power, and at entrances or exits. At stairways where there are six or more steps between floors, lighting outlets must be controlled wall-mounted switches at each floor. The Code contains the requirement for the location of the lighting

TABLE 22.2 TYPICAL INPUT WATTS FOR FLUORESCENT LAMP BALLASTS

Lamp type	Nominal lamp current	Nominal lamp, W	System input, W				Circuit type
			Standard ballasts		Energy-saving ballasts		
			One-lamp	Two-lamp	One-lamp	Two-lamp	
F20T12	0.380	20	32	53	—	—	Rapid start, preheat lamp
F30T12	0.430	30	46	81	—	—	Rapid start
F30T12, ES	0.460	25	42	73	—	—	Rapid start
F32T8	0.265	32	—	—	37	71	Rapid start
F40T12	0.430	40	57	96	50	86	Rapid start
F40T12, ES	0.460	34/35	50	82	43	72	Rapid start
F48T12	0.425	40	61	102	—	—	Instant start
F96T12	0.425	75	100	173	—	158	Instant start
F96T12, ES	0.455	60	83	138	—	123	Instant start
F48T12, -800 mA	0.800	60	85	145	—	—	Rapid start
F96T12, -800 mA	0.800	110	140	257	—	237	Rapid start
F96T12, ES, 800 mA	0.840	95	125	227	—	207	Rapid start
F48 -1500 mA	1.500	115	134	242	—	—	Rapid start
F96 -1500 mA	1.500	215	230	450	—	—	Rapid start

outlet, but does not require a specific location for the switch. The **NEC** does not prohibit installation of the switch behind a door; common sense does. Neither does the **NEC** require that you relocate the switch if you change the swing of the door.

Q8.   How do you design a basic lighting circuit?

A.   The basic rules appear in Sec. 220-11. For a standard residential or commercial circuit, take the wattage of the lamps and divide by the voltage. Take the current and multiply by 1.25 to get the wire and circuit breaker size. The demand factors listed in

TABLE 22.3 TYPICAL INPUT WATTS FOR HID LAMP BALLASTS

Lamp type	ANSI designation	Watts	Ballast type				
			Reactor	High-reactance auto-transformer (LAG)	Constant wattage auto-transformer (CWA)	Constant wattage regulated (CW)	High reactance regulated (regulated lag)
Mercury	H46	50	68	74	74	—	—
	H43	75	94	91–94	93–99	—	—
	H38/44	100	115–125	117–127	118–125	127	—
	H39	175	192–200	200–208	200–210	210	—
	H37	250	272–285	277–286	285–300	292–295	—
	H33	400	430–439	430–484	450–454	460–465	—
	H36	1000	1050–1070	—	1050–1082	1085–1102	—
Metal-halide	M57	175	—	—	210	—	—
	M58	250	—	—	292–300	—	—
	M59	400	—	—	455–465	—	—
	M47	1000	1050	—	1070–1100	—	—

	M48	1500	—	—	1610–1630	—	—
High-pressure sodium	S76	35	43	—	—	—	—
	S68	50	60–64	68	—	—	—
	S62	70	82	88–95	95	—	105
	S54	100	115–117	127–135	138	—	144
	S55	150 (55 V)	170	188–200	190	—	190–204
	S56	150 (100 V)	170	188	188	—	—
	S66	200	220–230	—	245–248	—	254
	S50	250	275–283	296–305	300–307	—	310–315
	S67	310	335–345	—	365	—	378–380
	S51	400	463–440	464–470	465–480	—	480–485
	S52	1000	1060–1065	—	1090–1106	—	—

Table 220-11 shall apply to that portion of the total branch-circuit load computed for general illumination. They must not be applied in determining the number of branch circuits for general illumination. Refer to Fig. 22.1.

When wiring a branch circuit, many designers simply allow 8 to 16 100-W fixtures on each branch circuit using a 15-A circuit with THHN wiring. Per **NEC** 210-70, there must be at least one wall switch-controlled lighting outlet in rooms, halls, stairways, attached garages, detached garages with electric power, and at outdoor entrances.

New to the 1999 **NEC** is the requirement [Sec. 210-23(a)] that lighting fixtures are not to be included in determining 50 percent of the branch-circuit rating where lighting fixtures and cord- and plug-connected utilization equipment are supplied by the same branch circuit. Designers must not include the VA of lighting fixtures when they apply the 50 percent rule to determine if fastened-in-place apparatus can be added to an existing branch circuit.

All of the foregoing must be preceded by a determination of how many fixtures are needed in order to provide the necessary illumination level. While the data contained in Fig. 22.2 may serve as a starting point, there is considerable "art" in the science of lighting. The reader should consult one of the many fine textbooks published by the Illumination Engineering Society.

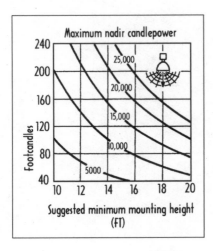

Figure 22.2
Suggested Minimum Mounting Height Versus Average Footcandle Illumination.

Q9.   *On what ampere basis is track light wiring designed?*

A.   Section 220-12 says that for load calculations, a maximum of 2 ft (609.6 mm) of lighting track or fraction thereof is considered to be 150-VA. Where multicircuit track is installed, the load requirement of this section shall be considered to be divided equally between the circuits. The 150-VA rating per 2 ft of track is for load calculations only and does not limit the length of track that can be run or the number of fixtures allowed. Some exceptions apply.

Q10.   *Does the NEC put limits on the size of a lighting fixture?*

A.   No limit on size, but there is a 50-lb limit on the weight of ceiling hung fixtures unless a listed outlet box is applied, designed, and listed for the purpose of carrying the extra weight. [Sec. 410-16(a)].

Q11.   *Who determines whether an occupancy is in a wet, damp, or dry location?*

A.   The AHJ. The 1999 **NEC** now provides language that grants the AHJ the authority to apply the definitions of Art. 100 to make this determination. It makes a difference because no one wants to see overhead fixtures corrode and fall from the mounting hardware. These definitions appear under the heading "Locations."

Q12.   *Can neon lights be installed in houses and multioccupancy buildings?*

A.   No. Section 410-80 says that equipment having an open-circuit voltage exceeding 1000 V cannot be installed in or on dwelling occupancies. Such lighting systems are often used as decorative lighting as well as for outline lighting and signs. Thus, the beer sign taken from the tavern and placed in a fraternity rumpus room is not permitted.

Q13.   *Can a motor and a lamp be supplied power on the same feeder circuit?*

A.   Yes, according to Sec. 430-63, where a feeder supplies a motor load and, in addition, a lighting or a lighting and appliance load, the feeder protective device is permitted to have a rating or setting sufficient to carry the lighting or the lighting and appliance load as determined in accordance with Arts. 210 and 220 plus, for a single motor, the rating permitted by Sec. 430-52 and, for two or more motors, the rating permitted by Sec. 430-62.

Q14. *Are there any **special** rules for overcurrent protection of lighting circuits?*

A. Some. Although conductors are required by Sec. 240-3 to be protected at their ampacities, there are many options. Parts (a) through (g) specify conditions that modify the basic rule and generally allow conductors to be protected by an overcurrent device with a rating or setting that exceeds conductor ampacity. Similarly, Sec. 240-4 allows fixture wires to be protected by overcurrent devices that exceed the ampacities given in Table 402-5. Section 240-4 allows a run of 50 ft of No. 18 or 100 ft of No. 16 to be protected at 20 A. No. 14 fixture wire can be used on 20-A branch circuits and No. 12 fixture wire can be connected to 40- or 50-A branch circuits. If fixture wires are smaller than permitted by the exception, supplementary overcurrent protection meeting the requirement in Sec. 240-10 is acceptable.

Q15. *Are there any special rules for grounding lighting circuits?*

A. Exposed conductive parts of lighting fixtures and equipment directly wired or attached to outlets supplied by a wiring method that provides an equipment ground must be grounded. Metal fixture housings, transformers, and their enclosures must be grounded where connected to circuits operating at over 150 V to ground. A means for connecting an equipment grounding conductor must be provided for all lighting fixtures with exposed metal parts. Lighting fixtures and associated components must be grounded by one of the methods mentioned in Sec. 250-118. Equipment grounding conductors must be sized to comply with Sec. and Table 250-122. Requirements for grounding appear in Secs. 410-17 through 410-21.

Q16. *What are the rules that apply to keeping lighting circuits fire safe?*

A. Where branch-circuit conductors are within 3 in of a ballast that they can contact, conductor insulation must be rated 90°C or higher. This applies to all insulation except type THW, which is permitted even though the temperature rating is 75°C. (see Sec. 410-31). Flush and recessed fixtures must be installed so that adjacent combustible material will not be subjected to temperatures in excess of 90°C. Recessed incandescent fixtures are required to have thermal protection and be so marked. Thermal protection is not required for fixtures that are identified for use in poured concrete or for fixtures designed so that the temperature performance characteristics are similar to thermally protected fixtures.

Section 410-66 requires that thermal insulation be kept at least 3 in away from recessed fixture enclosures, wiring compartments, and ballasts and is not permitted above lighting fixtures if the insulation would prevent free circulation of air. Recessed fixtures that are suitable for installation in direct contact with thermal insulation are marked "Type IC" or "Inherently Protected." Supply wiring to lighting fixtures must have temperature ratings suitable for the temperature encountered. A marking such as "For Supply Connections, Use Wire Rated for at Least 75°C" may appear near a knockout. Other temperature markings may specify 90 or 150°C insulation. Part (b) of Sec. 410-67 permits running branch-circuit conductors directly to a recessed or flush fixture, provided that the insulation is suitable for the temperature encountered.

Armored cable (type AC) and metal-clad cable (type MC) usually contain conductors with 90°C insulation. Fixture wire may also be used to supply these fixtures where the insulation is suitable for the temperature; the length of the raceway enclosing the tap conductors is at least 4 and not more than 6 ft; and the tap originates in an outlet box that is at least 1 ft from the fixture. A maximum lamp wattage marking must appear on all flush and recessed incandescent fixtures. Light blinking might occur when a thermally protected fixture is overlamped or mislamped. Where fluorescent fixtures are installed indoors, class P (thermally protected) ballasts are required by Sec. 410-73(e). Replacement ballasts for all fluorescent fixtures installed indoors must also have thermal protection.

Interior Lighting

Q17. *How many lighting circuits are required in a 2000 sq ft house?*

A. Use **NEC** Table 220-3(a), General Lighting Loads by Occupancies, assuming the square footage does not include the unfinished basement. Then 2000 sq ft/ × 3 VA/sq ft. = 6000 VA. Then 6000/120 V = 50 A. Then 50 A/15 A = 3.33 circuits. To determine the actual number of circuits, you must assume that the load is continuous. Thus, as a continuous load you may load the circuit to no more than 80 percent. In the case of a 20-A lighting circuit, the maximum permitted for a continuous load would be 16 A. More numerical examples appear in Chap. 9 of the **NEC**.

Q18. *What loads are permitted to be connected to the emergency generator?*

A. Emergency generation systems should be designed primarily for providing standby electric service during outages for life and fire safety loads, including emergency white and exit lights, fire-alarm and fire-detection systems, fire sprinkler booster pumps, security systems and door locks, public address and telecommunication systems, closed circuit and master antenna television systems, elevators.

Emergency white (EW) and exit lights are required at all egress points and changes in direction within corridors, as well as at main stairwell landings (nonintermediate), stairwell egress points, lobbies, and foyers. Exit light fixtures should have downlights, so they can be used in lieu of EW, and those exit lights delineating changes in egress direction require directional arrows.

Exterior Lighting

Q19. *What are the NEC basics for exterior lighting?*

A. For the exterior lighting such as roadways, parking lots, and athletic fields, the ballasts for electric discharge lamps can be connected to circuits not exceeding 600 V between conductors provided that the lighting units are mounted on poles at a minimum height of 22 ft. This permits these outdoor units to be connected to ungrounded 480-V systems. Per Sec. 210-70(a)(2), illumination is required for all outdoor entrances and exits. The floodlight is not required to be installed immediately adjacent to the entrance or exit but may be mounted in or below the overhang of the roof. The **NEC** is silent on the matter of a switch.

Q20. *We have a soccer field with night lighting. Where do we put the lighting disconnect?*

A. Disconnect for lighting fixtures poles can be remote from the pole. Section 225, Part B (new) clarifies the intent of the Code. Where more than one building, structure, or pole is on the same property and under single management, each building or structure must have a readily accessible disconnecting means at the separate building, structure, or pole. A new Exception 3 clarifies that the disconnecting means for poles used as lighting standards may be located remotely from the pole, but the disconnect must be readily accessible.

Q21. *What are the rules for wiring exterior lighting fixtures on poles?*

A. Article 225 says that when lighting fixtures are installed on poles and supplied by overhead conductors, the maximum distance between supports cannot be greater than 50 ft for No. 10 copper or No. 8 aluminum conductors. For spans exceeding 50 ft, the minimum conductor size is No. 8 copper or No. 6 aluminum. For a string of outdoor lights suspended between two points more than 15 ft apart (festoon lighting), the minimum conductor size is No. 12, unless supported by a messenger wire. Section 225-7 provides outdoor lighting branch-circuit requirements that are in addition to those found in Art. 210. Part(b) of this section allows a common neutral for multiwire branch circuits.

Similar safety rules for exterior lighting should be followed by the local utility. As shown in Fig. 22.3, public and private lighting systems frequently can, and should, merge. Refer also to the discussion in Question 25.

Q22. *What are the rules for installing lighting equipment outdoors?*

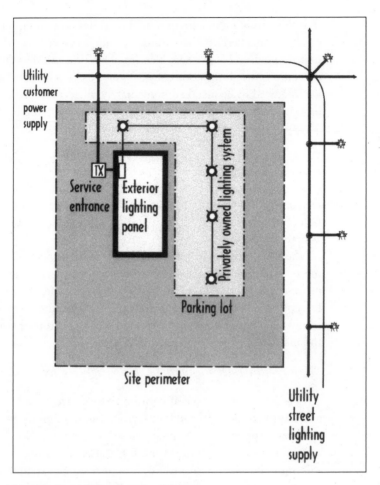

Figure 22.3 Merging of Utility and Private Street lighting system.

A.   One of the basics is that the ampacity of the neutral conductor cannot be less than the maximum net computed load current between the neutral and all ungrounded conductors connected to any one phase of the circuit. Refer to Sec. 225-7.

Circuits exceeding 120 V, nominal, between conductors and not exceeding 277 V, nominal, to ground are not permitted to supply lighting fixtures for illumination of outdoor areas of industrial establishments, office buildings, schools, stores, and other commercial or public buildings where the fixtures are not less than 3 ft from windows, platforms, fire escapes, and the like.

Where outdoor lampholders are attached as pendants, the connections to the circuit wires must be staggered. Where such lampholders have terminals of a type that puncture the insulation and make contact with the conductors, they must be attached only to conductors of the stranded type.

Q23.   *What are the rules for placing (locating) outdoor lamps?*

A.   Section 225-25 says that lamps for outdoor lighting should be below all energized conductors, transformers, or other electric utilization equipment, except where clearances or other safeguards are provided for relamping operations and except where equipment is controlled by a disconnecting means that can be locked in the open position. Vegetation such as trees cannot be used for support of overhead conductor spans.

Q24.   *How are luminaires protected from damage by water?*

A.   The bulk of the rules for the construction, location, wiring, and installation of lighting fixtures using incandescent lamps or electric discharge lamps are found in Art. 410. Lighting fixtures installed in wet locations, such as outdoors exposed to the weather, or vehicle washing areas, must be marked "Suitable for Wet Locations."

To prevent the entrance of water, caulking compound may have to be used at the outlet box, fixture housing interface. Fixtures installed outdoors under marquees, canopies, overhangs, or carports, roofed open porches, and the like must be marked "Suitable for Damp Locations" or "Suitable for Wet Locations." Where corrosive conditions are present, only lighting fixtures suitable for the environment are permitted.

Q25.   *What are the rules for parking lot lighting?*

A.   There has been some back and forth on this in the **NEC** technical panels and the answer falls somewhere in the gray area over which the authority having jurisdiction has the final word. It makes sense to think that **NEC** should apply to such lighting, however similar to roadway lighting and whether or not it is owned, installed, and maintained by a utility. The question arises: how utility-operated roadway lighting is safe and the same lighting in the parking lot is unsafe.

In much the same way overcurrent protection requirements for transformers vary according to the degree of supervision, the **NEC** installation requirements for parking lot lighting are based on whether qualified personnel are exclusively involved in the ongoing equipment usage. Very few jurisdictions will overturn their utility work practice regulations on the supply side of the service point just because the **NEC** thinks they should.

Q26.   *What special considerations should be given to hospital lighting systems?*

A.   Since hospitals require several branches of power, three will contain some amount of lighting: normal, critical, and life safety. Refer to **NEC** Art. 517 for the nomenclature of the various types of hospital power systems. You can derive 120- or 277-V lighting circuits with either step-up or step-down transformers. The choice of lighting-system voltage depends upon the load class mix.

Systems that operate on 120 V allow for complete flexibility in the selection of luminaires. Arcing ground faults, for example, are rarely sustained on 120-V systems but do occur with somewhat more frequency on 480-V systems. Both the building cost and the reliability of the system will be somewhat enhanced by selecting 120 V as the lighting system voltage (Fig. 22.4). Two dedicated branch-circuit panels are required.

For large hospitals, 277 V may be the optimal choice even though the lighting fixtures will require a duplication of panels. The higher voltage will save copper, copper losses, and possibly

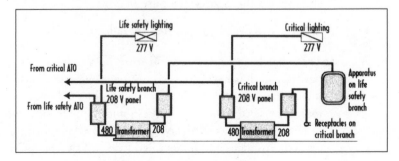

Figure 22.4 Small Health Care Facility with 120 V Life Safety and Critical Lighting.

workspace. This might be enough to offset the costs of providing the extra switchgear shown in Fig. 22.5. Four dedicated branch-circuit panels are required.

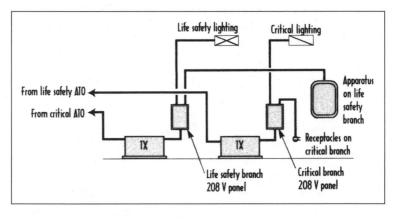

Figure 22.5 Large Hospital with 277 V Life Safety and Critical Lighting.

Lighting Accessories

Q27.   *What are the rules for lighting circuits operating at 30 V or less?*

A.   These are typically decorative lighting. They cannot be installed within 10 ft of pools. Where bare conductors are used, they cannot be installed less than 6 ft above the finished floor. They must be supplied from a 15-A maximum ampere circuit. A fixture that weighs more than 6 lb or exceeds 16 in in any dimension cannot be supported by the screw shell of a lampholder.

Q28.   *Can track lighting be installed over bathtubs?*

A.   Only if it is outside a zone 3 ft horizontally and 8 ft vertically from the top of the bathtub rim (Sec. 410-101).

Q29.   *What broad principles apply to lighting track?*

A.   Part R of Art. 410 covers the subject of track lighting. Section 410-101 says that lighting track must be permanently installed and permanently connected to a branch circuit. Only lighting track fittings must be installed on lighting track. Lighting track fittings must not be equipped with general-purpose receptacles. The connected load on lighting track cannot exceed the rating of the track. Lighting track must be supplied by a branch circuit having a rating not more than that of the track. Lighting track cannot be installed (1) where likely to be subjected to physical damage; (2) in wet or damp locations; (3) where subject to corrosive vapors; (4) in storage battery rooms; (5) in hazardous (classified) locations; (6) where concealed; (7) where extended through walls or partitions; (8) less than 5 ft above the finished floor except where protected from physical damage or track operating at less than 30 V rms open-circuit voltage.

Fittings identified for use on lighting track must be designed specifically for the track on which they are to be installed. They must be securely fastened to the track, maintain polarization and grounding, and must be designed to be suspended directly from the track.

Q30. *What are the design rules for track lighting?*

A. Lighting track is an assembly designed to support and energize lighting fixtures that can be moved along the track. For branch-circuit calculations, each 2 ft or fraction thereof of track is considered a 180-VA load. This limits the length of a 120-V, 20-A, single circuit track to 26 ft. This load calculation does not apply to track installed in dwelling units or the guest rooms of hotels and motels [see Sec. 220-12(b)]. Lighting track must be secured to the wall or ceiling in at least two places for lengths not exceeding 4 ft with additional supports at 4-ft intervals for longer track.

Q31. *What are the rules for underwater lighting?*

A. Article 680 says that lighting fixtures in walls of swimming pools must be installed so that the top of the fixture lens is not less than 18 in below the normal water level in the pool. They cannot be supplied from circuits operating at more than 150 V between conductors. And a ground-fault circuit interrupter is required in all branch circuits supplying fixtures that operate at more than 15 V.

Q32. *In an unfinished basement, would a porcelain light fixture with a receptacle built in be acceptable in meeting the Code?*

A.   No. Section 210-52(g) requires at least one receptacle outlet in addition to any provided for the laundry equipment in each basement. The plug in the light fixture is a lighting outlet. Section 210-8(a)(4) requires the receptacle in an unfinished basement to be protected with ground-fault protection for personnel.

Q33.   *What are the rules for emergency lighting?*

A.   Section 700-16 says that each internally illuminated exit sign must contain two or more light bulbs to prevent total darkness should one bulb burn out. All required egress lighting is subject to the same requirement, but single lamp fixtures closely spaced are acceptable, provided that the burning out of a bulb does not reduce the lighting level in the area to a value below the minimum required by the local Building Code or Life Safety Code (NFPA-101).

To reduce energy consumption, exit fixture conversion or retrofit kits are available to convert specific incandescent exit fixtures to fluorescent fixtures. For additional information on the application and installation restrictions that are placed on some lighting fixtures, see the 1994 edition of the white book published by UL which bears the title, "General Information for Electrical Construction, Hazardous Locations, and Electric Heating and Air Conditioning Equipment."

Q34.   *What are the general rules for locating lighting fixtures?*

A.   Section 410-4 gives the details. Fixtures installed in wet or damp locations must be so installed that water cannot enter or accumulate in wiring compartments, lampholders, or other electrical parts. All fixtures installed in wet locations must be marked, "Suitable for Wet Locations." All fixtures installed in damp locations must be marked, "Suitable for Wet Locations" or "Suitable for Damp Locations."

Installations underground or in concrete slabs or masonry in direct contact with the earth, and locations subject to saturation with water or other liquids, such as locations exposed to weather and unprotected, vehicle washing areas, and like locations, must be considered to be wet locations with respect to the above requirement.

Interior locations protected from weather but subject to moderate degrees of moisture, such as some basements, some barns, some cold-storage warehouses, and the like, the partially protected locations under canopies, marquees, roofed open porches, and the like, must be considered to be damp locations with respect to the above requirement.

Q35.   *How heavy can a bulb-supported fixture be?*

A.   Section 410-15 says that a fixture that weighs more than 6 lb or exceeds 16 in in any dimension must not be supported by the screw shell of a lampholder. Metal poles may be used to support lighting fixtures and enclose supply conductors, provided that a few access, rainproofing, and wiring requirements are met.

Q36.   *What is new to the 1999 NEC?*

A.   Several new requirements and improvements to the language:

■   All fixtures must be marked with the maximum watt requirement and the size of the supply wire [Sec. 410-35(a)].

■   Branch-circuit wiring must not pass through fixture outlet boxes unless the fixture is identified for through wiring. (This mimics a general rule that panelboards cannot be used as junction boxes.)

■   Type AC cable can be run in lengths up to 18 in when used as fixture tap conductors for lighting fixtures.

The subject of lighting will always be closest to the heart of architects and interior designers. The reader should consult the core references on 1999 NEC changes for comprehensive treatment of how changes in wiring safety practice will affect the way lighting systems are applied to new and/or existing space.

Solved Problem—Division 16500

SITUATION.

A 120- by 60-ft room requires bulk lighting branch circuits. A total of 70 luminaires is required, each with two 800-mA, 96-in, 113-W fluorescent lamps. From a manufacturer's catalog it is known that the ballast for each lighting unit is rated at 252 W with a line current of 0.92 A at 277 V.

REQUIREMENTS.

Determine the minimum number of branch circuits.

SOLUTION

1. The maximum allowable rating of each circuit is 20 A. The maximum loading of each circuit is 80 percent $\times$ 20 = 16 A.

2. Maximum number of units per circuit is 16/0.92 = 17.4 $\rightarrow$ 17.

3. Minimum number of circuits is 70/17 = 4.1 $\rightarrow$ 5.

The logical arrangement is one circuit for each row. The circuit loading is 14$\times$0.92=12.9 A.

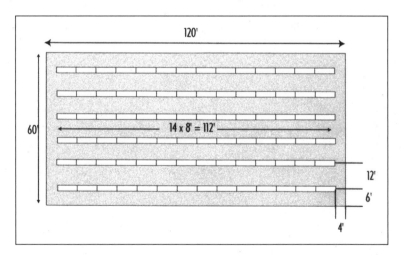

Figure 22.6 Layout for solved lighting problem.

REMARKS.

Bulk lighting at 277 V typically requires fewer branch circuits to wire the complete system.

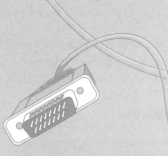

Division 16600—Lightning Protection

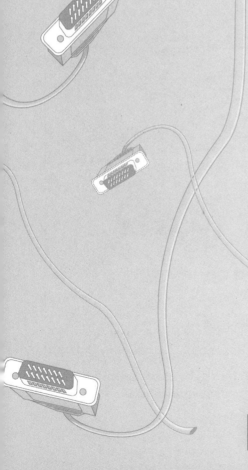

Q1.   *What* **NEC** *requirements commonly appear in Division 16600?*

A.   Article 280 contains the bulk of them, and very few changes have occurred in the 1999 **NEC** revision cycle. Related cross references appear in Arts. 230-82 and 230-209 and throughout Chap. 8. Section 800-13 requires a separation of at least 6 ft between communication wires and lightning conductors.

Factory Mutual Loss Prevention data sheets have been prepared on this subject, particularly on protecting cooling towers from lightning. NFPA-780 is a lightning protection code but, as a suggested design document, does not make enforceable requirements. It is worth noting that the **NEC** uses the term "surge," whereas the CSI uses the term "lightning" protection. Surge is a more general term that includes overvoltage phenomena associated with switching, among others.

Q2.   *How are the broad principles of lightning protection applied?*

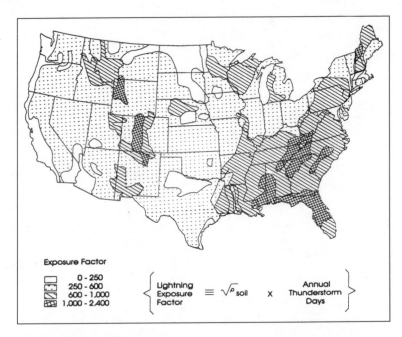

Figure 23.1 Lightning exposure factors. To use Fig. 23.1, you need to obtain the soil resistivity from Fig. 21.5. Lightning exposure factor = $\sqrt{\rho_{soil}} \times$ annual thunderstorm days.

A.   The basic idea is to provide a short-circuit path to ground for the lightning-induced current.

A lightning protection system is a system of multiple rooftop air (lightning) terminals, down conductors, equalizing conductors, and ground terminals that surround a building for the exclusive purpose of intercepting, diverting, and dissipating direct lightning strikes. Some systems are designed into the building structure so that the structural steel performs the down conductor and equalizing functions. Note that the zone of protection resembles the intercept of a parabola rather than a "cone." Section 250-106 cites a requirement to bond a lightning protection system to the grounding electrode system.

The 1999 **NEC** deletes the 6-ft side flash limits. Per Sec. 280-3, where used at a point on a circuit, *a surge arrester must be connected to each ungrounded conductor.* A single installation of such surge arresters can protect a number of interconnected circuits, provided that no circuit is exposed to surges while disconnected from the surge arresters.

Q3.   *How does lightning travel through a building?*

A.   Refer to Fig. 23.2.

Point 1: Assume a 210-kA lightning strike. The highest lightning

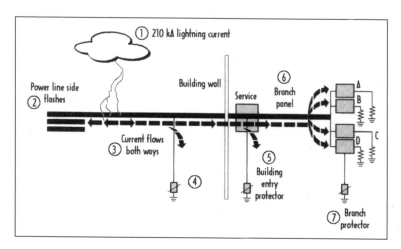

Figure 23.2 Building lightning protection system.

current occurs 0.5 percent of the time in high-exposure locations, so this is a near worst-case scenario.

Point 2: Power-line side flashes divide the 210 kA into three parts, 70 kA each.

Point 3: Current flows (in parallel) both ways, dividing 70 kA in half, 35 kA each way.

Point 4: The utility-owned distribution transformer arrester diverts 21 kA to ground. This assumes that the arrester has been specified with this capacity and that there is negligible ground resistance. A good preventive maintenance program by the local utility will ensure this.

Point 5: The customer-owned building entry protector diverts 10 kA to ground.

Point 6: At the branch panel 11 kA flows toward the branch protectors.

Point 7: The branch protector must be sized on the basis of the difference between the 11 kA and the tolerance of the utilization equipment.

The reader is referred to a more comprehensive treatment of this subject in Dion Neri, *EC&M Magazine,* March 1997.

Q4.   *How are surge arresters specified?*

A.   Section 280-4 says that on circuits of less than 1000 V the rating of the surge arrester must be equal to or greater than the maximum continuous *phase-to-ground* power frequency voltage available at the point of application. Surge arresters installed on circuits of less than 1000 V must be listed for the purpose. On circuits of 1 kV and over the rating of a silicon carbide–type surge arrester cannot be less than 125 percent of the maximum continuous phase-to-ground voltage available at the point of application.

Q5.   *What are the connecting details at the service entrance?*

A.   Section 280-21 says that line and ground connecting conductors cannot be smaller than No. 14 copper or No. 12 aluminum. The arrester grounding conductor must be connected to one of the following: (1) the grounded service conductor, (2) the grounding electrode conductor, (3) the grounding electrode for the service, or (4) the equipment grounding terminal in the service equipment. Similar rules apply to surge arresters installed on the load side services of less than 1000 V. The grounded conductor and the grounding conductor must be interconnected only by the normal operation of the surge arrester during a surge.

Q6. *Do underground distribution systems require lightning protection?*

A. It is a matter to be left to engineering judgment. At some point, underground circuits are joined with overhead power lines. Arresters should be installed at the junction point. Further, it is not uncommon for underground circuits to be struck by lightning, even though they may be 3 ft below grade. How much voltage gets to the service equipment depends upon the type of media, its reflectivity, and how many traveling waves get through the path. Refer to Fig. 23.3.

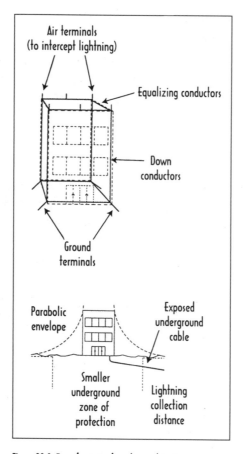

Figure 23.3 Zone of protection for underground circuits.

The building provides a zone of protection, but circuits outside the perimeter shown above—even though underground—can be damaged by lightning. Metal fences, installed underground shield conductors, and other conducting structures can affect this by diverting strikes away from or into protected zones.

Q7.   *Do the rules change for medium-voltage services?*

A.   Yes. Section 280-23 says that the conductor between the surge arrester and the line and surge arrester and the grounding connection cannot be smaller than No. 6 copper or aluminum. The grounding conductor of a surge arrester protecting a transformer that supplies a secondary distribution system must be interconnected as follows:

- A metallic interconnection must be made to the secondary grounded circuit conductor or the secondary circuit grounding conductor.

- Where the surge arrester grounding conductor is not connected but is otherwise grounded, an interconnection must be made through a spark gap.

- An interconnection of the surge arrester ground and the secondary neutral is permitted with permission from the AHJ.

As usual, some exceptions and special conditions apply. For example, surge arrester grounding connections must be built according to requirements stated in Art. 250. Grounding conductors cannot be run in metal enclosures unless bonded to both ends of such enclosure.

Division 16700—Communications

Remarks. The power industry is 250 percent larger than the telecommunications industry and, as such, is one of the largest customers of telecommunication companies in its use of telephony to control generation facilities, water supplies, transmission lines, and automated distribution switching stations. It wasn't very long ago when we thought of telecommunications and computers as different industries. Now the 15-A desktop computer has long since replaced the 1200-A mainframe installation, and many people are using their home computers to make long-distance telephone calls. There is a growing demand for communication with every piece of electrical equipment in the home or office. With electrotechnology continuing the covergence with the use of power lines as a communications medium, plain old telephone service will never be the same. It should be plain that the fun is just beginning.

Q1.   *What NEC topics commonly appear in Division 16700?*

A.   Fire alarms and related systems, telecommunication, data, public address, CATV, satellite, microwave, and other systems. Power consumed by these systems, while relatively low, still has the capability of producing enough heat to ignite combustible materials. These materials may be the insulation and related fittings of the transmission media itself. Thus, **NEC** requirements are concentrated upon which signal-carrying conductors can be run in the same raceway with other signal-carrying conductors. This is very much an economic optimization problem: How do you make a system economical to build while still retaining its safety? The ability to safely put all transmission media in the same raceway reduces costs substantially, but in many cases, it is not permitted because of the likelihood of fires that will produce toxic fumes. A related problem is the manner in which human beings interface with communication systems that are dedicated to life safety. One such arrangement is shown in Fig. 24.1.

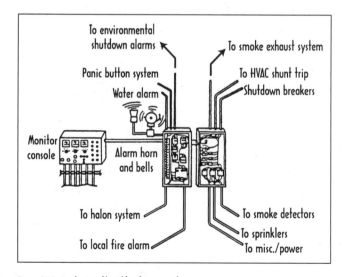

Figure 24.1 Coordination of basic life-safety system elements.

A single, readily accessible cabinet should be provided in the schematic design phase. With this approach, the requirements of a common interface for coordination of the various life-safety functions can be established. An emergency telephone with a separate line (not via PBX or switchboard) is needed near the exit and life safety indicators and controls. The audible alarm should not be so close to the telephone that it is impossible to listen and be heard.

Q2. *Where can I find communication-related requirements in the NEC?*

A. Chapter 8 is the first place to look. It has been overhauled because the existing **NEC** articles pertaining to telephony and cable television did not address significant issues that broadband applications present. You will find that new Art. 830 covers broadband technology that integrates cable television, telephony, and data transport, ultimately through the same transmission media. Article 645 covers information technology equipment. Chapter 7 contains requirements for signaling circuits that may or may not use telephony as a communication medium. The usual caveats with regard to scope apply: Chaps. 6, 7, and 8 material modifies the more general rules of Chaps. 1 through 4.

Much of the telecommunications standards are covered in UL 568 and EIA/TIA. In this text there will be some carryover from Art. 645; some issues will be covered in Division 13, Special Construction. We cover only the most common types of installations: fire alarms, telecommunications, power-limited circuits. Many signaling systems come as a listed assembly with substantial field installation.

Q3. *What is the difference between a signal, a remote control, and a power-limited circuit?*

A. A signal is any circuit which supplies energy to a device that gives a visual or audible signal. Examples of signal circuits are doorbells, buzzers to multiple dwelling units, code-call systems, signal lights, annunciators for fire or smoke detection, and fire or security alarms. A remote-control circuit is any circuit which has an operating coil as its load; with the operating coil controlling the opening and closing of the power-supply circuit to a (typically large) piece of utilization apparatus.

Common examples are a magnetic motor starter, a magnetic lighting or heating contactor, or a relay as part of a process control system. A "power-limited" circuit is a circuit that can operate as either a signal or a remote-control circuit but one in which the source of the energy supply is limited in its power to specified maximum levels. Low-voltage lighting that uses 12-V lamps in fixtures fed from 120/12-V transformers is a common example.

Common Terminology for Telecommunication Systems

Building Distribution Frame (BDF): The primary communication cable room in any building where cables external to the building meet cables internal to the building.

Communications Outlet: A connecting device located in a work area on which horizontal wiring system cable terminates and which can receive a mating connector.

Crosstalk: The introduction of signals between conductors in proximity to each other as a result of electromagnetic flux lines that are caused by the signal currents flowing in a conductor. The phenomenon results in the telephone conversation on an adjacent wire being faintly audible. Crosstalk is tolerable in plain old telephone service but devastating to digital signals.

Horizontal Twisted Pair Communication Cable: Low-voltage twisted pair copper communication cable placed from a wire concentration point (usually a local distribution frame room within a building) to a user location (a communications outlet usually located on the *same floor* as the LDF) for the purpose of providing the physical medium from a wire concentration point to a specific user for voice, data, and video services.

Local distribution frame (LDF): The secondary communications cables room(s) in a building where horizontal cables from individual user locations are terminated. LDFs typically service only the floor or part of the floor on which they are located. The total length of horizontal cable from the user location (communications outlet) to the LDF may not exceed 90 m.

RS232: The RS stands for "Recommended Standard," and it should be now indicated as EIA/TIA 2323-E, based upon definitions by Bell Telephone. It defines 1 as a negative voltage from +3 to −3 V.

Q4.   *In what ways do the general circuit design and construction rules established in the **NEC** differ for these kinds of circuits?*

A.   While all the applications under Art. 725 must follow the general power and light wiring methods of Chap. 3 of the **NEC** there are

several and they are not minor: overcurrent protection, conductor
selection and raceway sharing, mechanical protection. Wiring
must be done in a workmanlike fashion that does not invalidate
the fire rating of floors, ceilings, or wall penetrations. The wiring
methods for signaling circuits refer heavily back to the first principles of wiring methods in Chap. 3, largely because signaling circuits are hidden in plenums, suspended ceilings, and air-handling
spaces. That's where you want fire safety apparatus. In general
metal-clad cable is used in spaces where air is moving and at any
temperature, high or low. Exceptions exist for manufactured specialties known as plenum or duct rated. The overcurrent protection for class 1 conductors No. 14 remains in agreement with the
general ampacity determination rules of Sec. 310-15. The number
of class 1 conductors permitted in a raceway are the same as permitted in the fill percentages of Chap. 9, Table 1.

Q5.   *Are programmable logic controller circuits and power conductors
permitted in the same raceway?*

A.   There is an apparent contradiction. Class 1 conductors are permitted in the same raceway with other conductors if all of the
conductors are insulated for the maximum voltage of any conductor in the raceway. Class 1 conductors are permitted in the
same raceway with other system conductors only if the equipment being powered is functionally similar, for example, a group
of class 1 remote-control circuits installed in a wireway with
motor power conductors for the operation of a process. If one
section of the process motors develops a problem, the whole
process is affected. Should the different systems not be functionally associated, the installation of the class 1 conductors with the
other systems would violate this section, regardless of the voltage
rating of the conductor insulation. An exception is provided to
allow the separate systems in a factory- or field-assembled control center. Sometimes such systems are built incrementally.

Q6.   *Can class 2 and class 3 be built in the same raceway with power
circuits?*

A.   Not unless the different circuits are separated by a barrier, which
is sometimes available in wireways and always available in cable
tray. This barrier is not necessary if the circuits are in sheathed

cable assemblies. When you cannot get them in sheathed cable, there must be at least 2-in separation within the compartment or enclosure and the power conductors. You cannot even install class 1 with class 2 and class 3 conductors unless you build this barrier.

The performance characteristics of phone and data cable are divided into five categories, 1 through 5, in order of increasing transmission speed (Table 24.1).

TABLE 24.1 CATEGORIES OF COMMUNICATION CABLES

Category number	Service requirement
1	Voice and low speed data up to 20 kbits
2	Voice and data transmissions up to 1 Mbits/s
3	Voice and data transmissions up to 10 Mbits/s
4	Voice and data transmissions up to 16 Mbits/s
5	Voice and data transmissions up to 100 Mbits/s

The performance characteristics of phone and data cable are divided into five categories, with category 1 having the lowest transmission speed requirements and category 5 having the highest. UL will change its term "level" to "category" in the near future.

Fire Alarm System

Q7. *What are the basic types of fire alarm systems covered by the NEC?*

A. Fairly simple: power-limited and non-power-limited. When the class 1 alarm system is driven at no more than 1000 VA at 30 V, it is called a power-limited system. Some commercial bank or nurse call stations in health-care facilities operate at 30 V or below. When the class 1 alarm system is non-power-limited, it can be driven at any voltage up to 600 V. A remote-control circuit to a lighting contactor or motor starter is an example of a class 1 non-power-limited circuit. Class 2 circuits operate at even lower voltages than class 1 circuits so that these circuits may be serviced "hot" by virtue of their protection from electric shock. Class 3 circuits also operate at low levels of voltage and current but require protection from shock. Power sources may be provided by a transformer or from a dry cell battery.

Q8.   *What should we expect an NEC compliant fire alarm system to do?*

A.   Local building codes may well assert requirements much in harmony with the NFPA standards with respect to compartmentalization, fireproofing, sprinkler systems and smoke control to provide adequate fire protection. Fire alarm systems should provide early detection, accurate location of the alarm origin, fire department notification, and automatic control of the HVAC system, elevators, and other building systems that are necessary to make the building safer for its occupants. Fire alarm initiation devices include smoke and heat detectors, manual pull stations, water flow and tamper switches, and other fire-suppression systems. These devices are usually arranged by zone and monitored by a control panel. The control panel operates numerous devices including alarms, door holders, fan motor shutdown, elevator recall, and smoke control systems. Systems for high-rise buildings may include voice communication in stairwells, and fan and elevator override controls for fire fighters. Local signaling is provided by various types of annunciators. Automatic reporting to a listed monitoring agency or municipal fire department is recommended and often required.

These requirements are explained in full in NFPA 72, Protective Signaling Systems. The key provisions of this standard are as follows:

- Fire alarm systems must have two independent, reliable power supplies.

- In the home, smoke detectors are required outside each separate sleeping area and on each level of the living unit, including basements.

- In new construction, smoke detectors must be placed in each sleeping room. All detectors must be interconnected so that when one detector activates, all detectors sound the alarm.

- Fire alarm systems must produce an alarm signal that is different from all other signals. This signal may not be used for other purposes.

- The minimum visual signal appliance rating for daytime (nonsleeping) areas is 15 candela.

- Testing and maintenance of fire alarm systems is the responsibility of the property owner.

- Smoke detectors require an annual functional test and periodic sensitivity testing.

These standards originate in NFPA activity dating back to 1898. It establishes a basic framework for safety which may or may not be modified by local jurisdictions.

Q9. *What other standards apply to fire alarm systems?*

A. Quite a few. Starting with NFPA 70, The National Electric Code, there is:

NFPA 71—Central Station Signaling Systems—Protected Premises Unit
NFPA 72—Protective Signaling Systems
NFPA 72B—Auxiliary Protective Signaling Systems
NFPA 72C—Remote Station Protective Signaling Systems
NFPA 72D—Proprietary Protective Signaling Systems
NFPA 72E—Automatic Fire Detectors
NFPA 73—Public Fire Service Communications
NFPA 74—Household Fire Warning Equipment

In addition to these, there are UL 864, Control Units for Fire Protective Signaling Systems; UL 268, Smoke Detectors for Fire Protective Signaling Systems; UL268A, Smoke Detectors for Duct Applications; UL 217, Smoke Detectors, Single and Multiple Station; UL 521, Heat Detectors for Fire Protective Signaling Systems; UL 228, Door Closers-Holders for Fire Protective Signaling Systems; UL 464, Audible Signaling Appliances; UL 1638, Visual Signaling Appliances; UL 38, Manually Actuated Signaling Boxes; UL 346, Waterflow Indicators for Fire Protective Signaling Systems; UL 1481, Power Supplies for Fire Protective Signaling Systems. Some specifications will ask that all fire alarm control equipment be listed under UL Category UOJZ as a single control unit, with partial listings not acceptable.

Q10. Can a fire alarm system extend beyond one building?

A. Yes. Section 760-7 says fire alarm circuits that extend beyond one building must either meet the requirements of Art. 800 and be classified as communications circuits or must meet the requirements of Art. 225. One way to determine accepted industry practice is to refer to nationally recognized standards such as Commercial Building Telecommunications Wiring Standard, ANSI/EIA/TIA 568-1991; Commercial Building Standard for Telecommunications Pathways and Spaces, ANSI/EIA/TIA 569-1990; and Residential and Light Commercial Telecommunications Wiring Standard, ANSI/EIA/TIA 570-1991.

Q11. What are the basics for specifying smoke detectors in health facilities?

A. NFPA Standard 101, Life Safety Code, is the primary national standard addressing corridor smoke detectors in health-care facilities. Section 713-10 of the 1997 Uniform Building Code is another, related standard. The UBC requires a smoke detector in each patient sleeping room with annunciation at the local nurse station.

Since fire and health codes ensure that no area will be far from a corridor, all spaces in a health-care facility will be near a corridor. This standard requires, for example, that duct detectors be provided within 5 ft of smoke dampers. This requirement can be eliminated when dampers are located in a corridor wall or ceiling where the corridor is protected by a smoke-detection system.

If sprinkler systems are contemplated, consider the effect of an accidental discharge upon patients, upon ICUs and the expensive equipment typically found therein. There are various smoke-detection methods that should be considered and they may well be mandatory by the AHJ. Detectors selected for use in health-care facilities should be listed under the applicable edition of UL-268.

Telecommunication and Data Systems

Q12.   *How do you apply the general principles of the NEC to telecommunication and data systems?*

A.   The general principles of Sec. 110-3(b) are extended to telecommunications systems as indicated in the early sections of Art. 800. The notion of service, feeder, and branch circuit in the power industry bears some resemblance to the notion of network access point, local area network, and standard input/output unit in the telecommunications industry. Refer to Figs. 24.2 and 24.3. A few of the generic requirements are as follows:

- Equipment intended to be electrically connected to a telecommunications network must be listed for the purpose.

- Access to this equipment must not be denied by an accumulation of wires that prevents removal of panels.

- Equipment must be installed in a neat and workmanlike manner.

- Cables must be supported by the building structure so that the cable will not be damaged by normal building use.

- Equipment installed in a location that is classified locations must comply with the applicable requirements of NEC Chap. 5.

There are several terms that the telecommunications industry uses as an equivalent "service entrance" in the power industry. These may be referred to as *network access point, demarcation point,* or *point of minimum presence.* Each of these terms describes the beginning of the building distribution system. This is the point at which the telephone company cable first enters the premises and is terminated on cross-connect blocks. Figure 24.2 shows the principal components. The local telephone company will provide the service entrance cable, the primary protector, and the registered network interface block. All other components are provided by the owner.

New to the 1999 NEC is a controversial requirement that an underground communications entry be located no more than

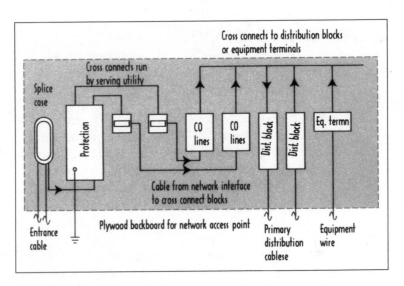

Figure 24.2 Network access point.

20 ft from the electric service. If this is impractical, a separate electrode can be used provided it is bonded to the power service. See Sec. 800-11.

In Fig. 24.3 we see that the layout of a multiple story building consists of the vertical distribution of various types of services over riser cables between the minimum point of presence and the distribution points on each floor. This network of riser primary distribution consists of riser cables connecting the apparatus closet of each floor with the minimum point of presence. From the apparatus closets, house distribution cables will feed tandem satellite closets.

NEC requirements are broken down into communications wiring on the exterior (Part B) and in the interior of the building (Part E).

*Q13. What requirements, if any, has the **NEC** adopted from the TIA/EIA and/or UL?*

A.   There is no shortage of standards in a rapidly developing industry. All of the following UL and EIA/TIA standards are referenced as fine-print notes. This means that they are not mandatory per the **NEC**, but compliance with them may be mandatory in a given construction contract:

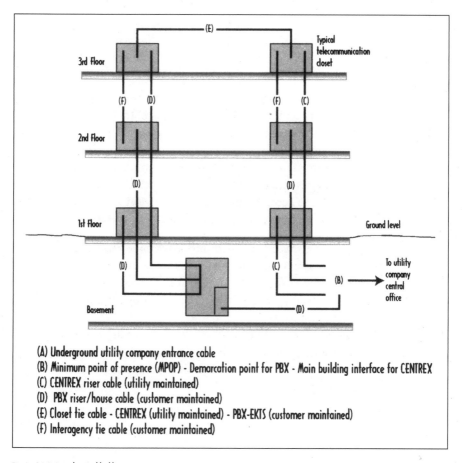

Figure 24.3 Typical vertical building system. (*Adapted from Maybin, Low Voltage Wiring, McGraw-Hill*)

1. Commercial Building Telecommunications Wiring Standard, ANSI/EIA/TIA 568

2. Commercial Building Standard for Telecommunications Pathways and Spaces, ANSI/EIA/TIA 569-1990

3. Residential and Light Commercial Telecommunications Wiring Standard, ANSI/EIA/TIA 570-1991

4. Standard for Telephone Equipment, UL 1459-1987, or Communication Circuit Accessories, UL 1863-1990

5. Standard for Protectors for Communications Circuits, ANSI/UL 497 (for primary protectors) and UL 497A (for secondary protectors)

Each of these standards refers to general wiring and ground methods described in Chaps. 2 and 3 of the **NEC.**

Q14.   *What are the working space rules for communications equipment?*

A.   In general the working clearances are the same as provided in Sec. 110-26 except as modified in Chap. 8. Table 24.3 is a recommended size based upon UL 569. These recommendations are modified somewhat by Sec. 800-5, which says that access to equipment must not be denied by an accumulation of wires and cables that prevents removal of panels, including suspended ceiling panels. It is accepted practice to provide 120-V power, lights, and service receptacles to all communication closets.

Usable ceiling space is divided into areas or zones for cables (Fig. 24.4). Cables may be pulled through a rigidly supported conduit from a nearby closet to the center of each zone. When plenum cable is used, conduit is not required, although it may be used to provide mechanical separation and protection. From the center of the zone, cables are run to nearby walls or utility columns and then to the floor below or above. At the center of the zone, cable terminates in an adapter that converts a large cable to a small one that terminates at each UIO.

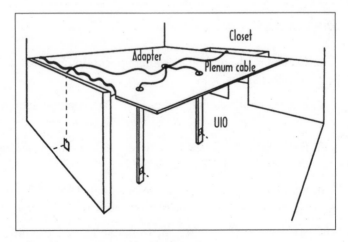

Figure 24.4 Zone method for horizontal overhead distribution.

Plenum cable runs directly from the equipment room or closet to the desired UIO (Fig. 24.5). This method is relatively economical and offers the most flexibility for distributing cables and wires in a ceiling. It also eliminates the possibility of signal interference caused by mixing analog and digital signals in the same cable sheath.

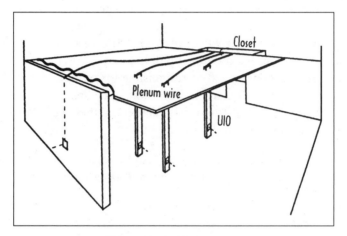

Figure 24.5 Home run method for horizontal overhead distribution.

Q15.   *What is good installation practice for communication system raceway and boxes?*

A.   The NECA installation standards generally follow BICSI and the **NEC**.

- Pull boxes must be installed on all runs over 90 ft in length.

- All empty conduit must have a fish tape installed.

- All horizontal cables must be placed in conduits and/or cable trays and left unterminated at the ends.

- All cables must be neatly pressed and coiled at the communications closet end. The length of slack cable at the closet end must equal the perimeter of the closet.

- The length of cable at the outlet end must be 12 in and coiled neatly in the outlet box.

All pretty commonsense stuff here. **NEC** Sec. 830-57 requires that if bends are made the bend should not damage it. This is particularly necessary for fiber bends which, unlike power cable which can be bent a total of 360°, can only bend *180°*.

Q16.   *Who builds the backboard and how do we build it?*

A.   In some specifications, contractors need only supply a backboard, an empty 1-in conduit and/or wireway. Some engineers will ask that no conduit run have more than the equivalent of one and one-half 90° bends(135°). If necessary to keep within this limitation, additional pull boxes may be installed. The telephone company will be responsible for termination of the cable at the station and at the LDF room. Backboards are typically plywood, 4 × 8 ft, 3/4 in thick. Sometimes these boards are supplied by the architectural trades.

It is not a bad idea to have a (land-line) telephone permanently installed in the main telecommunication room in order to assist service personnel.

Q17.   *What do I need to know about new Art. 830?*

A.   There are three power levels: low, medium, and high. A new table 830-4 defines them. The high-powered circuits are limited to 150 V, but they can carry up to 15 A (2250 VA) total. Such systems provide power to large buildings housing a number of repeaters, hubs, and switches. A disconnecting means is required on high-power circuits similar in location to service disconnects. A primary protector must be located near the point of entry. This assures that the shield will be grounded. The NIU, the network interface units, will be grounded such that an appreciable fraction of the return current on the shield will return over the pathways established by building steel, grounded service raceway, and the like.

Since it is so frequently the case that the telephone company will install the wiring, designers and electrical contractors need only design and install conduit for these systems and provide access to a ground. The number of riser sleeves for telephone service based upon total usable floor area may be estimated from this table. Keep in mind that this is only for telephone service. Where there will be integrated communication facilities, the overall space requirements may be more than double these figures. (Refer to Table 24.2.)

TABLE 24.2 RISER SLEEVES OR CONDUITS FOR TELEPHONE SERVICE

Total usable floor area to be served in thousands of sq ft	Number of sleeves or conduits (minimum size 4 in)
0–50	2
50–100	3
100–200	4
200–400	6

For every additional 200,000 sq ft or part thereof, add two additional sleeves or conduits.

Q18.   *Can communication circuits share the same raceway with power conductors within building interiors?*

A.   In general, No. The voltage difference is too great. The insulating and conduit material requirements may differ enough to make this practice uneconomical, if not unsafe. Section 800-52, however, contains a few exceptions to the general rule as long as specific separation requirements are met for the circuit type and cable class involved. There may be local modification of this rule, however. Some jurisdictions will not allow power and telecommunication cable in the same conduit but will allow them in shared boxes. Separation distances between power and telecommunication lines are given in Table 24.4.

Closets are part of the support structure for communication systems (Fig. 24.6). They may be arranged in a hierarchy with satellite closets, main closets, and central building telecommunications rooms. They should be located one identically above the other in a stacked arrangement. They should have enough working clearances for additional (or larger) equipment and have sufficient extra conduit to allow changes to occur without interrupting continuity of service.

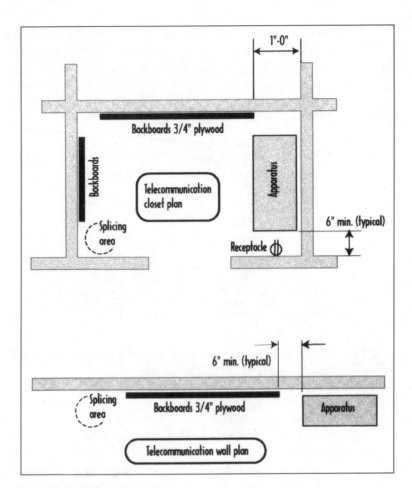

Figure 24.6 Typical apparatus closet layout.

TIA/EIA 569 recommends at least one telecommunications closet per floor in a building. All such closets should be dedicated to telecommunications and not shared with electrical power systems. Table 24.3 suggests an ideal correlation between the square foot area of the usable space and the size of the closet.

Q19.   *What are the rules for communications wiring in pipe chases?*

A.   Article 725 says that communications wires and cables run in the same shaft with conductors of electric light or power, class 1, or non-power-limited fire alarm circuits must be separated from

TABLE 24.3 CLOSET SIZE

Serving area		Closet size	
Sq m	Sq ft	mm	ft
1000	10,000	3000 × 3400	10 × 11
800	8000	3000 × 2800	10 × 9
500	5000	3000 × 2200	10 × 7

The 569 standard provides telecommunications closet sizing guidelines, based on the size of the floor the closet is intended to serve.

light or power conductors, Class 1, or non-power-limited fire alarm circuit conductors by not less than 2 in.

Q20.   *What are the general rules for installation of communications wiring within the building?*

A.   The generic rules appear in **NEC** Sec. 800-52, where there is considerable concentration on fire safety. Installers must follow some reasonable rules for separation from other conductors, keep the firestops intact, and provide mechanical support of the raceway.

The rules differ according to the type of wiring involved: twisted-pair, coaxial, or fiber optic. Copper (coaxial or twisted-pair) should be supported at least 2 ft from power cables unless the power cables are enclosed in conduit. Fiber-optic cables are permitted to be included in current-carrying cables as long as the voltage is less than 600 V. The metal current-carrying members of any fiber-optic cables should be grounded. Whenever possible, twisted-pair and coaxial cable should be run perpendicular to power cables.

Q21.   *Under what conditions will the **NEC** allow communications circuits to share the same raceway?*

A.   Only if they are power-limited circuits. Section 800-52 says that communications conductors must not be placed in any raceway, compartment, outlet box, junction box, or similar fitting with conductors of electric light or power circuits or class 1 circuits.

As always, however, there are exceptions, two rather common ones: where the power conductors are separated from the communications circuits by a barrier (some manufacturers have listed cable tray that complies under this exception), where the power conductors are introduced solely for power supply to communications equipment or for connection to remote-control equipment. The electric light or power, class 1, or non-power-limited fire alarm circuit conductors must be routed within the enclosure to maintain a minimum of 0.25 in separation from the communications circuit conductors. These two exceptions apply to conductors of electric light or power, class 1, or non-power-limited fire alarm circuits as well.

Q22.   *Can power raceway be used as "messenger" support for telecommunication circuits?*

A.   No. The prohibition against this for communications cables installed within buildings appears in Sec. 800-52(e). Communications cables or wires must not be strapped, taped, or attached by any means to the exterior of any conduit or raceway as a means of support. (Refer to Fig. 24.7.)

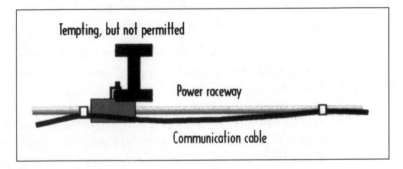

Figure 24.7

Communications cables are not permitted to use raceways as a means of support.

Q23.   *Why are there no voltage markings on communications cables?*

A.   Voltage markings on cables may be misinterpreted to suggest that the cables may be suitable for class 1, electric light, and power applications. Communications wires and cables installed as wiring within buildings must be listed as being suitable for the purpose, marked in accordance with Table 800-50, and installed in accordance with Sec. 800-52.

Q24.   *What is a primary protector and how does the **NEC** require that it be applied?*

A.   There are two basic types: fused and fuseless. (Refer to Fig. 24.8.) The air gap protector consists of carbon blocks inside a metal housing with a threaded cap. When the voltage on the protected conductor exceeds a predetermined value—typically 300 V—it is limited by arcing in the air gap between the carbon blocks. If the voltage is transient in nature, the protector returns to an open-circuit condition. When the voltage persists above 300 V, arcing continues long enough to melt a lead alloy spacer inside the units on the spring-loaded carbon block, causing it to move up against the other carbon block and establish a direct path to ground. The gas tube protector consists of a surge protector and a fusible disk, mounted inside a housing with a threaded brass cap. It functions like the air gap protector, except that it uses two metallic electrodes sealed in a glass envelope filled with an inert gas. This is one of the reasons communications cables are listed at 300 V.

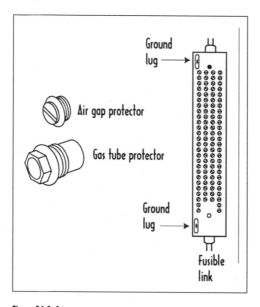

Figure 24.8 Primary protector.

Primary protectors, also known as voltage surge limiters, control the magnitude of high-voltage surges on building wiring. Their operating principle resembles that of varisters (or variable resistors) that are almost always applied on overhead power distribution lines. Figure 24.8 shows that these devices are resin-filled blocks embedded with an array of screw-in air gap or a gas tube elements, two ground lugs, and a 26-gauge input stub cable that serves as a fusible link. Primary protector terminals must be marked to indicate line and ground.

TABLE 24.4 SEPARATION OF TELECOMMUNICATIONS PATHWAYS FROM ≤480-V POWER LINES

Condition	Minimum separation distance		
	<2 kVa	2–5 kVA	>5 kVA
Unshielded power lines or electrical equipment in proximity to open or nonmetal pathways	127 mm (5 in)	305 mm (12 in)	610 mm (24 in)
Unshielded power lines or electrical equipment in proximity to a grounded metal conduit pathway	64 mm (2.5 in)	152 mm (6 in)	305 mm (12 in)
Power lines enclosed in a grounded metal conduit (or equivalent shielding) in proximity to a grounded metal conduit pathway	— —	76 mm (3 in)	152 mm (6 in)

The 569 standard indicates minimum separation distances between telecommunications pathways and power wiring to counter unwanted EMI.

Q25. *Where does the NEC say the primary protector should be applied?*

A.   In every facility where there is exposure to lightning or likelihood of a power line falling on a communication line, the primary protector must be located as practicable to the point at which the exposed conductors enter or attach to the owner's premises. (Per Sec. 800-2, an exposed circuit is one that, in the case of a failure of insulators, standoffs, or other support, can come into contact with another circuit.)

Section 800-30 says that a listed primary protector must be provided on each aerial circuit where accidental contact with electric light or power conductors is possible. In addition, where there exists a lightning exposure, each interbuilding circuit on a premises must be protected by a listed primary protector at each end of the interbuilding circuit.

Related material appears in Chap. 23.

Q26.   *What is a secondary protector?*

A.   It protects building interior apparatus that is not protected by the primary protector. Section 800-32 says that a secondary protector must provide means to safely limit currents to less than the current-carrying capacity of listed indoor communications wire and cable, listed telephone set line cords, and listed communications terminal equipment having ports for external wire line communications circuits. Any overvoltage protection, arresters, or grounding connection must be connected on the equipment terminals side of the secondary protector current-limiting means. Where a secondary protector is installed in series with the indoor communications wire and cable between the primary protector and the equipment, it must be listed for the purpose.

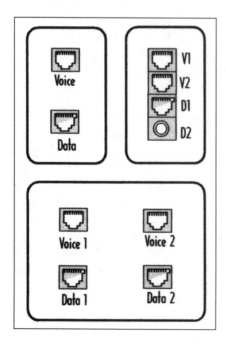

Figure 24.9
Multijack SIO units.

Q27.   *Is the telecommunication system ground the same as the power system ground?*

A.   The equipment ground for the power and the telecommunication system is the same, but not the power *system* ground. The building electrode system is described in Art. 250. There must be an identifiable common grounding electrode which is used as the voltage sink for all enclosures and equipment in or on the building. Figure 24.10 shows an arrangement that meets minimum requirements.

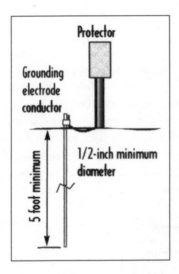

Figure 24.10 Communications grounding electrode. In the absence of any other grounding method listed in Sec. 800-40, the grounding electrode for communication equipment is permitted to be a pipe or rod not less than ½-in diameter and a minimum of 5 ft in length as shown here.

The telephone *switching ground,* however, is specified by the supplier of the switching or data equipment. The equipment room ground, which is the reference ground used to ground the telephone and data cables, should be installed according with the equipment manufacturer's instructions. For high-rise buildings the preferred cable is the telephone house cables. These cables are constructed with a corrugated aluminum or copper shield that services as the grounding conductor to the upper floors. If corrugated cable shields are not used, a No. 6 AWG or larger copper wires must be run parallel to the house cable to provide riser ground points. Grounding for communication equipment should be provided by an independent ground riser from the building's electrode network. It should terminate on an isolated ground bus and be sized on at least 2 kcmil ft of cable run. Single- or multiple-point grounding, or both, may be used depending on the communication equipment's frequency characteristics.

Q28.   *What listing agencies are relevant to new* **NEC** *Art. 830?*

A.   LAN cable performance testing agencies. They should evaluate local area network cable to industry performance specifications. Cable should be verified to one of five performance categories, based on recognized communications industry standards for performance, including Telecommunications Industry Association/Electronic Industries Association (TIA/EIA) standards. UL publishes a semiannual directory of all wire and cable products verified through either of these programs. A related program covers electrical connecting hardware for compatibility with LAN cable.

Effect of information system equipment on power systems.

Computers and telecommunications used to be in different rooms. One needed cooling, the other surge protection. Now each needs both. In addition, the cumulative effect of many low-voltage ac to dc power supplies indigenous to information system equipment poses a unique problem for power system designers. The harmonics generated by these power supplies generate high neutral currents. Supply conductors may need to be oversized by about one-third even though the current read by 50- to 60-Hz meters will not indicate it. Bus bars in panels and the supply transformers may need to be oversized.

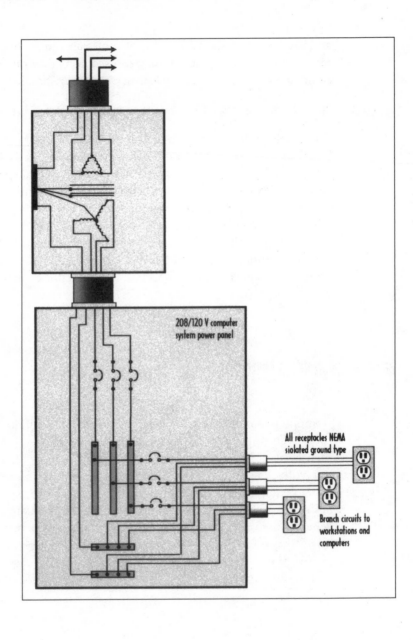

208/120 V computer system power panel

All receptacles NEMA siolated ground type

Branch circuits to workstations and computers

Division 16800—
Electric Heating Cables

Remarks. The Energy Policy Act of 1992 requires states to enforce commercial building energy efficiency codes that are at least as stringent as ASHRAE Standard 90.1-1989 or its successors. Since 1989, the ASHRAE 90.1 Project Committee has worked on revisions to Standard 90.1-1989. In 1996, ASHRAE published a draft for public review that received approximately 20,000 comments—most of which opposed ASHRAE's attempt to impose additional construction costs on buildings that used electric heat vis-à-vis all other forms of heating. In response to these public comments, ASHRAE removed the offending provisions and published the standard for a second public review in December 1997.

So goes the Edison Electric Institute's side of the story. It should be plain that energy policy will continue to be politically charged. Electric heating has its place, and politics aside, the National Electric Code is vital in its safe and economical application.

Q1.   *What **NEC** requirements commonly appear in Division 16800?*

A.   Electric heating is covered in Art. 424. Almost all of the **NEC** basics regarding workspace, disconnects, conductor sizing, grounding, and overcurrent protection that appear in the first three chapters of the **NEC** remain intact. The types of equipment covered include heating cable, unit heaters, boilers, central systems, or other approved fixed electric space-heating equipment. Article 424 does not apply to process heating and room air conditioning.

Q2.   *How do we keep electric baseboard heaters from thermally damaging appliances that need receptacle power?*

A.   The manufacturers have got a work-around for this. The Code permits permanently installed electric baseboard heaters equipped with factory-installed receptacle outlets, or outlets provided as a separate listed assembly, in lieu of a receptacle outlet(s) that is required by Sec. 210-50(b). Such receptacle outlets cannot be connected to the heater circuits. Listed baseboard heaters include instructions that may not permit their installation below receptacle outlets.

Q3.   *What are the rules for designing electric heating supply branch circuits?*

A.   Individual branch circuits are permitted to supply any size fixed electric space-heating equipment. Branch circuits supplying two or more outlets for fixed electric space-heating equipment must be rated 15, 20, or 30 A. (In nondwelling occupancies, fixed infrared heating equipment can be supplied from branch circuits rated not over 50 A.) The ampacity of the branch-circuit conductors and the rating or setting of overcurrent protective devices supplying fixed electric space-heating equipment consisting of resistance elements with or without a motor cannot be less than 125 percent of the total load of the motors and the heaters. See the Solved Problem at the end of this chapter.

Q4.   *What protective devices must be provided for snow-melting cables?*

A.   Section 426-28 says that all resistance snow-melting cables require supplementary protection against low-level arcing ground faults, whether located at dwelling or at nondwelling occupancies. This is not GFCI protection, however. GFCI devices are residual current devices that operate in the same way as ground fault protection equipment (GFPE) devices. A GFCI merely provides enhanced GFPE protection. Therefore, a GFCI could be used to establish GFPE; though, it may well nuisance trip and shouldn't be used except in very limited applications.

Q5.   *What are the basic principles to apply to heat-tracing equipment for water pipes in a residence?*

A.   Whether the piping is grounded or not, the tracing cables need their own grounded metal covering. GFPE needs to be provided so a fault to the metal shield will be detected and interrupted. The exception for an alarm indicator applies only to industrial occupancies with qualified maintenance and supervision.

Q6.   *How shall we install heating cables on dry board, in plaster, and on concrete ceilings?*

A.   Section 424-41 lays out all the rules that will be important to the architectural trades as well:

■   Cables must not be installed in walls.

■   Adjacent runs of cable not exceeding $2^3/4$ W per ft must be installed not less than $1^1/2$ in on centers.

■   Heating cables must be applied only to gypsum board, plaster lath, or other fire-resistant material. With metal lath or other electrically conductive surfaces, a coat of plaster must be applied to completely separate the metal lath or conductive surface from the cable.

■   All heating cables, the splice between the heating cable and nonheating leads, and 3 in minimum of the nonheating lead at the splice must be embedded in plaster or dry board in the same manner as the heating cable.

■   The entire ceiling surface must have a finish of thermally noninsulating sand plaster having a nominal thickness of

$1/2$ in or other noninsulating material identified as suitable for this use and applied according to specified thickness and directions.

- Cables must be secured at intervals not exceeding 16 in by means of approved stapling, tape, plaster, nonmetallic spreaders, or other approved means.

- In dry board installations, the entire ceiling below the heating cable must be covered with gypsum board not exceeding $1/2$ in thickness.

- The void between the upper layer of gypsum board, plaster lath, or other fire-resistant material and the surface layer of gypsum board must be completely filled with thermally conductive, nonshrinking plaster of other approved material of equivalent thermal conductivity.

- Cables must be kept free from contact with metal or other electrically conductive surfaces.

- In dry board applications, cable must be installed parallel to the joist, leaving a clear space centered under the joist of $2^{1}/2$ in (width) between centers of adjacent runs of cable. A surface layer of gypsum board must be mounted so that the nails or other fasteners do not pierce the heating cable.

- Cables must cross joists only at the ends of the room.

Cables must not be installed in walls except that isolated single runs of cable may be run down a vertical surface to reach a dropped ceiling (Fig. 25.1).

Solved Problem—Division 16800

SITUATION.

A 240-V, single-phase, fixed electric space heater has a nameplate rating of 20 kW. The fractional-horsepower blower motor requires 3 A, 120 V.

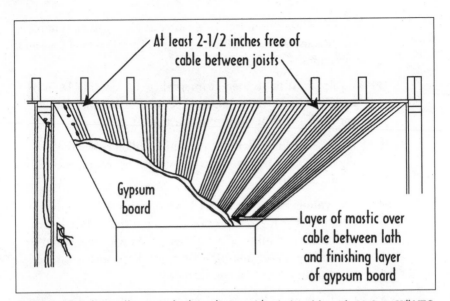

At least 2-1/2 inches free of
cable between joists

Gypsum
board

Layer of mastic over
cable between lath
and finishing layer
of gypsum board

Figure 25.1 Installation of Heating Cables on Dry Board, in Plaster, and on Concrete Ceilings. (*Adapted from The McGraw-Hill NEC Handbook.*)

REQUIREMENTS.

Select supply circuit elements.

SOLUTION.

1. Determine full-load amperes of heater [Sec. 424-3(b)]
   (20 kW × 1000)/240 V = 83 A

2. Determine total load (83 A + 3 A) × 125% = 107.5 A

3. Select conductors per Table 310-16. Select No. 2 THWN Cu.

4. Per Secs. 240-6(a) and 240-3(b) select next higher size above
   107.5 A. Select 125 A.

REMARKS.

The circuit breaker may serve as the disconnect if it can be locked in the
open position, per Sec. 422-33. When the elements in resistance-type heating
units exceed 48 A, they must be subdivided by the manufacturer.

Division 16900—Testing and Acceptance

*Remarks. Up until the early 1970s the **NEC** contained information about electrical acceptance testing. We present here a few of the most important acceptance testing tables to provide you with some sense of what tests are necessary, which are optional, and what to look for in test results. The retainer of a NETA certified testing contractor will ensure that the tests will be conducted appropriately. In many specifications, this division contains a requirement that the owner's representative and, in some cases, the AHJ, be present to witness the tests.*

Q1.   *What NEC requirements typically appear in Division 16900?*

A.   The generic requirements for the use of listed equipment—installed in accordance with manufacturer requirements—is the most direct linkage between the **NEC** and this CSI division. In most variants of the CSI *Masterspec©* format, acceptance testing is labeled Division 16950.

The inspection and testing performed on an electrical power system and its components prior to initial energization is perhaps the most important preventive maintenance that a given system and/or components will receive during their operational life. Acceptance tests are valuable in several different ways:

- They are used to determine whether the system or equipment is in compliance with the purchase specifications and design intent, and to verify that the equipment has been installed properly.

- They are used to determine whether equipment has been subjected to damage during shipment or installation. For example, if the conditions of construction required that a wall be replaced in order to install a large item of electrical apparatus, you would want to have the electrical equipment tested before you rebuilt the wall.

- They establish benchmarks which can be referred to during subsequent tests performed over the life of the equipment.

When failures of components of power systems fail, the bulk of them fail during the first year of operation. Similarly, there are many unnecessary outages during the initial period of operation of a power system due to a variety of human errors. These failures and outages can be significantly reduced by means of a comprehensive program of acceptance inspection and testing prior to the initial system energization.

Q2.   *What should we look out for when a new item of electrical apparatus arrives?*

A.   At the very least, a good specification should include some language—a punchlist—about how the electrical apparatus is accepted at the loading dock or at the job site. For the acceptance of a substation, for example, a few of the punchlist items might be:

- Check all enclosure sheets for dents and nicks and other damage.

- Check all exposed bolted connections for tightness.

- Check all control connections. Continuity of instrument transformer circuits.

- Check nameplate data to confirm compliance with specifications.

- Check the operation of the cooling fans and alarms, if any.

- Provide heat as required until the transformer is energized.

- Inspect all high- and low-voltage bushings for cracks.

- Inspect all gauges for broken glass, base for dents and deformation.

- Remove the enclosure sheets and inspect the core and coils, insulators, and all bus work for damage.

If any of these are overlooked and damaged equipment is installed before it is energized, it will not comply with the **NEC** purpose of ensuring the safe and practical use of electrical energy.

Q3.   *What is "megger" testing?*

A.   The term "megger" is electrical slang for a test that requires a "megohmmeter" or an instrument that can measure insulation resistance on the order of 100,000 ohms. It is not the same as a "hypot" test (electrical slang for "high potential")

A megger test will result in measurements in units of megaohms rather than microamps. It is typically applied to determine the insulation resistance of electric cables, insulators, buses, motors, and other electrical components. It consists of either a hand-cranked or motor-driven dc generator and resistance indicator and is available in different voltage ratings, usually 500, 1000, or 2500 Vdc. Note that the test voltage is relatively low.

A resistance test of the electrical insulation, prior to placing equipment in service or during routine maintenance, will indicate the condition of the insulation. Wet or defective insulation can be detected very readily. A high megger reading, however, does not necessarily mean that the equipment's insulation can withstand rated potential, since the megohmmeter's voltage is normally not equal to the equipment's rated potential.

Q4.   *How do you know when an electrical component "meggers well"?*

A.   Consult with the equipment manufacturer. After this, *periodic* testing and plotting the resistance readings will show trends that indicate possible problems. That means that you will need a baseline.

The insulation resistance of electrical apparatus is of doubtful significance as compared with the dielectric strength. It is subject to wide variation with design, temperature, dryness, and cleanliness of the parts. When the insulation resistance falls below prescribed values, it can, in most cases of good design and where no defect exists, be brought up to the required standard by cleaning and drying the apparatus. The insulation resistance, therefore, may afford a useful indication as to whether the apparatus is in suitable condition for application of the dielectric test.

The significance of values of megger tests generally requires some interpretation, depending on the design and the drying and cleanliness of the insulation involved. If a user decides to make megger tests, it is recommended that megger test values and that these periodic values be plotted. Substantial variation in the plotted values of megger should be investigated for cause. Insulation resistances may vary with applied voltage, and any comparison must be made with measurements at the same voltage.

Q5.   *How do you prepare for a megger test?*

A.   First of all, it should only be done by an approved testing company with qualified personnel. The International Electrical Testing Association provides the industry with guidelines for test voltages, a few of which appear in this chapter. The manufacturers will provide information about how to confirm a successful test result. A few of the salient points:

■   Windings should be in their normal insulation environment.

■   All bushings or terminals should be in place.

■   Transformer temperature should be at about ambient.

In this test, as in all the other tests described in this chapter the humidity of the equipment ambient environment is an important factor in measurements. A barometer should be used to record humidity on test results.

Q6.   *How is the megger test performed?*

A.   Assuming all safety precautions have been applied, all the circuits of equal voltage above ground are connected together. Appropriate guard circuits should be utilized over all bushings. Circuits or groups of circuits of different voltages above ground are tested separately, for example, high voltage to low voltage and ground, low voltage to high voltage and ground.

■   The dc voltage applied for measuring megger to ground should not exceed a value equal to the rms low-frequency applied voltage allowed.

■   Partial discharges should not be present during megger tests, since they can damage a transformer and may also result in erroneous values of megger.

■   When measurements are to be made using dc voltage exceeding the rms operating voltage of the winding involved (or 1000 V for a solidly grounded wye winding), a relief gap may be employed to protect the insulation.

■   Voltage should be increased in increments (usually 1 to 5 kV), holding each step for 1 min. The test should be discontinued immediately in the event the current begins to increase without stabilizing.

Test duration recommendations are 10 min with resistances tabulated at 30 s, 1 min, and 10 min. Dielectric absorption ratio and polarization index may be calculated.

Q7.   *What is "hypot" testing?*

A.   It is not the same as a megger testing. The units of a hypot test are
        microamps of leakage current (rather than megaohms). A hypot
        test is commonly performed *after* the equipment has passed the
        megometer test. Properly conducted dc hypot testing will indicate
        such faults as cracks, discontinuity, thin spots or voids in the insula-
        tion, excessive moisture or dirt, faulty splices, faulty potheads, etc.

Q8.   *What is the purpose of hypot testing?*

A.   You can use test results to estimate future operating life. The use
        of direct current has several important advantages over ac test-
        ing. For instance, the test equipment itself may be much smaller,
        light in weight, and lower in price. When properly used and
        interpreted, the dc test will give much more information than is
        obtainable with ac testing. With dc testing there is far less chance
        of damage to equipment and less ambiguity in interpreting
        results. The high-capacitance current frequency associated with
        ac testing is not present to mask the true leakage current, nor is it
        necessary to actually break down the material being testing to
        obtain information regarding its condition. Though dc testing
        may not simulate the operating conditions as closely as ac test-
        ing, the many other advantages indicated above make it well
        worthwhile including operating practice.

Q9.   *What are the basic types of hypot tests?*

A.   *Design tests.* These are tests usually made in the laboratory to
        determine proper insulation levels prior to manufacturing.
        Factory tests. These are tests made by the manufacturer to deter-
        mine compliance with design or production requirements.
        Acceptance tests. These are the tests made immediately after
        installation but prior to putting the equipment or cables in ser-
        vice. *Proof tests.* There are the HV tests made soon after the
        equipment and during the guarantee period. *Maintenance tests.*
        Maintenance tests are those performed during normal mainte-
        nance operations or after servicing or repair of equipment or
        cables. *Fault locating.* Fault-locating tests are made to determine
        the location of a specific fault in a cable installation.

        The maximum test voltage, the testing techniques, and the inter-
        pretation of the test results will vary somewhat depending upon
        the particular type of test. It should be plain that an acceptance
        test must be much more severe than a maintenance test, while

a test made on equipment that is already faulted would be conducted in a manner different from a test being conducted on a piece of equipment in active service.

Q10. *At what voltage do you perform hypot tests?*

A. Practice in the field varies widely—especially when it involves cable. The idea is not to blow a hole through the insulation. One rule of thumb—though by no means the only rule of thumb—has it that *acceptance tests* are made at or near the values shown in Table 26.1. *Proof tests* are usually made at about 60 percent of the factory test voltage. The maximum voltage used *in maintenance testing* depends on the age, previous history, and condition of the equipment, but an acceptable value would be approximately 50 to 60 percent of the factory test voltage. It cannot be emphasized enough, however, that you should consult with the cable manufacturer.

TABLE 26.1 MEDIUM-VOLTAGE CABLES MAXIMUM FIELD ACCEPTANCE TEST VOLTAGES (kV, DC)

Insulation type	Rated cable voltage, kV	Insulation level, %	Test voltage kV, dc
Elastomeric: butyl and oil base	5	100	25
	5	133	25
	15	100	55
	15	133	65
	25	100	80
Elastomeric: EPR	5	100	25
	5	133	25
	8	100	35
	15	133	45
	15	100	55
	15	133	65
	25	100	80
	25	133	100
	28	100	85
	35	100	100

TABLE 26.1 MEDIUM-VOLTAGE CABLES MAXIMUM FIELD ACCEPTANCE TEST VOLTAGES (kV, dc) (*Continued*)

Insulation type	Rated cable voltage, kV	Insulation level, %	Test voltage kV, dc
Polyethylene	5	100	25
	5	133	25
	8	100	35
	8	133	45
	15	100	55
	15	133	65
	25	100	80
	25	133	100
	35	100	100

Table 26.1 appears in the 1995 NETA Acceptance Testing Specification Handbook. It is derived from ANSI/IEEE Standard 141 and by factoring applicable ICEA/NEMA standards.

Q11. *What broad principles apply to the testing of cables?*

A. An acceptance test on a power cable installation is one of the most important tests that can be performed on a new electrical power circuit. All other major components of a system are factory-built and inspected, shipped to the site, and then installed. Little is done to this kind of equipment at the site except possibly some reassembly of parts removed for shipping. A cable installation, though, is something entirely different. The cable is shipped to the site on reels. Then it is installed—either in conduit or directly buried. It is pulled into conduit; it is cut; it is shaped around manholes or switchgear; it is spliced; it is terminated.

For all practical purposes we can say with confidence that the cable system is truly fabricated on the job site. The acceptance test, performed after installation but prior to energization, is the factory test of the cable system. The cable itself was production tested prior to shipment—but that cable bears little resemblance to the installed cable system.

Q12. *What are the most common types of tests performed on cable systems?*

A. The megger and hypot tests. The dc hypot test uses a test voltage above the dc equivalent of the crest of the normal ac operating voltage. In the step-voltage test the test voltage is applied in a number of equal increments with each being held for an equal time interval. A leakage current reading is taken at each level immediately before the voltage is increased to the next level. When performed as an acceptance test, the maximum test voltage for either of the above two tests is listed in the Insulated Cable Engineers Association Standards (ICEA).

There are two refinements of the insulation resistance test—the dielectric absorption test and the polarization index test. Both of these tests involve leaving the test voltage on the cable insulation for an extended period of time, 1 to 10 min. Evaluation of the results of the tests is based on changes in the apparent insulation resistance of the cable with respect to time.

Q13. *What items should a good specification for cable testing include?*

A. A great deal of generic text about cable testing has been cut and pasted in U.S. construction industry electrical specifications over the years. Some of the items that are common to most variants of the Masterspec format are as follows:

- Exercise suitable and adequate safety measures prior to, during, and after the high-potential tests, including placing warning signs and preventing people and equipment from being exposed to the test voltages.

- Tests should be made after the installation is complete with all splices and terminations but before connection to the equipment.

- Each phase should be tested individually.

- Where new cable is connected to the existing cable, the new cable should be tested prior to its connection to the existing. If testing of the combined circuit is required, the new cable should be tested prior to testing of the combined circuit.

■ Prior to the high-potential test the contractor should ensure that the cable and shields are continuous, that the cable is free of shorts and grounds, and that the shields are grounded at splices and terminations.

■ Measure the leakage current from each conductor to the insulation shield. Corona shields guard rings, taping, mason jars, or plastic bags are often used to prevent corona current from influencing the readings. Unprepared cable shield ends should be trimmed back 1 in or more for each 10 kV of test voltage.

■ At terminations, look for evidence of corona or excessive heat, shield grounds, porcelain cracking, pothead compound leaks, condition of connections, and physical damage.

■ In manholes, look for cracks in rigid fireproofing that may indicate excessive stress on the cable sheath, sharp bends, excessive cable movement, oil or compound leaks, sheath wear at cable supports and duct mouths, swollen or collapsed splices, pitting, and ground connections.

You should consult NETA, NEMA, and the cable manufacturer for particulars on how to apply test voltages and test methods and how to interpret results.

In the absence of consensus standards dealing with insulation-resistance tests, the NETA Technical Committee suggests the

TABLE 26.2 SWITCHGEAR INSULATION-RESISTANCE TEST VOLTAGE

Voltage rating	Minimum dc test voltage	Recommended minimum insulation resistance, megohms
0–250	500	50
251–600	1,000	100
601–5,000	2,500	1,000
5,001–15,000	2,500	5,000
15,001–25,000	5,000	20,000
25,001–35,000	15,000	100,000

representative values in Table 26.2 for switchgear insulation test voltages.

Q14.   *What is a typical testing requirement for transformers?*

A.   Transformer inspection, testing, and maintenance can be divided into two general categories—work that can be performed with the transformer energized, and work which requires that the unit be deenergized. There isn't a lot of testing that is done to a transformer while it is energized other than testing for sound and testing the temperature controller.

Several inspections can be safely performed only when the unit is deenergized. These include the inspection of connections for overheating or corrosion, the inspection of bushings for cracks, chips, or leaks, and the inspection of pressure relief devices for leaks, corrosion, and diaphragm integrity. If an internal inspection is to be performed, this also must be performed with the transformer deenergized, but this procedure is recommended only when there is a need indicated or when the transformer is to be opened for some other purpose.

Certain optional tests such as no load loss, exciting current, phase relation, and polarity tests are not normally required on utilization-level substations, i.e., substations that are specified at less than 1500 kVA for use in buildings.

A turns ratio and polarity test, for example, determines the number of turns in one winding of a transformer in relation to the number of turns in the other winding of the same transformer. It is used as an acceptance test to verify nameplate voltage ratios and polarity of the transformer and is also used as a maintenance test to detect shorted turns in the transformer or defective tap changers.

Another optional test for transformers is the *Doble test*. This is an ac qualitative test of the overall insulation system of the transformer. It detects changes in measurable characteristics of insulation systems caused by the effects of heat, moisture, ionization, and other destructive agents which reduce the breakdown strength of insulation. This test is performed as an acceptance test but should not be performed on transformers where insulation damage is suspected unless contingencies are in place for a complete loss of the transformer.

Insulation power factor is the ratio of the power dissipated in the insulation in watts to the product of the effective voltage and current in voltamperes when tested under a sinusoidal voltage and prescribed conditions. While the real significance that can be attached to the insulation power factor of dry-type transformers is still a matter of opinion, experience has shown that insulation power factor is helpful in assessing the probable condition of the insulation when good judgment is used.

In interpreting the results of the insulation power factor test values, the comparative values of tests taken at periodic intervals are useful in identifying potential problems rather than an absolute value of insulation power factor. A factory insulation power factor test will be of value for comparison with field insulation power factor measurements to assess the probable condition of the insulation.

It has not been feasible to establish standard insulation power factor values for dry-type transformers because experience has indicated that little or no relation exists between insulation power factor and the ability of the transformer to withstand the prescribed dielectric tests. Table 26.3 should serve as a guideline in the absence of specific manufacturer information.

TABLE 26.3 TRANSFORMER INSULATION-RESISTANCE ACCEPTANCE TEST VOLTAGE AND MINIMUM RESULTS

Transformer coil rating type, volts	Minimum dc test voltage	Recommended minimum insulation resistance, megohms	
		Liquid filled	Dry
0–600	1,000	100	500
601–5,000	2,500	1,000	5,000
5001–15,000	5,000	5,000	25,000

In the absence of consensus standards, the NETA Technical Committee suggests these reference values. Since insulation resistance depends upon insulation rating and winding capacity, values obtained should be compared to the manufacturer's test data.

Q15.   *What tests are common for secondary switchgear buses?*

A.   The completely installed busway run should be electrically tested prior to being energized. The testing procedure should first verify that the proper phase relationships exist between the busway and associated equipment. This phasing and continuity test can be performed in the same manner as similar tests on other pieces of electric equipment on the job.

All busway installations should be tested with a megometer or high-potential voltage to be sure that excessive leakage paths between phases and ground do not exist. Megometer values depend on the busway construction, type of insulation, size and length of busway, and atmospheric conditions. Acceptable values for a particular busway should be obtained from the manufacturer.

Q16.   *Should low-voltage breakers be tested?*

A.   Typically, single-phase branch circuit breakers are not. The cost of testing does not justify the expense. However, many molded feeder breakers above, say 400 A, may require testing. The cutoff point (in terms of ampere rating) should be determined jointly by the owner and the engineer (see Tables 26.4 and 26.5).

You should always test steel-frame breakers (600 to 800 A and above)—especially those controlled by microprocessor based trip units. Even though the breakers may test well—and they usually do—it is a good time to confirm that breaker interrupting ratings, sensors, plugs and trip units have been specified and built correctly. This may require shipping the breakers to the testing facility or bringing the testing equipment to the site. Wherever the tests are performed, the following tests should be performed and the tripping functions observed:

■   High-level phase fault with instantaneous tripping of the breaker

TABLE 26.4 MOLDED-CASE CIRCUIT BREAKERS VALUES FOR INVERSE TIME TRIP TEST AT 300% OF RATED CONTINUOUS CURRENT OF CIRCUIT BREAKERS*

Range of rated continuous current amps	Maximum trip time, for each maximum frame rating	
	≤250 V	251–600 V
0–30	50	70
31–50	80	100
51–100	140	160
101–150	200	250
151–225	230	275
226–400	300	350
401–600	—	450
601–800	—	500
801–1000	—	600
1001–1200	—	700
1201–1600	—	775
1601–2000	—	800
2001–2500	—	850
2501–5000	—	900

*For integrally fused circuit breakers, trip times may be substantially longer if tested with fuses replaced by shorting bars (1995 NETA Acceptance Testing Specifications for Electrical Power Distribution Equipment and Systems). The original source is Table 5-3 from NEMA publication AB4-1991.

TABLE 26.5 INSTANTANEOUS TRIP SETTING TOLERANCES FOR FIELD TESTING OF MARKED ADJUSTABLE TRIP CIRCUIT BREAKERS*

Ampere rating	Tolerances of high and low settings, %	
	High	Low
≤250	+40	+40
	−25	−30
>250	±25	±30

*For circuit breakers with nonadjustable instantaneous trips, tolerances apply to the manufacturer's published trip range, i.e., +40 percent on high side, −30 percent on low side (1995 NETA Acceptance Testing Specifications for Electrical Power Distribution Equipment and Systems). The original source is Table 5-4 from NEMA publication AB4-1991.

- Moderate phase overload with the breaker tripping within a prescribed time period

- Moderate level ground fault with the breaker tripping instantaneously

- Moderate level ground fault with the breaker tripping after a slight time delay

These are functional tests only and are not intended as a check of the actual calibration of the breaker. Testing the overcurrent calibration of an electromechanical device can only be performed by primary injection that is actually imposing an overcurrent or fault current on the breaker and observing its tripping characteristics. On breakers with solid-state trip devices, this testing can be performed either by primary injection or by putting a low-level current into the electronic circuitry (secondary injection).

Appendix

This appendix is provided to the reader as a sampler and a starting point for the assembly of a professional library. There is no substitute for reading the original sources from which most of the material in this book has been taken. Grateful acknowledgment is made to the various professional organizations for having granted permission to reprint illustrations and tables that appear in this text. Note that graphical items appear before tabular material, followed by the list of core references and the bibliography.

A1. *Errata to the First Printing of the 1999 National Electric Code.*

A2. Example of a Building and Safety Submittal Checklist. City of Las Vegas. Reprinted courtesy of City of Las Vegas Development Services Center, 731 S. Fourth Street, Las Vegas, NV 89101.

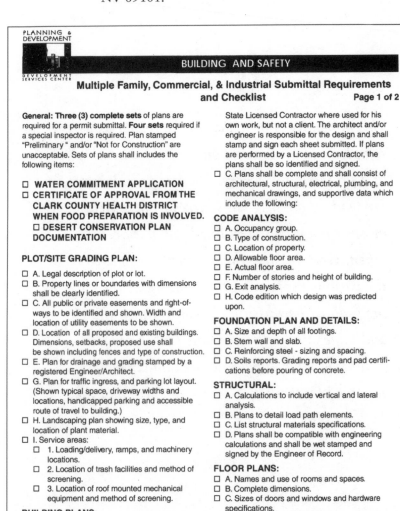

PLANNING &
DEVELOPMENT

BUILDING AND SAFETY

DEVELOPMENT
SERVICES CENTER

Multiple Family, Commercial, & Industrial Submittal Requirements and Checklist   Page 1 of 2

General: Three (3) complete sets of plans are required for a permit submittal. **Four sets** required if a special inspector is required. Plan stamped "Preliminary " and/or "Not for Construction" are unacceptable. Sets of plans shall includes the following items:

☐ **WATER COMMITMENT APPLICATION**
☐ **CERTIFICATE OF APPROVAL FROM THE CLARK COUNTY HEALTH DISTRICT WHEN FOOD PREPARATION IS INVOLVED.**
☐ **DESERT CONSERVATION PLAN DOCUMENTATION**

PLOT/SITE GRADING PLAN:

☐ A. Legal description of plot or lot.
☐ B. Property lines or boundaries with dimensions shall be clearly identified.
☐ C. All public or private easements and right-of-ways to be identified and shown. Width and location of utility easements to be shown.
☐ D. Location of all proposed and existing buildings. Dimensions, setbacks, proposed use shall be shown including fences and type of construction.
☐ E. Plan for drainage and grading stamped by a registered Engineer/Architect.
☐ G. Plan for traffic ingress, and parking lot layout. (Shown typical space, driveway widths and locations, handicapped parking and accessible route of travel to building.)
☐ H. Landscaping plan showing size, type, and location of plant material.
☐ I. Service areas:
   ☐ 1. Loading/delivery, ramps, and machinery locations.
   ☐ 2. Location of trash facilities and method of screening.
   ☐ 3. Location of roof mounted mechanical equipment and method of screening.

BUILDING PLANS:

☐ A. Plans shall be submitted to Building and Safety for Plan check and only a licensed con tractor of the State of Nevada can be issued a permit for construction.
☐ B. Plans must be drawn by a Nevada State Registered Architect or Engineer, or Nevada

State Licensed Contractor where used for his own work, but not a client. The architect and/or engineer is responsible for the design and shall stamp and sign each sheet submitted. If plans are performed by a Licensed Contractor, the plans shall be so identified and signed.

☐ C. Plans shall be complete and shall consist of architectural, structural, electrical, plumbing, and mechanical drawings, and supportive data which include the following:

CODE ANALYSIS:

☐ A. Occupancy group.
☐ B. Type of construction.
☐ C. Location of property.
☐ D. Allowable floor area.
☐ E. Actual floor area.
☐ F. Number of stories and height of building.
☐ G. Exit analysis.
☐ H. Code edition which design was predicted upon.

FOUNDATION PLAN AND DETAILS:

☐ A. Size and depth of all footings.
☐ B. Stem wall and slab.
☐ C. Reinforcing steel - sizing and spacing.
☐ D. Soils reports. Grading reports and pad certifications before pouring of concrete.

STRUCTURAL:

☐ A. Calculations to include vertical and lateral analysis.
☐ B. Plans to detail load path elements.
☐ C. List structural materials specifications.
☐ D. Plans shall be compatible with engineering calculations and shall be wet stamped and signed by the Engineer of Record.

FLOOR PLANS:

☐ A. Names and use of rooms and spaces.
☐ B. Complete dimensions.
☐ C. Sizes of doors and windows and hardware specifications.
☐ D. Wall and ceiling finish materials and specifictions.

FRAMING PLANS AND DETAILS:

☐ A. Plans, sections, details, and schedules showing:
   ☐ 1. All Beams, supports, and structural details.
   ☐ 2. Roof construction, venting, openings, and materials.

PLANNING &
DEVELOPMENT

BUILDING AND SAFETY

DEVELOPMENT
SERVICES CENTER

☐ 3. Exterior walls and bearing partitions.
☐ 4. Joist and rafter size, spacing, and layout.
☐ 5. Type and thickness of floors.
☐ 6. Truss layouts and wet stamped calculations.
☐ B. Roofing type, class, and manufacturer.

ELEVATIONS AND SECTIONS:
☐ A. Exterior elevations to include all weather resistive construction.
☐ B. Cross sections sufficient to reflect structural systems.
☐ C. Occupancy and area separation walls including hourly ratings.
☐ D. Rated corridors.
☐ E. Interior wall lateral support.

FIRE RESISTIVE CONSTRUCTION:
☐ A. All fire resistive construction is to be shown in section view.
☐ B. Opening or penetrations of fire resistive constuction are to be detailed in section view with applied references.
☐ C. Closure construction between fire resistive floors and walls and structural or exterior wall components shall be detailed in section view.
☐ D. Fire resistive assemblies shall be identified by their listings.

MISCELLANEOUS DETAILS AND MATERIALS:
☐ A. Details of construction features such as stairs, balconies, retaining walls, ramps, etc., including specifications of all materials.

ELECTRICAL PLANS:
☐ A. Single line diagram.
☐ B. Service and load calculations, to include all short circuit and fault current calculations.
☐ C. Panels schedules and descriptions of circuits with connected loads, panel ratings, and feeder size.
☐ D. All outlets, smoke detectors, equipment and feeders shown on plan with appropriate panel and circuit numbers at devices.
☐ E. Show emergency power system, type, and model.
☐ F. Show voltage drop catculations for all feeders to sub-panels, panels, area lighting, free stand ing signs, and air conditioning units.

MECHANICAL PLANS:
☐ A. Show model and type of equipment.
☐ B. Show AFUE/SEER rating.

☐ C. Energy demand, input, and BTU.
☐ D. Location, access and working space for mechanical equipment.
☐ E. Combustionair, flue sizes, and materials.
☐ F. Dampers -- type, size, and locations.
☐ G.Sizes of supply/terurn air dicts and grilles shown in plan view. CFM capacity of ducts, grilles and diffusers.
☐ H. Location, material, and insulation of mechanical pipes and ducts.
☐ I. Size, location, and piping material of all air conditioning condensate drains.
☐ J. Size, location, and ducting of all smoke control systems.

PLUMBING PLANS:
☐ A. Plan view of all water, drainage, waste and vent piping, with location, size, and material.
☐ B. Drainage, waste, vent, and water supply plans. Indicate types of fixtures with symbols.
☐ C. Location and size of gas, fuel oil, or LPG piping with appliance demands.
☐ D. Size, location, and materials of P/T water relief valve.
☐ E. Location of all water heaters and/or boiler combustion air flues.
☐ F. Show location of cleanouts, backwater valves, and water shutoff valves.

MODEL ENERGY CODE CALCULATIONS FIRE PROTECTION CODES:
☐ A. Location of fire hydrants, fire department access roads, fire department hook-ups, etc.
☐ B. Fire flow calculations (in civil engineering package).
☐ C. State if building will include fire protection systems (and any relevant design details) including, but no limited to:
  ☐ 1. Fire alarm diagrams.
  ☐ 2. Layout, size, location, material, and calculations of fire sprinkler systems.
  ☐ 3. Halon system.
  ☐ 4. Kitchen protection.
  ☐ 5. Specialized system.
  ☐ 6. Smoke control design and operation.
  ☐ 7. Standpipe systems.
  ☐ 8. Flammable/combustible liquid tanks/lines.
  ☐ 9. Medical gas system design.
  ☐ 10. Cut sheets for above systems (Catalog).
☐ D. Smoke control operation/design description.

A2 (continued)

A3. Application for Electrical Permit. Building Department, City of Ann Arbor, Michigan.

A4. Typical Electrical Permit Fees. The structure of the permit requires that the electrical contractor have accurate information on the blueprints to apply for the permit.

APPLICATION FOR ELECTRICAL PERMIT

BUILDING DEPARTMENT • 100 N. Fifth Ave. • P.O. Box 8647
Ann Arbor, Michigan 48107-8647 • (734) 994-2674

PERMIT NO. _____

Date Issued _____

Location Address _____

Owner _____

Address _____

Phone _____

Date _____

Contractor _____

Business Address _____

Phone _____

NUMBER	ITEM	FEE	Ad. To
	EACH INSPECTION		
	SERVICE FEEDERS/PER AMP		
	CIRCUITS & FEEDERS, EACH		
	20 AMPS AND UNDER		
	OVER 20 TO 50		
	OVER 50 TO 100		
	OVER 100 TO 200		
	OVER 200 TO 400		
	OVER 400 AMPS		
	OVER 800 AMPS		
	MISC WIRING/CODE REPAIRS		
	TEMPORARY SERVICE		
	MINIMUM PERMIT FEE		
	IF A PERMIT IS NOT OBTAINED BEFORE THE WORK IS STARTED, THE APPLICANT SHALL BE SUBJECTED TO DOUBLE THE PERMIT FEE.		
	EXTRA INSPECTION ON JOB WITH PERMIT		
	TOTAL FEE		

Existing Building _____   New Building _____

Residential _____   Commercial _____

Name of Business or Building _____

APPLIANCES

Dryer _____   Dishwasher _____

Range _____   Food Disposal _____

Water Heater _____   Air Cond. Outlet _____

Washer _____   Other (Specify) _____

INSPECTOR'S REPORT

Rough _____

Date Apr. _____   Inspector _____

Final _____

Date Apr. _____   Inspector _____

Remarks _____

License No. _____

E-63 Revised 7/92

A3

A5. Data from IEEE Reliability Survey of Industrial Plants. From IEEE Standard 493-1990. The 1997 edition of this standard contains an updated and more comprehensive version of this table.

A6. State Electric Rate Information. Source: Energy Information Administration, Form EIA-861, "Annual Electric Utility Report"; copyright 1997 by Edison Electric Institute.

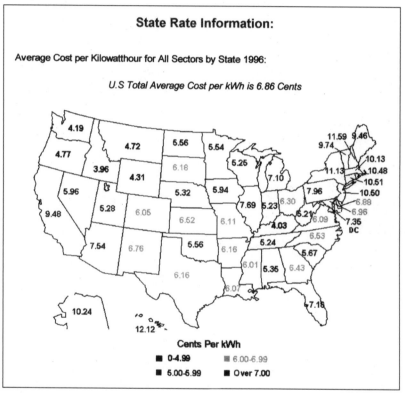

State Rate Information:

Average Cost per Kilowatthour for All Sectors by State 1996:

U.S Total Average Cost per kWh is 6.86 Cents

Cents Per kWh

- 0-4.99
- 5.00-5.99
- 6.00-6.99
- Over 7.00

A6

A7. NESC Table 234-1. Clearance of Wires, Conductors, Cables and Unguarded Live Parts Adjacent but Not Attached to Buildings and Other Installations Except Bridges.

A8. Resistivity of Soils and Resistance of Single Rods. From IEEE Standard 142.

A9. Smoke Detector Spacing. From NFPA 72-1996.

Table 234-1   **FT**

Clearance of Wires, Conductors, Cables, and Unguarded Rigid Live Parts
Adjacent but Not Attached to Buildings and Other Installations Except Bridges ⑩

(Voltages are phase to ground for effectively grounded circuits and those other circuits where all ground faults are
cleared by promptly de-energizing the faulted section, both initially and following subsequent breaker operations.
See the definitions section for voltages of other systems. Clearances are with no wind displacement
except where stated in the footnotes below. See Rules 234C1a, 234C2, and 234H4.)

Clearance of	Insulated communication conductors and cables; messengers; surge-protection wires; grounded guys; neutral conductors meeting Rule 230E1; supply cables meeting Rule 230C1 (ft)	Supply cables of 0 to 750 V meeting Rules 230C2 or 230C3 (ft)	Unguarded rigid live parts, 0 to 750 V; noninsulated communication conductors (ft)	Supply cables over 750 V meeting Rules 230C2 or 230C3; open supply conductors, 0 to 750 V (ft)	Open supply conductors, over 750 V to 22 kV (ft)	Unguarded rigid live parts, over 750 V to 22 kV (ft)
1. Buildings						
a. Horizontal						
(1) To walls, projections, and guarded windows	4.5⑦	5.0	5.0	5.5①②⑨	7.5①②⑨⑪	7.0
(2) To unguarded windows⑧	4.5	5.0	5.0	5.5①②⑨	7.5⑨⑪	7.0
(3) To balconies and areas readily accessible to pedestrians③	4.5	5.0	5.0	5.5⑨	7.5⑨⑪	7.0
b. Vertical						
(1) Over or under roofs or projections not readily accessible to pedestrians③	3.0	3.5	10.0	10.5	12.5	12.0
(2) Over or under balconies and roofs readily accessible to pedestrians③	10.5	11.0	11.0	11.5	13.5	13.0
(3) Over roofs accessible to vehicles but not subject to truck traffic⑥	10.5	11.0	11.0	11.5	13.5	13.0
(4) Over roofs accessible to truck traffic⑥	15.5	16.0	16.0	16.5	18.5	18.0
2. Signs, chimneys, billboards, radio and television antennas, tanks, and other installations not classified as buildings or bridges						
a. Horizontal④	3.0	3.5	5.0	5.5①②⑨	7.5①②⑨⑪	7.0
b. Vertical over or under④	3.0	3.5	5.5	6.0①	8.0	7.5

① Where building, sign, chimney, antenna, tank, or other installation does not require maintenance such as painting, washing, changing of sign letters, or other operations that would require persons to work or pass between supply conductors and structure, the clearance may be reduced by 2 ft.

② Where available space will not permit this value, the clearance may be reduced by 2 ft provided the conductors, including splices and taps, have covering that provides sufficient dielectric to prevent a short circuit in case of momentary contact between the conductors and a grounded surface.

③ A roof, balcony, or area is considered readily accessible to pedestrians if it can be casually accessed through a doorway, ramp, window, stairway, or permanently mounted ladder by a person on foot who neither exerts extraordinary physical effort nor employs special tools or devices to gain entry. A permanently mounted ladder is not considered a means of access if its bottom rung is 8 ft or more from the ground or other permanently installed accessible surface.

④ The required clearances shall be to the closest approach of motorized signs or moving portions of installations covered by Rule 234C.

⑤ This footnote not used in this edition.

⑥ For the purpose of this rule, trucks are defined as any vehicle exceeding 8 ft in height.

⑦ This clearance may be reduced to 3 in for the grounded portions of guys.

⑧ Windows not designed to open may have the clearances permitted for walls and projections.

⑨ The clearance at rest shall be not less than the value shown in this table. Also, when the conductor or cable is displaced by wind, the clearance shall be not less than 3.5 ft; see Rule 234C1b.

⑩ The clearance at rest shall be not less than the value shown in this table. Also, when the conductor or cable is displaced by wind, the clearance shall be not less than 4.5 ft; see Rule 234C1b.

⑪ Where available space will not permit this value, the clearance may be reduced to 7.0 ft for conductors limited to 8.7 kV to ground.

⑫ The clearance values shown in this table are computed by adding the applicable Mechanical and Electrical (M&E) value of Table A-1 to the applicable Reference Component of Table A-2b of Appendix A.

A10. Voltage Drop of Conductors in Cable or Conduit. From IEEE Standard 142. Tables like this are available in various standards and in various manufacturer catalogs. This version offers some insight into how conduit and cable combinations (often called circuit "makeup") affect voltage.

A11. Fault Current Table—240 V. Tables like this are available in various standards and in various manufacturer catalogs.

Sometimes the data is shown in graphical form. Such tables and graphs are good for a first-cut estimate but should be backed up by more rigorous calculation in borderline design situations (on the cusp of circuit breaker interrupting ratings, for example). The fault currents listed are maximum available symmetrical rms values based on liquid-filled transformers, with nominal impedances of 4.5 percent for ratings up to and including 500 kVA and 5.5 percent for ratings above 500 kVA, and include motor contribution based on 100 percent motor load.

A12. Fault Current Table—480 V. See A11.

A13. Typical Appliance/General-Purpose Receptacle Loads (excluding plug-in-type A/C and heating equipment). From IEEE Standard 241.

A14. Typical Apartment Loads. Adapted from IEEE Standard 241.

A15. Connected Load and Maximum Demand by Tenant Classification. From IEEE Standard 241.

A16. Typical Loads in Commercial Kitchens. From IEEE Standard 241.

A17. Typical Ratings of Engine-Driven Generator Sets. From IEEE Standard 493.

A18. Summary of Possible or Typical Emergency or Standby Power Needs for the *Health-Care* Industry. From IEEE Standard 446.

A19. Summary of Possible or Typical Emergency or Standby Power Needs for the *Commercial Building* Industry. From IEEE Standard 446.

A20. Preferred Ratings for Medium Voltage Indoor Oiless Circuit Breakers. From ANSI/IEEE 37.06.

A21. ANSI Device Function Numbers for Protective Devices.

A22. Transformer Approximate Loss and Impedance Data. From Westinghouse/Cutler-Hammer Consulting Guide.

A23. Typical (Ambient) Transformer Sound Levels. From Westinghouse/Cutler-Hammer Consulting Guide.

A24. Maximum Average Sound Levels. From Westinghouse/Cutler-Hammer Consulting Guide.

TABLE A.7 Preferred Ratings ‡ for Medium-Voltage Indoor Oilless Circuit Breakers

| Rated maximum voltage[1] kV, rms | Rated voltage range factor K[2] | Rated continuous current at 60 Hz[3] amperes, rms | Rated short-circuit current* (at rated maximum kV[4,5,6]) kA, rms | Transient Recovery Voltages | | | Rated interrupting time[7] cycles[9] | Rated maximum voltage divided by K kV, rms[1] | Maximum symmetrical interrupting capability and rated short-time current[4,5,8] kA, rms | Closing and latching capability 2.7K times rated short-circuit current,[4] kA, crest |
				Rated time to point P T_2† μ sec	Rated rate R kV/μ sec	Rated delay time T_1 μ sec				
4.76	1.36	1200	8.8	—	—	—	5	3.5	12	32
4.76	1.24	1200, 2000	29	—	—	—	5	3.85	36	97
4.76	1.19	1200, 2000, 3000	41	—	—	—	5	4.0	49	132
8.25	1.25	1200, 2000	33	—	—	—	5	6.6	41	111
15.0	1.30	1200, 2000	18	—	—	—	5	11.5	23	62
15.0	1.30	1200, 2000	28	—	—	—	5	11.5	36	97
15.0	1.30	1200, 2000, 3000	37	—	—	—	5	11.5	48	130
38.0	1.65	1200, 2000, 3000	21	—	—	—	5	23.0	35	95
38.0	1.0	1200, 3000	40	—	—	—	5	38.0	40	108

* For the related required capabilities associated with the rated short-circuit current of the circuit breaker, see Note 4.

† These rated values are not yet standardized. Work is in progress.

‡ For service conditions, definitions, and interpretation of ratings, tests, and qualifying terms, see ANSI/IEEE C37.04-1979, ANSI/IEEE C37.09-1979, and ANSI/IEEE C37.100-1981.

§ The interrupting ratings are for 60-Hz systems. Applications on 25-Hz systems should receive special consideration.

⁹ Current values have been rounded off to the nearest kiloampere (kA) except that two significant figures are used for values below 10 kA.

¹ The voltage rating is based on ANSI C84.1-1982, where applicable, and is the maximum voltage for which the breaker is designed and the upper limit for operation.

476

[2] The rated voltage range factor, K, is the ratio of rated maximum voltage to the lower limit of the range of operating voltage in which the required symmetrical and asymmetrical current interrupting capabilities vary in inverse proportion to the operating voltage.

[3] The 25-Hz continuous current ratings in amperes are given herewith following the respective 60-Hz rating: 600–700; 1200–1400; 2000–2250; 3000–3500.

[4] Related required capabilities. The following related required capabilities are associated with the short-circuit current rating of the circuit breaker. (a) Maximum symmetrical interrupting capability (kA, rms) of the circuit breaker is equal to K times rated short-circuit current. (b) 3-sec short-time current carrying capability (kA, rms) of the circuit breaker is equal to K times rated short-circuit current. (c) Closing and latching capability (kA, rms) of the circuit breaker is equal to 1.6 K times rated short-circuit current. If expressed in peak amperes, the value is equal to 2.7 K times rated short-circuit current. (d) 3-sec short-time current carrying capability and closing and latching capability are independent of operating voltage up to and including rated maximum voltage.

[5] To obtain the required symmetrical current interrupting capability of a circuit breaker at an operating voltage between 1/K times rated maximum voltage and rated maximum voltage, the following formula shall be used:

Required symmetrical current interrupting capability

$$= \text{rated short-circuit current} \times \frac{(\text{rated maximum voltage})}{(\text{operating voltage})}$$

For operating voltages below 1/K times rated maximum voltage, the required symmetrical current interrupting capability of the circuit breaker shall be equal to K times rated short-circuit current.

[6] With the limitation stated in 5.10 of ANSI/IEEE C37.04-1979, all values apply for polyphase and line-to-line faults. For single phase-to-ground faults, the specific conditions stated in 5.10.2.3 of ANSI/IEEE C37.04-1979 apply.

[7] The ratings in this column are on a 60-Hz basis and are the maximum time interval to be expected during a breaker opening operation between the instant of energizing the trip circuit and interruption of the main circuit on the primary arcing contacts under certain specified conditions. The values may be exceeded under certain conditions as specified in 5.7 of ANSI/IEEE C37.04-1979.

[8] Current values in this column are not to be exceeded even for operating voltages below 1/K times rated maximum voltage. For voltages between rated maximum voltage and 1/K times rated maximum voltage, follow (5) above.

A20

Approximate Loss and Impedance Data

Table R4: 15 kV Class Oil Liquid-Filled Transformers

65°C Rise

kVA	No Load Watts Loss	Full Load Watts Loss	%Z	%R	%X	X/R
112.5	550	2470	5.00	1.71	4.70	2.75
150	545	3360	5.00	1.88	4.63	2.47
225	650	4800	5.00	1.84	4.65	2.52
300	950	5000	5.00	1.35	4.81	3.57
500	1200	8700	5.00	1.50	4.77	3.18
750	1600	12160	5.75	1.41	5.57	3.96
1000	1800	15100	5.75	1.33	5.59	4.21
1500	3000	19800	5.75	1.12	5.64	5.04
2000	4000	22600	5.75	0.93	5.67	6.10
2500	4500	26000	5.75	0.86	5.69	6.61

Table R5: 15 kV Class Primary – Dry-Type Transformers Class H

150°C Rise

kVA	No Load Watts Loss	Full Load Watts Loss	%Z	%R	%X	X/R
300	1600	10200	4.50	2.87	3.47	1.21
500	1900	15200	5.75	2.66	5.10	1.92
750	2700	21200	5.75	2.47	5.19	2.11
1000	3400	25000	5.75	2.16	5.33	2.47
1500	4500	32600	5.75	1.87	5.44	2.90
2000	5700	44200	5.75	1.93	5.42	2.81
2500	7300	50800	5.75	1.74	5.48	3.15

80°C Rise

kVA	No Load Watts Loss	Full Load Watts Loss	%Z	%R	%X	X/R
300	1800	7600	4.50	1.93	4.06	2.10
500	2300	9500	5.75	1.44	5.57	3.87
750	3400	13000	5.75	1.28	5.61	4.38
1000	4200	13500	5.75	0.93	5.67	6.10
1500	5900	19000	5.75	0.87	5.68	6.51
2000	6900	20000	5.75	0.66	5.71	8.72
2500	7200	21200	5.75	0.56	5.72	10.22

For 600-volt class dry-type transformers, see pages 116 and 117.

Table R6: 600-Volt Primary Class Dry-Type Transformers

150°C Rise

kVA	No Load Watts Loss	Full Load Watts Loss	%Z	%R	%X	X/R
3	33	231	7.93	6.60	4.40	0.67
6	58	255	3.70	3.28	1.71	0.52
9	77	252	3.42	1.94	2.81	1.45
15	150	875	5.20	4.83	1.92	0.40
30	200	1600	5.60	4.67	3.10	0.66
45	300	1900	4.50	3.56	2.76	0.78
75	400	3000	4.90	3.47	3.46	1.00
112.5	500	4900	5.90	3.91	4.42	1.13
150	600	6700	6.20	4.07	4.68	1.15
225	700	8600	6.40	3.51	5.35	1.52
300	800	10200	7.10	3.13	6.37	2.03
500	1700	9000	5.50	1.46	5.30	3.63
750	2200	11700	6.30	1.27	6.17	4.87
1000	2800	13600	6.50	1.08	6.41	5.93

Table R7: 600-Volt Primary Class Dry-Type Transformers

115°C Rise

kVA	No Load Watts Loss	Full Load Watts Loss	%Z	%R	%X	X/R
15	150	700	5.20	3.67	3.69	1.01
30	200	1500	4.60	4.33	1.54	0.36
45	300	1700	3.70	3.11	2.00	0.64
75	400	2300	4.60	2.53	3.84	1.52
112.5	500	3100	6.50	2.31	6.08	2.63
150	600	5900	6.20	3.53	5.09	1.44
225	700	6000	7.20	2.36	6.80	2.89
300	800	6600	6.30	1.93	6.00	3.10
500	1700	6800	5.50	1.02	5.40	5.30
750	1500	9000	4.10	1.00	3.98	3.98

Table R8: 600-Volt Primary Class Dry-Type Transformers

80°C Rise

kVA	No Load Watts Loss	Full Load Watts Loss	%Z	%R	%X	X/R
15	200	500	2.30	2.00	1.14	0.57
30	300	975	2.90	2.25	1.83	0.81
45	300	1100	2.90	1.78	2.29	1.29
75	400	1950	3.70	2.07	3.07	1.49
112.5	600	3400	4.30	2.49	3.51	1.41
150	700	3250	4.10	1.70	3.73	2.19
225	800	4000	5.30	1.42	5.11	3.59
300	1300	4300	3.30	1.00	3.14	3.14
500	2200	5300	4.50	0.62	4.46	7.19

A22

A25. Average Efficiencies and Power Factors for Polyphase Squirrel-Cage Induction Motors.

A26. Specifying Motor Overloads.

A27. Spectrums of Several Standard Frequency Bands. From BICSI Training Manual 1998.

A28. Twisted-Pair Cable Standards. From BICSI Training Manual 1998.

A1. Errata to NEC First Printing*

1. Page 70-31. In Secs. 110-31(a)(2) and 110-31(b)(2), change 110-35 to 490-24.

---

*Taken from the web site
www.iaei.com/news/1999necerrata.htm
Copyright © 1999 International Association of Electrical Inspectors. Last modified January 19, 1999.

2. Page 70–64. In Sec. 230-40, Exception 5, change 230-82(3) to 230-82(4).

3. Page 70–75. Section 240-21(c)(4)(b) should read as follows: "(b) The conductors terminate at a single circuit breaker or a single set of fuses that will limit the load to the ampacity of the conductors. This single overcurrent device shall be permitted to supply *any number of additional* overcurrent devices on its load side."

4. Page 70–93. In Sec. 250-997, in the last sentence of the first paragraph, Section 250-94 (1) through (4) should be changed to Section 250-94, except for (1).

5. Page 70-109. In Sec. 300-7 (b), FPN, change Table 347-9 to Table 347-9(A).

6. Pages 70-114 and 70-115. Exceptions 1 through 6 that follow Sec. 300-50(e) should directly follow Table 300-50; they are exceptions to the table.

7. Page 70-140. In the title of Table 310-86, change "copper" to "aluminum."

8. Page 70-172. In Table 347-9(B), "Length Change of PVC Conduit" should be changed to "Length Change of Fiberglass Conduit" in both columns.

9. Page 70-174. Section 349-1 should be deleted and Section 349-2 should be renumbered to 349-1.

10. Page 70-186. In Sec. 362-23, FPN, change Table 347-9 to Table 347-9(A).

11. Page 70-199. In Sec. 370-70(2), change 370-23(f) to 370-23(g).

12. Page 70-238. In Sec. 422-16(b)(1)(a). Exception, the first sentence, "A listed kitchen disposer distinctly marked..." should be changed to "A listed kitchen waste disposer distinctly marked.....".

13. Page 70-271. Add the following notes to the bottom of Table 430-72(b) Notes:

   ■ Value specified in Sec. 310-15 as applicable.

■   400 percent of value specified in Table 310-17 for 60°C conductors.

■   300 percent of value specified in Table 310-16 for 60°C conductors.

14. Page 70-275. In Sec. 430-102(b), delete the word "separate."

15. Page 70-291. In the title of Table 450-3(a), change "Transformer-related Current" to "Transformer-Rated Current."

16. Page 70-307. In Sec. 490-41(b), delete the exception.

17. Page 70-354. Section 516-(a)(5) should read as follows: "Sumps, pits, or below-grade channels within 25 ft. (7.625m) horizontally of a vapor source. If the sump, pit, or channel extends beyond 15 ft. (7.625m) from the vapor source, it shall be provided with a vapor stop or it shall be classified as Class I, Division 1 for its entire length."

18. Page 70-354. Section 516-2(a)(6) should read as follows: "The interior of any enclosed dipping or coating process or apparatus."

19. Page 70-420. In Sec. 551-77(e), change 110-16 to 110-26.

20. Page 70-434. In Sec. 600-7, line 15, add the words "or less" after 100 Hz.

21. Page 70-486. Change Sec. 680-42 to read as follows: "680-42 Protection. The outlet(s) that supplies:

(a) A self-contained spa or hot tub, or

(b) A packaged spa or hot tub equipment assembly, or

(c) A field-assembled spa or hot tub with a heater load of 50 amperes or less shall be protected by a ground-fault circuit interrupter. A listed self-contained unit or listed packaged equipment assembly marked to indicate that integral ground-fault circuit-interrupter protection is provided for all electrical parts within the unit or assembly (pumps, air blowers, heaters, lights, controls, sanitizer generators, wiring, etc.) shall not require that the outlet supply be protected by a ground-fault circuit interrupter.

"A field-assembled spa or hot tube or spa assembly

commonly bonded need not be protected by a ground-fault circuit interrupter .FPN: See Sec. 680-4 for definitions of *self-contained spa or hot tub,* and for *packaged spa or hot tub equipment assembly.*"

22. Page 70-614. In Example D8, under Conductor Ampacity, the calculation for 25 hp motor should be "34A × 1.25 = 42.5A" and the first calculation for 30 hp motors should be "40A × 1.25 = 50A." Under Branch-Circuit Short-Circuit and Ground-Fault Protection, the calculation under non time-delay fuse, should be "300% × 34A = 102A." The text should read as follows: "The next larger standard fuse is 110A....If the motor will not start with a 110A non time-delay fuse...." The calculation under time delay fuse should be "175% × 34A = 59.5A."

A4. Typical Electrical Permit Fees

In some jurisdictions if a permit is not obtained before work is started, the applicant may be subject to a penalty equal to the permit fee.

Each inspection	$25
Service feeders	$0.15/A
Circuits/feeders, each:	
20 A under	$1.75
Over 20–50 A	$4.75
Over 50–100 A	$6.00
Over 100–200 A	$7.25
Over 200–400 A	$8.50
Over 400–800 A	$11.00
Over 800 A	$17.00
Misc. wiring/code repairs	$30
Temporary service	$25

Industrial and commercial inspections:

Existing buildings	$40 per hour or fraction thereof
Annual inspection fee	$400
Minimum permit fee	$35
Special or overtime inspection	$40 per hour or fraction thereof with a minimum fee of $80
Extra inspection	$35

A5. Data from IEEE Reliability Survey of Industrial Plants

Equipment category	λ, Failures per year	r, Hours of downtime per failure	$\lambda \cdot r$, Forced hours of downtime per year
Protective relays	0.0002	5.0	0.0010
Metal-clad drawout circuit breakers			
0–600 V	0.0027	4.0	0.0108
Above 600 V	0.0036	83.1*	0.2992
Above 600 V	0.0036	2.1†	0.0076
Power cables (1000 circuit ft)			
0–600 V, above ground	0.00141	10.5	0.0148
601–15,000 V, conduit below ground	0.00613	26.5*	0.1624
601–15,000 V, conduit below ground	0.00613	19.0†	0.1165
Cable terminations			
0–600 V, above ground	0.0001	3.8	0.0004

Equipment category	λ, Failures per year	r, Hours of downtime per failure	$\lambda \cdot r$, Forced hours of downtime per year
601–15,000 V, conduit below ground	0.0003	25.0	0.0075
Disconnect switches enclosed	0.0061	3.6	0.0220
Transformers			
601–15,000 V	0.0030	342.0*	1.0260
601–15,000 V	0.0030	130.0†	0.3900
Switchgear bus— bare			
0–600 V (connected to 7 breakers)	0.0024	24.0	0.0576
0–600 V (connected to 5 breakers)	0.0017	24.0	0.0408
Switchgear bus— insulated			
601–15,000 V (connected to 1 breaker)	0.0034	26.8	0.0911
601–15,000 V (connected to 2 breakers)	0.0068	26.8	0.1822

*Repair failed unit.
†Replace with spare.

A8. Resistivity of Soils and Resistance of Single Rods

Soil description	Group symbol*	Average resistivity, ohm·cm	Resistance of 5/8 in (16 mm) × 10 ft (3 m) rod, ohms
Well-graded gravel, gravel-sand mixtures, little or no fines	GW	60,000–100,000	180–300
Poorly graded gravels, gravel-sand mixtures, little or no fines	GP	100,000–250,000	300–750
Clayey gravel, poorly graded gravel, sand-clay mixtures	GC	20,000–40,000	60–120
Silty sands, poorly graded sand-silts mixtures	SM	10,000–50,000	30–150
Clayey sands, poorly graded sand-clay mixtures	SC	5,000–20,000	15–60
Silty or clayey fine sands with slight plasticity	ML	3,000–8,000	9–24
Fine sandy or silty soils, elastic silts	MH	8,000–30,000	24–90
Gravelly clays, sandy clays, silty clays, lean clays	CL	2,500–6,000†	17–18†
Inorganic clays of high plasticity	CH	1,000–5,500†	3–16†

*The terminology used in these descriptions is from the Unified Soil Classification and is a standard method of describing soils in a geotechnical or geophysical report.
†These soil classification resistivity results are highly influenced by the presence of moisture.

A9. Smoke Detector Spacing

Min/air change	Air changes/h	ft² (m²)/detector
1	60	125 (11.61)
2	30	250 (23.23)
3	20	375 (34.84)
4	15	500 (46.45)
5	12	625 (58.06)
6	10	750 (69.68)
7	8.6	875 (81.29)
8	7.5	900 (83.61)
9	6.7	900 (83.61)
10	6	900 (83.61)

A10. Voltage Drop of Conductors in Cable or Conduit

Conductor size	Cable or conduit	DC resistance ohms/100 ft	Computed IR drop V/1000 A/ 100 ft	Measured drop V/1000 A/ 100 ft
500 kcmil	3/C VCI (steel armor)	0.0383	38.3	35
500 kcmil	3/C VCI (steel armor with internal grounding conductor)	—	—	5
No. 4/0	3/C VCI (aluminum armor)	0.286	286.0	151
No. 4/0	3/C VCI (aluminum armor with internal grounding conductor)	—	—	12

A10. Voltage Drop of Conductors in Cable or Conduit (Continued)

Conductor size	Cable or conduit	DC resistance ohms/100 ft	Computed IR drop V/1000 A/ 100 ft	Measured drop V/1000 A/ 100 ft
No. 4/0	3/C VCI (steel armor)	—	—	55
No. 4/0	3/C VCI (steel armor with internal grounding conductor)	—	—	11
No. 4/0	3/C lead sheath (15 kV)	0.00283	2.83	11
No. 4/0	4″ rigid steel conduit	0.0025	2.5	1
No. 2/0	2″ rigid steel conduit	0.0095	9.5	6
No. 1/0	3/C VCI (steel armor)	0.0458	45.8	51
No. 1/0	3/C VCI (steel armor with internal grounding conductor)	—	—	19
No. 2	Aluminum sheath (solid sheath M/C cable)	0.01	10.0	9
No. 2	1¼ in rigid steel	0.0108	10.8	11
No. 2	1¼ in EMT	0.0205	20.5	22
No. 2	1¼ in greenfield (flexible metal conduit)	0.435	435.0	436

Conductor size	Cable or conduit	DC resistance ohms/100 ft	Computed IR drop V/1000 A/ 100 ft	Measured drop V/1000 A/ 100 ft
No. 8	¾ in rigid steel	0.02	20.0	21
No. 8	¾ in EMT	0.0517	51.7	48
No. 8	¾ in greenfield (flexible metal conduit)	1.28	1280.0	1000
No. 10	Aluminum sheath (solid sheath MC cable)	0.015	15.0	16

Conductor size	Cable or conduit	DC resistance ohms/100 ft	Computed IR drop V/1000 A/ 100 ft	Measured drop V/1000 A/ 100 ft
No. 12	½ in rigid steel	0.0223	22.3	25
No. 12	½ in EMT	0.0706	70.6	70
No. 12	BX without ground (ac cable)	1.79†	1790.0	1543

*Value read from bar chart (numeric values not published).
†Does not meet current Underwriter's Laboratories' listing requirements.

A11. Fault Current Available (Symmetrical RMS Amperes), 240 V

kVA rating of transformer	Conductor size per phase	Distance from transformer to point of fault, ft								
		0	5	10	20	50	100	200	500	1,000
150	No. 4	9,980	9,520	9,000	8,000	5,580	3,440	1,900	800	400
	No. 0	9,980	9,700	9,450	9,000	7,600	5,850	3,900	1,800	950
	250 MCM	9,980	9,820	9,660	9,350	8,500	7,220	5,550	3,200	1,900
	2–250 MCM	9,980	9,900	9,800	9,650	9,200	8,400	7,200	4,900	3,200
225	No. 4	14,940	13,800	12,800	10,600	6,500	3,800	2,000	800	450
	No. 0	14,940	14,500	14,000	12,900	10,100	7,100	4,300	2,000	1,000
	250 MCM	14,940	14,600	14,300	13,600	11,800	9,500	6,800	3,500	1,800
	2–250 MCM	14,940	14,700	14,500	14,300	13,200	11,700	9,400	6,000	3,500
	2–500 MCM	14,940	14,800	14,700	14,500	13,600	12,500	10,600	7,500	5,000
300	No. 4	19,970	18,000	16,000	12,700	7,000	4,000	2,000	800	400
	No. 0	19,970	19,100	18,100	16,200	11,800	7,800	4,500	2,000	1,000
	250 MCM	19,970	19,300	18,700	17,500	14,500	11,200	7,500	3,600	2,000
	2–250 MCM	19,970	19,500	19,300	18,700	17,000	14,500	11,200	6,400	3,600
	2–500 MCM	19,970	19,600	19,400	19,000	17,600	15,600	13,000	8,200	5,200

500	No. 4	33,100	28,000	22,900	15,900	7,800	4,200	2,200	900	500
	No. 0	33,100	30,800	28,000	23,100	14,800	9,000	4,900	2,000	1,000
	250 MCM	33,100	31,500	30,000	27,000	20,300	14,200	8,800	4,000	2,000
	2–250 MCM	33,100	32,300	31,400	29,800	25,300	20,100	14,000	7,000	3,900
	2–500 MCM	33,100	32,600	32,000	30,700	22,200	22,500	17,000	9,600	5,500
750	No. 4	40,900	33,000	26,000	17,000	8,000	4,000	2,000	900	500
	No. 0	40,900	37,400	33,900	27,000	15,900	9,200	5,000	2,000	1,000
	250 MCM	40,900	38,300	36,000	32,000	23,000	15,000	8,900	3,900	2,050
	2–250 MCM	40,900	39,800	38,500	36,000	30,000	22,900	15,000	7,300	4,000
	2–500 MCM	40,900	40,100	39,100	37,100	32,000	26,100	19,000	10,100	5,600
1,000	No. 4	54,400	41,000	29,500	18,000	8,200	4,200	2,100	950	400
	No. 0	54,400	48,800	42,200	32,100	17,900	9,900	5,000	2,050	1,000
	250 MCM	54,400	50,100	46,300	39,900	27,000	17,000	9,500	4,000	2,050
	2–250 MCM	54,400	52,100	50,000	46,000	36,800	26,900	17,000	8,000	4,050
	2–500 MCM	54,400	52,800	51,000	48,000	40,300	31,800	22,000	11,200	6,000
1,500	No. 4	80,100	53,200	35,500	20,500	9,900	4,800	2,500	1,200	900
	No. 0	80,100	66,500	55,000	40,000	20,000	10,500	5,800	2,800	1,800
	250 MCM	80,100	72,000	64,500	52,000	32,000	19,500	10,100	4,500	3,000

A11. Fault Current Available (Symmetrical RMS Amperes), 240 V (Continued)

kVA rating of transformer	Conductor size per phase	Distance from transformer to point of fault, ft									
		0	5	10	20	50	100	200	500	1,000
	2–250 MCM	80,100	76,000	72,000	64,000	47,000	32,000	19,500	8,500	4,800
	2–500 MCM	80,100	77,500	74,000	68,000	53,500	40,000	25,500	12,000	6,500
	No. 4	105,600	60,500	38,000	21,000	8,800	4,300	2,200	800	—
	No. 0	105,600	83,000	64,000	42,000	20,000	10,300	5,500	2,500	1,200
2,000	250 MCM	105,600	90,500	79,000	60,000	34,500	19,800	10,200	4,500	2,400
	2–250 MCM	105,600	97,500	91,000	78,000	54,000	34,000	19,000	8,500	4,600
	2–500 MCM	105,600	100,000	94,500	84,000	62,500	43,500	2,700	12,000	6,200

A12. Fault Current Available (Symmetrical RMS Amperes), 480 V

kVA rating of transformer	Conductor size per phase	Distance from transformer to point of fault, ft								
		0	5	10	20	50	100	200	500	1,000
150	No. 4	4,990	4,930	4,880	4,770	4,420	3,800	2,800	1,480	790
	No. 0	4,990	4,940	4,920	4,880	4,700	4,400	3,850	2,650	1,680
	250 MCM	4,990	4,960	4,930	4,910	4,800	4,600	4,250	3,350	2,500
	2–250 MCM	4,990	4,970	4,940	4,920	4,900	4,800	4,600	4,050	3,350
225	No. 4	7,470	7,380	7,240	7,000	6,140	4,880	3,300	1,600	840
	No. 0	7,470	7,400	7,320	7,200	6,800	6,200	5,100	3,180	1,860
	250 MCM	7,470	7,420	7,360	7,300	7,040	6,640	5,900	4,400	3,000
	2–250 MCM	7,470	7,440	7,400	7,350	7,220	7,000	6,600	5,580	4,300
	2–500 MCM	7,470	7,460	7,450	7,400	7,300	7,100	6,800	6,000	5,000
300	No. 4	9,985	9,800	9,600	9,100	7,600	5,600	3,560	1,620	840
	No. 0	9,985	9,840	9,750	9,520	8,800	7,650	5,900	3,400	1,920
	250 MCM	9,985	9,880	9,800	9,660	9,240	8,500	7,300	5,000	3,240
	2–250 MCM	9,985	9,920	9,825	9,790	9,580	9,200	8,450	6,800	5,020
	2–500 MCM	9,985	9,950	9,850	9,800	9,660	9,400	8,820	7,500	5,880

A12. Fault Current Available (Symmetrical RMS Amperes), 480 V (Continued)

		Distance from transformer to point of fault, ft								
kVA rating of transformer	Conductor size per phase	0	5	10	20	50	100	200	500	1,000
500	No. 4	16,550	16,000	15,400	14,000	10,250	6,800	3,800	1,600	800
	No. 0	16,550	16,200	15,950	15,250	13,250	10,500	7,400	3,500	1,900
	250 MCM	16,550	16,300	16,050	15,700	14,500	12,700	10,000	5,900	3,500
	2–250 MCM	16,550	16,350	16,250	16,100	15,450	14,400	12,500	9,000	6,000
	2–250 MCM	16,550	16,400	16,350	16,300	15,700	14,800	13,400	10,500	7,500
750	No. 4	20,450	19,700	18,700	16,800	11,700	7,500	4,000	1,600	800
	No. 0	20,450	20,000	19,500	18,700	16,000	12,400	8,100	3,800	2,000
	250 MCM	20,450	20,200	19,800	19,250	17,500	15,000	11,500	6,600	3,800
	2–250 MCM	20,450	20,250	20,200	19,700	19,000	17,500	15,000	10,500	6,600
	2–500 MCM	20,450	20,400	20,250	19,900	19,300	18,200	16,300	12,000	8,400
1,000	No. 4	27,200	26,000	24,200	21,000	13,400	7,900	4,400	1,800	800
	No. 0	27,200	26,700	25,900	24,300	20,000	14,400	9,000	4,100	2,200
	250 MCM	27,200	26,900	26,400	25,300	22,400	18,600	13,600	7,200	4,000

kVA	Conductor Size									
	2–250 MCM	27,200	27,000	26,700	26,200	24,500	22,200	18,500	12,100	7,200
	2–500 MCM	27,200	27,100	26,800	26,500	25,300	23,300	20,300	14,500	9,500
1,500	No. 4	40,050	37,000	33,100	26,000	14,400	8,200	4,000	1,400	600
	No. 0	40,050	38,800	36,800	33,200	24,500	16,000	9,200	4,000	2,000
	250 MCM	40,050	39,100	37,800	35,600	29,900	23,000	15,200	7,500	4,000
	2–250 MCM	40,050	39,600	39,000	37,900	34,100	29,000	22,500	13,000	7,400
	2–500 MCM	40,050	39,700	39,200	38,200	35,500	31,600	25,900	16,400	10,100
2,000	No. 4	52,800	47,400	40,700	30,000	15,100	8,200	4,200	1,900	1,000
	No. 0	52,800	50,200	47,000	41,200	28,000	17,000	9,700	4,200	2,400
	250 MCM	52,800	51,000	49,000	45,400	36,200	26,500	16,500	8,000	4,200
	2–250 MCM	52,800	51,800	50,900	48,900	43,100	36,000	26,700	14,000	8,000
	2–500 MCM	52,800	52,100	51,300	49,900	45,100	39,200	30,800	18,500	11,000

The fault currents listed are maximum available symmetrical rms values based on liquid filled transformers, with nominal impedances of $4^{1}/_{2}\%$ for ratings up to and including 500 kVA and $5^{1}/_{2}\%$ for ratings above 500 kVA, and include motor contribution based on 100% motor load.

*For equivalent metric sizes, see Table 6.

A13. Typical Appliance/General-Purpose Receptacle Loads

	Unit load, (VA/sq ft)		
Type of occupancy	Low	High	Average
Auditoriums	0.1	0.3	0.2
Cafeterias	0.1	0.3	0.2
Churches	0.1	0.3	0.2
Drafting rooms	0.4	1.0	0.7
Gymnasiums	0.1	0.2	0.15
Hospitals	0.5	1.5	1.0
Hospitals, large	0.4	1.0	0.7
Machine shops	0.5	2.5	1.5
Office buildings	0.5	1.5	1.0
Schools, large	0.2	1.0	0.6
Schools, medium	0.25	1.2	0.7
Schools, small	0.3	1.5	0.9

Other unit loads:

Specific appliances—ampere rating of appliance

Supplying heavy-duty lampholders— 5 A/outlet

A14. Typical Apartment Loads

Type	Load
Lighting and convenience outlets (except appliance)	3 VA/sq ft
Kitchen, dining appliance circuits	1.5 kVA each
Range	8 to 12 kW
Microwave oven	1.5 kW
Refrigerator	0.3–0.6 kW

Type	Load
Freezer	0.3–0.6 kW
Dishwasher	1.0–2.0 kW
Garbage disposal	0.33–0.5 hp
Clothes washer	0.33–0.5 hp
Clothes dryer	1.5–6.5 kW
Water heater	1.5–9.0 kW
Air conditioner (0.5 hp/room)	0.8–4.6 kW
Personal computer	0.6 kW (600 W)
Monitor	0.23 kW (230 W)
Printer	0.025 kW (25 W)
Fax	0.045 kW (45 W)

A15. Connected Load and Maximum Demand by Tenant Classification

Classification	Connected load, W/sq ft	Maximum demand, W/sq ft	Demand factor
10 Women's wear	7.7	5.9	0.75
3 Men's wear	7.2	5.6	0.78
6 Shoe store	8.5	6.9	0.79
2 Department store	6.0	4.7	0.74
2 Variety store	10.5	4.5	0.45
2 Drugstore	11.7	6.7	0.57
5 Household goods	5.4	3.9	0.76
10 Specialty shop	8.1	6.8	0.79
4 Bakery and candy	17.1	12.1	0.71
3 Food store (supermarkets)	9.9	5.9	0.60
5 Restaurant	15.9	7.1	0.45

Connected load includes an allowance for spares.

A16. Typical Loads in Commercial Kitchens

	Number served	Connected load, kW
Lunch counter (gas ranges, with 40 seats)		30
Cafeteria	800	150
Restaurant (gas cooking)		90
Restaurant (electric cooking)		180
Hospital (electric cooking)	1200	300
Diet kitchen (gas cooking)		200
Hotal (typical)		75
Hotel (modern, gas ranges, three kitchens)		150
Penitentiary (gas cooking)		175

A17. Typical Ratings of Engine-Driven Generator Seats

Nominal rating, kW	Prime power rating, kW	Standby rating, kW	Power factor	Prime mover			Speed, rpm
				Gasoline	Diesel	Natural gas/ LP gas	
5	5	5	1.0	x		x	3600
10	10	10	1.0	x		x	1800
25	25	30	0.8		x	x	1800
100	90	100	0.8		x	x	1800
250	225	250	0.8		x		1800
750	665	750	0.8		x		1800
1000	875	1000	0.8		x		1800

A18. Summary of Possible or Typical Emergency or Standby Power Needs for the Health-Care Industry

Power use	Application	Purposes	Range of tolerable outage time	Duration of need	Power sources	Remarks
Life safety equipment			10 s		4	Legally required emergency power
Critical equipment			10 s		4	Legally required emergency power
Essential equipment			1 min		4	Legally required standby power

These facilities have some of the greatest and the most regulated needs for emergency power for the preservation of human life. This chapter recognizes this industry as a significant user of emergency power, but refers the reader to ANSI/IEEE Standard 602-1986 [5], where complete coverage of this topic is provided, and also to Arts. 517 and 700 in the NEC, plus any local codes or ordinances, for detailed requirements.

A19. Summary of Possible or Typical Emergency or Standby Power Needs for the Commercial Building Industry

Power use	Application	Purposes	Range of tolerable outage time	Duration of need	Power sources	Remarks
Emergency lighting	Hotels, apartments, stores, theaters, offices, assembly halls	1, 3, 4, 5	10–60 s or statutory requirement	1½ h or statutory requirement or as needed	2, 4, 5*	Building management or safety authorities to determine *as needed* requirements
Elevators	Hotels, apartments, stores, offices	3, 5	10 s to 5 min	Evacuation completion	4†	Other methods of evacuation may be available
Phones in elevators	Signaling, switching, talking	3	10–60 s	Evacuation completion or extended occupancy	2, 4, 5	Many PBX or PAX systems no longer have batteries
Other phones or PAs	Signaling, switching, talking	3	10–60 s	Evacuation completion or extended occupancy	2, 4, 5	Many PBX or PAX systems no longer have batteries
HVAC	Space, process, or material	4, 5	1–30 min	Duration of outage	4	Negotiable with tenants

Tenants' computers	Various	4, 5	Negotiable	Negotiable	2, 3, 4	Negotiable with tenants
Fire fighting and smoke control	Fire pumps and ventilation control	1, 4	10 s	2 h or more	4	Pumps needed when high-rise exceeds water pressure capability
Electric security locks	Access controls	4	0–10 s	Until manual locking accomplished	1, 2, 3, 4	Loss of power should not prevent egress
Emergency lighting	Store lighting	4	10–60 s	Completion of evacuation	2, 4,* 5†	Protection from theft of stock and evacuation
Emergency lighting and HVAC	Casino lighting and environment	3, 4, 5	1–10 s	Duration of outage	2, 4	Protection from theft and continuing operation is financially significant

*Where applicable, NEC (National Electrical Code, ANSI/NFPA 70-1987), Art. 700-12, requires $1\frac{1}{2}$ h battery supply.
†Where applicable, NEC, Art. 700-12, requires 2 h generator fuel supply.

A21. ANSI Device Function Numbers for Protective Devices

2. Time-delay starting or closing relay

3. Checking or interlock relay

6. Starting circuit breaker

8. Control power-disconnecting device

12. Overspeed device

14. Underspeed device

15. Speed- or frequency-matching device

18. Accelerating or decelerating device

19. Starting-to-running transition contractor

21. Distance relay

23. Temperature-control device

25. Synchronizing or synchronism-check device

27. Undervoltage relay

30. Annunciator relay

32. Directional power relay

36. Polarity device

37. Undercurrent or underpower relay

40. Loss of excitation (field) relay

41. Field circuit breaker

42. Running circuit breaker

43. Manual transfer or selector device

46. Reverse-phase or phase-balance current relay

47. Phase-sequence voltage relay

48. Incomplete-sequence relay

49. Machine or transformer thermal relay

50. Instantaneous overcurrent or rate-of-rise relay

51. AC time overcurrent relay

52. AC circuit breaker

55. Power-factor relay

56. Field-application relay

59. Overvoltage relay

60. Voltage- or current-balance relay
62. Time-delay stopping or opening relay
64. Ground-protective relay
67. AC directional overcurrent relay
68. Blocking relay
69. Permissive control device
72. DC circuit breaker
74. Alarm relay
76. DC overcurrent relay
78. Phase-angle measuring or out-of-step protective relay
79. AC reclosing relay
81. Frequency relay
82. DC reclosing relay
85. Carrier or pilot-wire receiver relay
86. Locking-out relay
87. Differential protective relay
91. Voltage directional relay
92. Voltage and power directional relay
94. Tripping or trip-free relay

A22. Transformer Typical Ambient Sound Levels

	dB
Ratio, recording and TV studios	25–30
Theaters and music rooms	30–35
Hospitals, auditoriums, and churches	35–40
Classrooms and lecture rooms	35–40
Apartments and hotels	35–45
Private offices and conference rooms	40–45
Stores	45–55
Residence (radio, TV Off) and small offices	53
Medium office (3 to 10 desks)	58
Residence (radio, TV On)	60

A23. Transformer Typical Ambient Sound Levels (*Continued*)

	dB
Large store (5 or more clerks)	61
Factory office	61
Large office	64
Average factory	70
Average street	80

A24. Maximum Average Sound Levels

	Dry-type		Liquid-filled	
kVA	Self-cooled rating (AA)	Forced-air cooling (FA)	Self-cooled rating (OA)	Forced-air cooling (FA)
0–50	50	—	—	—
51–150	55	—	—	—
151–300	58	67	55	67
301–500	60	67	56	67
501–700	62	67	57	67
701–1000	64	67	58	67
1001–1500	65	68	60	67
1501–2000	66	69	61	67
2001–2500	68	71	62	67
2501–3000	70	71	63	67
3001–4000	71	73	64	67
4001–5000	72	74	65	67
5001–6000	73	75	66	68
6001–7500	—	76	67	69
7501–10000	—	76	68	70

A25. Average Efficiencies and Power Factors for Polyphase Squirrel-cage Induction Motors

hp	Efficiency			Power factor		
	One-half load	Three-fourths load	Full load	One-half load	Three-fourths load	Full load
1/2	60.0	67.0	69.0	45	56	65
3/4	64.0	68.0	69.0	48	58	65
1	75.0	77.0	76.0	57	69	76
1 1/2	75.0	77.0	78.0	64	76	81
2	77.0	80.0	81.0	68	79	84
3	80.0	82.0	81.0	70	80	84
5	80.0	82.0	82.0	76	83	86
7 1/2	83.0	85.0	85.0	77	84	87
10	83.0	85.0	85.0	77	86	88
15	84.0	86.0	88.0	81	85	87
20	87.0	88.0	87.0	82	86	87
25	87.0	88.0	87.5	82	86	87
30	87.5	88.5	88.0	83	86.5	87
40	87.5	89.0	89.5	84	87	88
50	87.5	89.0	89.5	84	87	88
60	88.0	89.5	89.0	84	87	88
75	88.5	89.5	89.5	84	87	88
100	89.0	90.0	90.5	84	88	88
125	90.0	90.5	91.0	84	88	89
150	90.0	91.5	92.0	84	88	89
200	90.0	91.5	92.0	85	89	90
250	91.0	92.5	93.0	84	89	90
300	92.0	93.5	94.0	84	89	90

A26. Specifying Motor Overloads

Motor full-load current, A	Thermal unit No.	Maximum fuse rating, A	Motor full-load current, A	Thermal unit No.	Maximum fuse rating, A
0.28–0.30	JR 0.44	0.6	2.33–2.51	JR 3.70	5
0.31–0.34	JR 0.51	0.6	2.52–2.99	JR 4.15	5.6
0.35–0.37	JR 0.57	0.6	3.00–3.42	JR 4.85	6.25
0.38–0.44	JR 0.63	0.8	3.43–3.75	JR 5.50	7
0.45–0.53	JR 0.71	1.0	3.76–3.98	JR 6.25	8
0.54–0.59	JR 0.81	1.125	3.99–4.48	JR 6.90	8
0.60–0.64	JR 0.92	1.25	4.49–4.93	JR 7.70	10
0.65–0.72	JR 1.03	1.4	4.94–5.21	JR 8.20	10
0.73–0.80	JR 1.16	1.6	5.22–5.84	JR 9.10	10
0.81–0.90	JR 1.30	1.8	5.85–6.67	JR 10.2	12
0.91–1.03	JR 1.45	2.0	6.68–7.54	JR 11.5	15
1.04–1.14	JR 1.67	2.25	7.55–8.14	JR 12.8	15
1.15–1.27	JR 1.88	2.5	8.15–8.72	JR 14.0	17.5
1.28–1.43	JR 2.10	2.8	8.73–9.66	JR 15.5	17.5
1.44–1.62	JR 2.40	3.2	9.67–10.5	JR 17.5	20
1.63–1.77	JR 2.65	3.5	10.6–11.3	JR 19.5	20
1.78–1.97	JR 3.00	4.0	11.4–12.7	JR 22	25
1.98–2.32	JR 3.30	4.0	12.8–14.1	JR 25	25

Motor overloads are sized according to the actual full-load current stamped on the motor nameplate. When you have this data you need to obtain an electrical equipment manufacturer's catalog with a schedule that correlates the overload (or heater) with the motor full-load current. A typical manufacturer schedule appears in the table.

Suppose you need to provide protection for a 240-V motor with a 1.5-A full-load amps stamped on the nameplate. According to the schedule, the appropriate overload (heater) to apply is a number JR 2.40. The **NEC** allows this heater to be applied at 125 percent of the motor nameplate full-load current provided that (1) the service factor is 1.15 or larger, and (2) the temperature rise is not greater than 40°C. The manufacturer has already taken this into consideration when setting up the schedule. If the service factor is less than 1.15 or the temperature rise is greater than 40°C, then the heater will be 10 percent oversized, thus making the motor vulnerable to burnout. If the motor service factor is less than 1.15, multiply the motor full-load current by 0.9 and use this new value to size the overload. In our example, multiplying 1.5 A by 0.9 results in 1.35 A. The overload corresponding to 1.35 A is the JR 2.10 unit.

A27. Spectrums of Several Standard Frequency Bands

Band	Symbol	Description	Range
Audio	VLF	Very low frequency	3–30 kHz
Audio	LF	Low frequency	30–300 kHz
Radio (RF)	MF	Medium frequency	300–3,000 kHz
Radio (RF)	HF	High frequency	3,000–30,000 kHz
Video (TV)	VHF	Very high frequency	30–300 MHz
Video (TV	UHF	Ultra high frequency	300–3,000 MHz
Radar	SHF	Super high frequency	3,000–30,000 MHz
Radar	EHF	Extremely high frequency	30–300 GHz

A28. Twisted-Pair Cable Standards

Parameter	EIA	IBM	UL®	NEMA	Bellcore	ICEA
Published specification	ANSI/TIA/EIA-568-A	GA27-3773-1	200-131A	WC63	TA-NWT-000133	S-80-576
Conductor sizes (AWG)	22, 24	22, 24, 26	22, 24	22, 24, 26	24	22, 24, 26
Impedance (ohms)	100	150	100	100, 150	100	Not specified
Cable sizes (pairs)	4 25-pair subunits	2 to 6	25 or less	6 or less	Any	3,600 or less
Shielding	UTP	STP	STP\UTP	STP\UTP	UTP*	STP\UTP
Performance	Category: 1–5	Type: 1–9	Category: 1–5	Standard; low loss; low loss extended frequency	Category: 1–5	Not specified
Equivalence to ANSI/TIA\EIA-568-A	1	None	1	None	1	None
	2	Type 3	2	None	2	None
	3	None	3	Standard	3	None
	4	None	4	Low loss	4	None
	5	None	5	Low loss, extended frequency	5	None

*The technical advisory does not preclude STP.

Core References

Analysis of the 1999 National Electric Code, International Association of Electrical Inspectors, Richardson, TX, 1998 (www.iaei.com).

Anthony, Michael A.: *Electrical Power System Protection and Coordination*, McGraw-Hill, New York, 1995 (www.mcgraw-hill.com).

Bierals, Gregory P.: *Applying the 1993 National Electric Code*, Fairmont Press, Inc., 700 Indian Trail, Lilburn, GA, 1994.

IEEE Standard 141 Recommended Practice for Electric Power Distribution for Industrial Plants, The Institute of Electrical and Electronic Engineers, Inc., New York, 1993 (www.ieee.org).

IEEE Standard 241 Recommended Practice for Electric Power Systems in Commercial Buildings, 1997.

Hartwell, F., Illustrated Changes in the 1999 National Electric Code, *EC&M Magazine*, Intertec Publishing Corp., Overland Park, KS, 1999.

Inspector's Guide to Low Voltage Systems, National Electrical Contractors Association, Bethesda, MD.

McPartland, B., *McGraw-Hill's National Electrical Code Handbook*, McGraw-Hill, New York, 1999.

Mike Holt Training Program, Mike Holt Enterprises, Ft. Lauderdale, FL (www.mikeholt.com).

National Electrical Code, NFPA 70.

National Fire Protection Association, *National Electrical Code Handbook*, Quincy, MA, 1999.

Stallcup, James: *Electrical Calculations Simplified*, Gray Boy Publishing, Grayboy & Associates, Fort Worth, TX, 1993.

Stallcup, James: *Stallcup's Illustrated Code Changes*, Gray Boy Publishing, Grayboy & Associates, Fort Worth, TX, 1998.

Surbrooke, Truman C.: *Interpreting the National Electric Code*, Delmar Publishers, Inc., Albany, NY, 1990.

PERIODICALS

Consulting-Specifying Engineer, Cahner's Publishing Company, Cahner's Plaza, Des Plaines, IL (www.csemag.com).

EC&M Magazine, Intertec Publishing Corp., Overland Park, KS (www.ecmweb.com).

Design Magazine, Square-D Company, Palatine, IL.

IEEE: *Industry Applications Society Transactions*.

Bibliography

Chapter 1

CEGS, The Corps of Engineers Guide Specifications, U.S. Army Corps of Engineers, Washington, DC.

Traits of a Good Electrical Inspector, Cox, Phil: *IAEI News,* International Association of Electrical Inspectors, Richardson, TX, May/June 1997.

DeDad, John: Educating the Owner/End User, *Electrical Construction & Maintenance,* Intertec Publishing Corp., Overland Park, KS, December 1995.

Lovorn, Kenneth L.: Holding to Electrical Specs, *Consulting-Specifying Engineer,* Cahner's Publishing Company, Des Plaines, IL, October 1998.

MASTERSPEC, American Institute of Architects, Washington, DC.

Miller, Mark E.: Avoiding Liability in the Event of Electrical Fire, *IAEI News,* International Association of Electrical Inspectors, Richardson, TX, January/February 1998.

Means, R.S.: *HVAC Handbook.*

NFGS, U.S. Naval Facilities Engineering Command Guide Specifications, U.S. Navy, Washington, DC.

Shapiro, Dave E.: Spreading Safety from the Podium, *IAEI News,* International Association of Electrical Inspectors, Richardson, TX, January/February 1998.

SPECTEXT, Construction Specifications Institute, Arlington, VA.

Stauffer, Brooke: Eliminating Electrical Inspections: Budget Cutbacks Threaten Critical Safety Function, *IAEI News,* November/December 1996.

Stetson, LaVerne E.: *Electrical Codes and Standards in Rural Applications,* IEEE Industry Applications Society, The Institute of Electrical and Electronic Engineers, New York, November/December 1998.

Walker, Nathan: *Legal Pitfalls in Architecture, Engineering and Building Construction,* McGraw-Hill, New York, 1979.

Chapter 2

Course Notes for Statistical Methods in Power Systems Operation and Planning, University of Michigan College of Engineering, July 1976.

Hartwell, Fred: Whose Code Is This? *Electrical Construction & Maintenance,* Intertec Publishing Corp., Overland Park, KS, August 1997.

IEEE Standard 493 Recommended Practice for Design of Reliable Industrial and Commercial Power Systems; The Institute of Electrical and Electronic Engineers, New York, NY.

Quiter, James R.: Is a Single Model Code Possible? *Consulting-Specifying Engineer,* Cahner's Publishing Company, Des Plaines, IL, April 1997.

Reliability Assessment of Large Electric Power System, Billinton and Allan, Kluwer Academic Publishers, 1988.

The National Electric Code: 100 Years of Quality in the Interest of Public Safety. Wells, Jack, Pass and Seymour/Legrand: *IAEI News,* International Association of Electrical Inspectors, Richardson, TX, November/December 1995.

Chapter 3

National Electric Safety Code, ANSI/IEEE Standard C2.

Liquid-Filled Transformer Installation Requirements, *Electrical Construction & Maintenance,* Primedia Intertec, Overland Park, KS, June 1996.

Young, David C.: Using the NESC, *IAEI,* International Association of Electrical Inspectors, Richardson, TX, July/August 1997.

1992 Survey Examines Three-Year Trend in Underground Distribution, Staff Report, *Transmission & Distribution Magazine,* Primedia Intertec, Overland Park, KS, July 1992.

Unearthing Underground Distribution Trends, Staff Report, *Transmission & Distribution Magazine,* Intertec Publishing Corp., Overland Park, KS, July 1994.

Mailman, Mitchell: Plowing-in Underground Cable, *Transmission & Distribution Magazine,* Intertec Publishing Corp., Overland Park, KS, August 1995.

Chapter 9

Devlin, John E.: Rating Fire Resistance, *Consulting-Specifying Engineer,* Cahner's Publishing Company, Des Plaines, IL, July 1998.

Knisley, Joseph R.: Establishing an Electrostatic Discharge Control Program, *Electrical Construction & Maintenance,* Intertec Publishing Corp., Overland Park, KS, December 1995.

Chapter 10

Guideline on Electrical Power for ADP Installations, *Federal Information Processing Standards Publication 94.*

Chapter 11

Hartwell, Frederic P.: Illustrated Changes in the 1996 NEC, *Electrical Construction & Maintenance,* Intertec Publishing Corp., Overland Park, KS, December 1995.

Chapter 13

IEEE Standard 602 Recommended Practice for Electric Systems in Health-Care Facilities, The Institute of Electrical and Electronic Engineers, New York, NY.
Johnson, Gordon S.: *Integrity of Emergency Systems,* IEEE Industry Applications Society, July/August 1998.

Chapter 14

Stallcup, James: *Electrical Calculations Simplified,* Grayboy & Associates, Fort Worth, TX, Gray Boy Publishing.

Chapter 15

Cutler-Hammer Consulting Application Guide for Adjustable Frequency Drives.
Heydt, G. T.: *Electric Power Quality,* Stars-in-a-Circle Publications, West Lafayette IN, 1991.
Lawrie, Robert J.: Understanding Modern Motors and Controllers, *EC&M Magazine,* Intertec Publishing Corp., Overland Park, KS, March 1995.
Milne, L. James: *Fire Protection of Critical Circuits—A Life and Property Preserver,* IEEE Industry Applications Society, The Institute of Electrical and Electronic Engineers, New York, NY, July/August 1998.
Peeran, Syed M.: A Reason for Rules, *Consulting-Specifying Engineer,* Cahner's Publishing Company, Des Plaines, IL, November 1998.
Simmons, J. Phillip: Air Conditioning and Heat Pump Equipment, *IAEI News,* International Association of Electrical Inspectors, Richardson, TX, September/October 1995.
Turkel, Solomon S.: Understanding Variable Speed Drives, *EC&M Magazine,* Intertec Publishing Corp., Overland Park, KS, March 1995.

CHAPTER 17

Hartwell, Fred: The 1999 NEC: Will You Recognize It? *Electrical Construction & Maintenance,* Intertec Publishing Corp., Overland Park, KS, April 1997.
Salimando, Joe: Plug into the New NEC, *Consulting-Specifying Engineer,* Cahner's Publishing Company, Des Plaines, IL, November 1998.

CHAPTER 18

Simmons, Phillip: Analysis of the 1996 NEC, *IAEI News,* International Association of Electrical Inspectors, Richardson, TX, September/October 1995.
Toy, John H.: Get on the Bus, *Consulting-Specifying Engineer,* Cahner's Publishing Company, Des Plaines, IL, July 1998.

CHAPTER 19

Prime Power: On-Site Power Generation for a Competitive Future, Supplement to *Consulting-Specifying Engineer,* Cahner's Publishing Company, Des Plaines, IL, November 1998.
Garrett, Robert E.: The Case of a Hospital's Emergency Power Failure, *Electrical Construction & Maintenance,* Intertec Publishing Corp., Overland Park, KS, July 1998.
Zackrison, Harry B., Jr.: Power Standing By, *Consulting-Specifying Engineer,* Cahner's Publishing Company, Des Plaines, IL, May 1998.
IEEE Standard 446, Recommended Practice for Emergency and Standby Power Systems for Industrial and Commercial Applications, The Institute of Electrical and Electronic Engineers, New York, NY.

CHAPTER 20

Enabnit, Elgin G.: 15 kV Is Still Preferred Distribution Voltage for Larger Utilities, *Transmission & Distribution Magazine,* Primedia Intertec, Overland Park, KS, January 1991.
Gonen, Turan: *Electric Power Distribution System Engineering,* McGraw-Hill, New York, 1986.

CHAPTER 21

Simmons, J. Phillip: Selecting Conductors for Services, Feeders and Branch Circuits, *IAEI News,* International Association of Electrical Inspectors, Richardson, TX, March/April 1997.

CHAPTER 21, DIVISION 16450

Haman, Richard: A Closer Look: Article 250, *IAEI News,* International Association of Electrical Inspectors, Richardson, TX, September/October 1995.
Simmons, J. Phillip: *Soare's Book on Grounding,* International Association of Electrical Inspectors, Richardson, TX.
IEEE Standard 142 Recommended Practice for Grounding of Industrial and Commercial Power Systems, The Institute of Electrical and Electronic Engineers, New York, NY.

CHAPTER 21, DIVISION 16460

Pauley, Jim: Transformers, Tap Conductors and Overcurrent Protection, *IAEI News,* International Association of Electrical Inspectors, Richardson, TX, March/April 1997.

CHAPTER 21, DIVISION 16470

Pauley, Jim: Conductors and Continuous Loads, *IAEI News,* International Association of Electrical Inspectors, Richardson, TX, November/December 1995.

CHAPTER 21, DIVISION 16480

Farrell, George, and Frank Valvoda: Integrating Power and Control, *Consulting-Specifying Engineer,* Cahner's Publishing Company, Des Plaines, IL, August 1998.
Gregory, George D., and Lorraine Padden: *Application Guidelines for Instantaneous Trip Circuit Breakers in Combination Motor Starters,*

IEEE Industry Applications Society, The Institute of Electrical and Electronic Engineers, New York, NY, July/August 1998.

Lawrie, Robert J.: Answering Twenty Key Questions about Premium Efficiency Motors, *Electrical Construction & Maintenance,* October 1994.

Hartwell, Fred: Applying the New 1996 NEC Rules for Motor Circuits, *EC&M,* Intertec Publishing Corp., Overland Park, KS, March 1996.

Lawrie, Robert J., Ed.: *Electric Motor Manual,* McGraw-Hill, New York, 1987.

Revelt, Jean J.: Don't Let the New Design I Motor Throw You a Curve, *EC&M,* Intertec Publishing Corp., Overland Park, KS, March 1996.

NEMA Standard MG-1, Revision 1, Park 12.58. Motor efficiency test. The test technique, called IEEE 112-A-Method B, provides a consistent efficiency measurement standard for those who use it.

CSA Standard 390.

ANSI/UL 508 Safety Standard for Industrial Control Equipment.

Waggoner, Ray: Faulty Drive Operation? Check Grounding and Wiring First, *Electrical Construction & Maintenance,* Intertec Publishing Corp., Overland Park, KS, August 1997.

CHAPTER 22

Hughes, S. David: *Electrical Systems in Buildings,* PWS-Kent Publishing Company, Boston, MA, 1988.

Page, Terry: Leased Lighting Leads to Utility Profits, *Transmission & Distribution Magazine,* April 1992.

Vandel, Kathi S.: Updating Exterior Lighting, *Consulting-Specifying Engineer,* Cahner's Publishing Company, Des Plaines, IL, May 1998.

CHAPTER 23

Neri, Dion: Surge Protection: Where and How Much? *EC&M,* Intertec Publishing Corp., Overland Park, KS, March 1997.

Valvoda, Frank, and George Farrell: Keeping Surges at Bay, *Consulting-Specifying Engineer,* Cahner's Publishing Company, Des Plaines, IL, August 1997.

Chapter 24

Maybin, Harry B.: *Low Voltage Wiring Handbook,* McGraw-Hill, New York, 1995.
McElroy, Mark: Fiber to the Desk, *Electrical Construction & Maintenance,* Intertec Publishing Corp., Overland Park, KS, November 1997.
Stauffer, Brooke, and Mike Holt: Inspecting Low Voltage Systems, *IAEI News,* International Association of Electrical Inspectors, Richardson, TX, March/April 1998.
Federal Information Processing Standard.
Commercial Building Telecommunications Wiring Standard, ANSI/EIA/TIA 568-1991.
Commercial Building Standard for Telecommunications Pathways and Spaces, ANSI/EIA/TIA 569-1990.

Chapter 25

Byres, Eric, and Ed Hanschke: *High Speed Fiber Optics Backbones,* IEEE Industry Applications Society, The Institute of Electrical and Electronic Engineers, New York, NY, July/August 1998.
Simmons, Phillip J.: Fixed Electric Space Heating Equipment, *IAEI News,* International Association of Electrical Inspectors, Richardson, TX, November/December 1995.

Chapter 26

Electrical Testing Specifications, International Electrical Testing Association, Morrison, Co. (www.netaworld.org).

Index

About the Author

Michael A. Anthony is an electrical engineer for University of Michigan Business Operations and is responsible for the engineering and planning of the world's largest campus bulk electric distribution grid. His writing on electrical industry issues appears in both technical and general interest publications. Mr. Anthony is also the author of McGraw-Hill's *Electric Power System Protection and Coordination*.